EXPERIMENTAL MAN

What One Man's Body Reveals
about His Future, Your Health,
and Our Toxic World

DAVID EWING DUNCAN

WILEY

John Wiley & Sons, Inc.

Published by John Wiley & Sons, Inc., Hoboken, New Jersey
Published simultaneously in Canada

The chart on page 296 is courtesy of Olivier Boss and Sirtris Pharmaceuticals.

Limit of Liability/Disclaimer of Warranty: While the publisher and the author have used their best efforts in preparing this book, they make no representations or warranties with respect to the accuracy or completeness of the contents of this book and specifically disclaim any implied warranties of merchantability or fitness for a particular purpose. No warranty may be created or extended by sales representatives or written sales materials. The advice and strategies contained herein may not be suitable for your situation. You should consult with a professional where appropriate. Neither the publisher nor the author shall be liable for any loss of profit or any other commercial damages, including but not limited to special, incidental, consequential, or other damages.

For general information about our other products and services, please contact our Customer Care Department within the United States at (800) 762-2974, outside the United States at (317) 572-3993 or fax (317) 572-4002.

Wiley also publishes its books in a variety of electronic formats. Some content that appears in print may not be available in electronic books. For more information about Wiley products, visit our web site at www.wiley.com.

Library of Congress Cataloging-in-Publication Data:

Duncan, David Ewing.
    Experimental man: what one man's body reveals about his future, your health, and our toxic world / David Ewing Duncan.
        p.   cm.
    Includes bibliographical references and index.
    ISBN 978-0-470-17678-8 (cloth)
    1. Health risk assessment.   2. Environmental health.   3. Medical innovations.
4. Medical genetics.   I. Title.
    RA427.3.D86   2009
    362.1—dc22

                                                        2008047040

Printed in the United States of America
10   9   8   7   6   5   4   3   2

To my family

I love fools' experiments.
I am always making them.

—CHARLES DARWIN, LETTER, 1887

CONTENTS

Experimental Man Index

Amount of blood in the human body, in liters: 5.6

Amount of blood drawn by author for this book at different times, in liters: ~1.4

Hours author spent in an MRI: 22

Number of gigabytes of data author produced: 100+

Number of chemical toxins author tested for: 320

Number of toxins detected in author's body: 165

Number of toxins detected in author's body at levels over the national mean: 155

Number of gene markers the author was tested for, in millions: 7–10

Number of gene markers tested for in author's family, in millions: 5

Number of gene markers tested in the author that are associated with a trait: 7,000+

Number of high-risk gene markers* (approximate): 122

Author's favorite gene marker: Caffeine fast metabolizer gene

Number of author's high-risk gene markers for heart attack: 7

Online genetic company's stated risk factor for the author's risk of a heart attack: high, medium, and low (see "I'm doomed. Or not.")

Rise in mercury levels after author eats two meals of big fish, in parts per billion, before and after: 4–14

Flame retardant levels (PBDE-47) detected in the author's blood, percentage above U.S. mean: 1,200

Predominant side of author's brain: Right

Status of the size of author's brain: Shrinking

Average background radiation exposure per person from the sun and other sources, per year, in millisievers: 3–4

Author's radiation exposure from a single full-body CT scan, in millisievers: 6–8

Average life span of an American in 1900, in years: 47.3

Average life span of an American in 2005, in years: 77.8

Author's maximum age, range of predictions, in years: 59–122

Author's production of a biomarker associated with longevity, percentage above average: 1,000

Cost of all the author's tests, including those conducted pro bono: $150,000–$500,000

---

* Risk factors of 1.5 times normal and above.

INTRODUCTION

All Life is an experiment. The more
experiments you make, the better.

—RALPH WALDO EMERSON

A fish and mercury story

When the halibut on my hook broke the surface, writhing in a splash of seawater, I was thinking less of this fish's fate than of my own. Considering that I planned to kill and eat it, this might seem cruel. Yet this flat, odd-looking creature had tucked inside its fat and muscle cells a substance as poisonous to me, if this fish becomes my meal, as it could be to him—methyl mercury, the most common form of mercury that builds up inside people (and fish). At the right dose and duration of exposure, methyl mercury can impair a person's memory, ability to learn, and behavior. Even in small doses, methyl mercury can cause birth defects in fetuses exposed to it in the womb and in breast-fed newborns whose mothers' milk is laced with tiny amounts of this heavy metal.

Scientists have assured me, however, that a single halibut contains nowhere near a dosage that might cause harm. These are the same scientists who admit that no one knows for sure what the threshold dose is that causes this toxic metal to subtly poison cells in the brain and the liver, two organs where it tends to accumulate.

I'm fishing from the bow of the *Osprey*, a tiny, open-decked trawler, as part of an experiment that started a few days earlier when I gave up 9 millimeters of blood and enough pee to fill a tiny cup. The idea is to test my normal level of methyl mercury—the background level that I typically have in my body from living on Earth in the twenty-first century, in the city of San Francisco, California—and then to give up more bodily fluids after I have eaten today's catch for lunch and a store-bought swordfish on the same day for dinner. Would my level rise?

I live just a few miles south of Bolinas, California, where we are now bobbing just off the coast above a field of kelp where halibut troll for food. The kelp is also a repository of lost hooks that landlubbers like me tend to lose when we lamely snag them on the leathery plants and the hooks have to be cut loose—so says the *Osprey*'s skipper and sole member of her crew, Josh Churchman. Fifty-something and stubbly bearded, with graying hair and a faded baseball cap, Churchman has been fishing solo for more than three decades on this twenty-four-foot vessel with a steering cabin large enough to fit a single person. He let me come along because he's worried about mercury in the fish he catches and regularly eats and because he loves the company. He hasn't stopped talking since we motored out here from the Bolinas Lagoon a couple of hours ago.

Methyl mercury gets into these fish from coal-burning power plants that rim the northern Pacific, from the United States and Mexico to Japan and China. Expelled from tall stacks, mercury is carried in the upper atmosphere until it rains down over the eastern Pacific Ocean, where microorganisms transform it into methyl mercury. It then moves up the food chain after being absorbed by plankton that is eaten by small fish. They are gobbled up by larger predators, as each bigger fish accumulates more mercury with every meal—including the halibut that was now tiring and allowing itself to be reeled in as Churchman leaned far over the gunwales with a net.

In my "before" test for mercury, I had registered a level of 4 micrograms per liter (μg/l), safely below the Environmental Protection Agency (EPA) threshold of 5.8 μg/l. (This is the same as saying parts per billion, a very small amount.) In a previous "before and after" test, I had registered 5 μg/l, the difference coming because mercury, like all toxins, stays inside the average person only as long as it takes the body to expel it—a process that depends on the chemical and on people's different physiologies. For instance, phthalates, which among other things make plastic soft and malleable, dissipate in the human body in only a few hours. Other chemicals, once on board, stay for years, such as perfluorinated acids (PFOAs), the hard, nonstick material in Teflon and formerly in Scotchgard fabric protector. Mercury stays in most people's bodies for 30 to 40 days, although constant exposure to the toxin means that levels re-up each day, keeping a steady background level of the sort that I had registered. Since I had not consumed fish in several weeks when I had my first round of blood and urine collected—fish are the major source of mercury exposure in humans—my level most likely came from the air and the water and from eating other food.

I planned to do much more than a simple before-and-after test. In a first-time experiment for an individual, I was going to investigate how well my body defends itself from the harmful effects of mercury. Not everyone responds to methyl mercury in the same way. Some people take more than the usual 30 to 40 days to expel this noxious metal from their bodies, with a few taking up to 190 days, which greatly increases their chances of suffering permanent physical and mental damage. These differences suggest a strong genetic component, says Karin Broberg, an environmental toxicologist who specializes in mercury exposure at Sweden's Lund University. This led me to investigate what is known about genes that make one susceptible to methyl mercury and to see whether I could track down genes inside my body that might confer a higher-than-average risk.

My methyl mercury tests are the opening salvos in a large battery of tests I plan to take for this book, exploring four major areas: genes, environment, brain, and body. In essence, I hope to answer two questions: how healthy am I, and what can the seemingly endless profusion of new high-tech tests for various diseases and traits tell me about my health in the present and for the future?

Wondering when we might get sick or die, which is at the heart of question one, has always been part of the human condition. This goes for people with no obvious ailments as well as for the ill. The second question is made possible only by a new precision in acquiring information about ourselves that was unavailable until recently, when technology began to produce its contemporary wonders, from the X-ray machine, invented more than a century ago, to the cracking of the human genome,* which was completed in 2003.

These discoveries and many others have revolutionized modern medicine, allowing physicians to give more accurate diagnoses and researchers to have better tools to delve into the secrets of how the body works. Technological advances also usher us closer to the long-sought goal of truly personalized medicine, where diagnoses and treatments—everything from diet and exercise to pills and surgery—will be customized to individuals based on their own genes, environmental input, and other factors specific to them. This is a radical departure for a medical system that has been largely based on lumping people into groups by age and sex and by such broad factors as whether one smokes or drinks. These categones are important but do not offer truly individual insight.

---

*For a definition of "genome" and other scientific terms, please consult the glossary.

We are witnessing the beginning of a new era of medicine, when healthy people will get a personal snapshot of their bodies—organs, cells, DNA, proteins, and a whole molecular universe of other tiny structures—cross-referenced with environmental input that ranges from heavy metals and other pollutants to stress, food, gravity, and ultraviolet rays from the sun. Changes in diet and lifestyle, in medications, and in other treatments will be tailored to an individual's specific profile, instead of physicians resorting to the one-size-fits-all medicine that is often practiced today.

A raft of new technologies will help make this happen, creating a world where a doctor's exam will include a quick scan of our bodies that reveals hundreds or thousands of bits of data seamlessly integrated by a computer into a health scorecard. (You Trekkies, think of the sick bay on the starship *Enterprise*). Or maybe we'll have our own handheld devices—let's call it an iHealth (with apologies, or perhaps a suggestion, to Apple)—that will keep track of our genomes plus the most recent scans of our brain and body, while inputting real-time environmental data about what we are exposed to as we walk around, eat, and work: levels of mercury and benzene, say, and exposure to UV rays. This information will be synced up at home with sophisticated biomonitors that record daily levels of thousands of chemicals, proteins, and other substances inside us. Our iHealth will download the data, calculate our current health, and determine up-to-the-minute probabilities of acquiring various diseases and exposures, while assessing risks all day long for everything from walking out of our front door into an outside environment teeming with hidden chemicals to eating a steaming piece of halibut. While we're waiting for all of the data to sync, we can play a game, check our e-mail, or watch a video on a futuristic version of YouTube.

The impact on humanity of such a device, and the intimate information it will provide, cannot be fathomed, much as people in the mid-twentieth century had only an inkling of the affects antibiotics, when they were invented in the 1940s, would have on future generations. They could not have imagined that diseases such as tuberculosis and whooping cough that terrified them and cut lives short would largely disappear in the West, and that millions of people would remain alive and vibrant (and receiving Social Security) well into their seventies and eighties. I suspect some people already inclined to fret over their health will be terrified to go outside while others will love having the information. Everyone will worry about health hackers and new forms of identity theft that make today's fears seem quaint by comparison.

My experiment aims to explore clues hidden deep inside a typical, healthy person—me—that might flag a future heart attack, diabetes, or dementia. It explores the biology behind behaviors in my head, such as risk taking, greed, anxiety, and even religious belief—and what in the environment might trigger certain actions and feelings or some dreadful neural malady. Most of the tests I have taken are in the early stages of reliability and offer an incomplete assessment of an individual's health. But the potential is there. Most of them are also unavailable to the public. I had to persuade dozens of labs and companies to allow me to partake. A few did so only after I agreed to underscore that their research is still preliminary; a very few said no thanks because of cost or a lack of time in their schedules, or because their experimental protocols required that their subjects, and results, stay anonymous. A few worried that erroneous or preliminary results might cause me unnecessary anxiety or even harm if I am compelled to act based on a sketchy finding—such as taking a drug that has side effects.

I can reveal enough of my findings to confirm that I am not a lunatic for wanting to subject myself to this investigation, inasmuch as modern medical science is able to determine such things. Yet I seriously wonder whether I possess an undiscovered Pandora gene that makes me insatiably curious in a manner that might get me in trouble. I do have a DNA marker on the DRD4 gene that supposedly predisposes me to taking risks, although the study that made the association between this gene and risky behavior is preliminary and needs to be confirmed. Another gene that investigators might want to look for is a narcissism gene—a hypothetical patch of DNA that predisposes humans to gaze back lovingly at their own reflections, as the original Narcissus did in the ancient Greek myth. My investigation, however, is not intended to be an exercise in self-absorption—though for some, indulging in a high-tech perusal of their body might be. I plan to peek beneath my reflection, an exercise that I suspect will reveal unpleasant realities about how I might get sick or die and undoubtedly a blemish or two.

Another motivation I have for writing this book is frustration. As a journalist trying to explain science to nonscientists like myself, I've noticed a tendency for readers' attention to stray when stories get overly technical or abstract. *Experimental Man* is an effort to humanize science by having a real person with a family and children intimately participate in leading-edge technologies, using the experience both as a tool to better understand the science and also to assess the usefulness of tests that will be available to everyone in the next few years.

We are about to acquire profound new powers of knowledge about ourselves, possibly more than we want to know. This data will impact and change lives; it will alter personal and family dynamics; it will confront us with ethical dilemmas, such as how we keep information private and who gets to use new medicines and other goodies that might extend the human life span or cure cancer. Like Pandora, it might also unleash plagues and monsters, although I don't think it has to.

Back on the *Osprey*, Josh Churchman scoops up my halibut in his net and drops it on the deck. After he stabs it and drains some of its blood, we fish for another hour or so amid whitecaps and a steady, chilly wind before heading back to Bolinas Lagoon with a second halibut and a rockfish. As the little boat rides up and down over the waves and the twin outboard engines roar, I wonder what my tests will reveal about my susceptibility to mercury. Do I have a super-gene deep inside me to fend off heavy metals or not? I know that I will die of something, someday, but I seldom give it much thought. Yet I wonder, Have I been born with the best genetic and physiological armor that evolution can provide?

A few days later I eat the halibut I caught, cooked with melted butter and basil, and later the same day the swordfish steak I bought, grilled with lemon juice. The next morning I have another 9 mm of blood drawn and give up another container of urine at the University of California at San Francisco, where my internist, Josh Adler, is the physician overseeing the testing for this book. I also put uneaten chunks of both fish on ice and mail them in a cooler to environmental biologist Robert Taylor at Texas A&M to measure how much mercury each has inside its body.

Soon after, Josh Adler e-mails me the test results. With just those two meals, my mercury level (most of which is probably methyl mercury) has spiked upward from 4 µg/l to 14 µg/l, more than doubling to well over the EPA's recommended level of 5.8 µg/l. These results were even more dramatic than when I ran the same test on store-bought fish caught in the Pacific for a 2006 *National Geographic* story on environmental toxins. That before-and-after test took me from

5 μg/l to 12 μg/l, a bump up that prompted pediatrician and mercury expert Leo Trasande of the Mount Sinai Medical Center in New York City to scold me about running a "fish gorge" experiment.

"No amount of mercury is really safe," Trasande says, although my results as a fifty-year-old white male are far less significant than they would be for children or for women who are of childbearing age. "Children have suffered losses in IQ at 5.8 micrograms."

After my first "gorge," Trasande had advised me to avoid repeating the experiment. I didn't tell him that I did it again.

The fishes' exposure levels came later from Texas A&M's Robert Taylor. My halibut carried more than three times the average level recorded in a 1997 FDA investigation. The swordfish registered a level that is two and a half times above the average for swordfish in the same study.

Armed with my onboard methyl mercury data, I go hunting for genes tucked into my cells that might influence whether I will follow Leo Trasande's advice on eating large fish in the future. This search begins with an e-mail to Trasande, who tells me that as a clinician he is not aware of human genes that are affected by mercury or of tests to determine a patient's genetic proclivities for coping with heavy metals. So I turn to animal toxicologists, who have identified several relevant genes in rodents, fish, dogs, dolphins, chickens, and fruit flies. Matthew Rand, a mercury toxicologist at the University of Vermont, has shown in flies that mercury binds to cells, including neurons, and interferes with receptors that send signals to the cell controlling how it develops, replicates, and dies. Rand says that Trasande is right about the lack of studies on mercury-gene interactions in humans, although he later corrects himself when he recalls a study by Karin Broberg and her lab in Sweden. In 2004, her team ran a study on 365 people to see whether ethyl mercury, a close cousin of methyl mercury, caused a reaction in several target genes that are involved with ridding the body of unwanted chemicals. She concluded that mutant variations of two genes did affect a critical system for flushing cells of toxic metals, such as mercury, cadmium, and arsenic.

Called GCL and GSTP1, these genes help produce enzymes such as glutathione-S-transferase that maintain levels of glutathione, a first line of defense in cells for expelling metals, among other things. Too little glutathione causes metals to stay in cells longer, according to Broberg, adding to the potential for damage, particularly in neurons. "These findings suggest that GCL polymorphisms [variatons] that affect glutathione production also affect methyl mercury retention," she wrote me in an e-mail, "and that GSTP1 may play a role in conjugating [chemically joining] methyl mercury with glutathione." Broberg's lab has identified a specific spot within the DNA sequence of GSTP1 that when mutated signals a slower elimination of ethyl mercury. In genetic-speak, this means that a single letter in the genetic code—in the Gs, Ts, Cs, and As that are used like an alphabet to describe the DNA inside us—is probably different for people who produce less glutathione. They have an "A" where most people have a "G" and presumably expel mercury at the normal rate.

Fortunately, when I checked my genetic results, I came up with the normal "G," meaning that my cells most likely expel methyl mercury in the 30- to 40-day range, rather than in the more dangerous 190-day end of the spectrum.

Even if my genes did contain the "bad" variation for GSTP1, there is a simple solution to avoiding this toxin: limit my consumption of large, predatory fish. According to experts, small fish have less mercury and are probably safer. Older fish also have higher concentrations of mercury. Unfortunately, eliminating or reducing the source isn't possible for most other pollutants that we breathe, eat, drink, and absorb through our skin, whether we want to or not, including new manmade chemicals such as the phthalates I mentioned and the PFOAs in Teflon These and thousands of other manufactured compounds we use in everyday products do not appear in nature and have entered our environment so recently that our genes, cells, brains, and bodies have not yet evolved specific mechanisms for coping with them. I suspect that some of us will develop adaptive defenses for these chemicals. In the meantime, we need to better understand the genes, the neural architecture, and the organs each of us has been born with: our strengths and vulnerabilities as our bodies and minds are daily faced with environmental input that ranges from trace levels of metals to the stresses and joys of life that make us happy, blue, crazy, angry, and ecstatic.

This is my journey in this book: to discover what one person can find out about his body as it interacts with the world, including, as I have my DNA scrutinized and my brain scanned, why I might be crazy enough to want to take all of these tests in the first place.

Your host

I'd like to introduce you to the main character in this book: me. Not *me* in the sense that I am going to bore you with details of my modest accomplishments. I have not conquered countries, held high office, or founded a multibillion-dollar company. The *me* here is the physical person who is not too different from you, the parts that one might disassemble like a machine to see how the programming and all of the pieces are working as I turn fifty years old.

Your host—the human guinea pig telling this story—is a father of three children, ages fourteen, twenty, and twenty-two; he is an older brother to his only sibling and the son of a mother and a father in their late seventies who are both in good health. This book is in part their story, too, the connection to a continuum of generations that stretches far back in history and will project forward into the future through my daughter and two sons. To make this link for *Experimental Man*, I have had genetic profiles run of my parents; my forty-nine-year-old brother, Donald; and my twenty-year-old daughter, Danielle.

But this physiological and hereditary "me"—the bundle of tissue, organs, sensors of environmental cues, cells, and genes—isn't the full story I plan to tell. This is the functional part, the mechanistic me. This is also an experiment about what happens to my mind (as opposed to my physical brain—the tissue that will be tested) and my emotions and to my conception of self as I learn the results of my investigation—if I have a reaction at all. This is the mental "me" that gets up each morning with an innate sense of who I am, how I feel about myself, and my place in the world: that package of hopes, fears, euphoria, loves, hates, needs, and resolutions that shifts and changes and underlies my days and nights as I interact with the world. As Fyodor Dostoyevsky wrote in *Notes from Underground*: "There are . . . things which a man is afraid to tell even to himself, and every decent man has a number of such things stored away in his mind."

Underneath this fluctuating interface with our environment and other people is a multilayered self-conception that each of us has deep down, aspects that we like and hate about ourselves: weaknesses, anxieties, and hidden reserves of strength. We are healthy or sickly, risk takers or not, shy or outgoing, generous or miserly. The outer layer of our self-conception is what we prefer to present to the world—our cool and confident public persona. The next layer down consists of the "selves" we share with our families and friends. These can include behaviors that we are proud of but also those that we view as deficits, such as a quick temper or a lack of humor, that others either like about us or tolerate or loathe. Deepest of all are the selves that we share with no one—our secret selves, our ids that ruminate about failures, expectations dashed, loves lost, flaws, and proclivities that we don't want to admit even to ourselves.

One of my hidden secrets since I was a boy is an intense and occasionally debilitating social anxiety, a shyness and a fear that I'll say or do something that will make me seem like an idiot—or that will upset or disappoint people I care about. In my early teens, I was terrified to raise my hand in class, and several times I vomited from nerves before or after piano recitals and plays in which I was acting. When I was a young adult, I fumbled and fidgeted around women I was attracted to and men whom I wanted to impress, and I felt the blood rushing through my head when I needed to be smart and articulate with teachers, my parents' friends, and, later, as a young journalist, with editors and subjects. Possibly, these anxieties were normal, but to me, they were sometimes terrible. And yet I also have extrovert tendencies and have spent my life taking risks professionally as a reporter and physically as a foreign correspondent who has covered conflicts and lived in dangerous areas of the world. In my twenties, I bicycled around the world and then from Cape Town to Cairo in Africa. This seeming contradiction of the shy risk taker has been reconciled over years of forcing my timid side to give way to the extrovert. Today I regularly give lectures, and I love nothing more than visiting tumultuous areas of the world or hurtling down a steep mountain on skis. I have learned to overcome my anxieties, yet they are still there, buried deep inside, along with a now low-grade fear that they will emerge at some inopportune moment that will cause me great embarrassment.

The one person who knows the truth is my mother, who suffers from more severe anxiety and occasionally has panic attacks that require medication. We will investigate whether her genes are the source of my own milder condition.

I relate this not because anyone should care about my details, but because we all have inner secrets that we closely guard, fearing they will be discovered. I'm mentioning it now because, as I contemplate my tests, I am rather absurdly anxious about being anxious. I feel old fears rising that the tests might expose an abnormality in my genes or my brain that will disturb or shatter the carefully wrought self-conception—or self-deception—that my anxious days are over.

Another thought as I launch this project is my oft-stated contention that I am a fatalist who gives little thought to how and when I will get ill or die. But is this true? Am I really so cavalier about my future health, figuring that when my time comes, it comes, and there is no point in worrying? Or is this more self-delusion? I come from a mostly long-lived family that suffers from few obvious genetic diseases. My grandmothers died well into their eighties. My father's father died in 1972 at age sixty-seven, after having been misdiagnosed with non-Hodgkin's lymphoma. He was subjected to the crude and highly toxic chemotherapy of that day and died of a heart attack, probably exacerbated by the treatments. My mother's father died of a rare form of small bowel cancer at age sixty-eight. Several aunts and uncles in my grandparents' generation lived past eighty and some past ninety, with a brother of each grandfather living to age ninety-three. This was when life expectancy was less than it is today. A great-great-aunt named Effie on my father's side lived to be 102 years old—I visited her in a retirement home when I was small, and she gave me homemade cupcakes. I had a great-aunt who lived into her nineties who suffered from dementia that was probably Alzheimer's disease and one ancestor in my great-grandparents' generation who died young of cancer, but she lived long ago.

Other than my brother—the sole member of my family afflicted by a genetic disorder that grew serious only in recent years, as I'll explain later in this book—the general good health of my loved ones, especially my parents, has enabled me to form a powerful self-belief about my physical well-being and an anticipated long life. If I had come from a family with a history of grave illness or relatives who died young, I would have a very different outlook. I have a friend who tells me he wakes up every morning convinced he will die that day or perhaps the next, or that he will be diagnosed with a fatal ailment.

Both of his parents died young—his father of a heart attack and his mother of breast cancer. He says he would be very selective about the tests I'm taking for *Experimental Man*. "I might be okay with taking tests for diseases that have good treatments," he told me. "But tests for diseases that have no treatment? Forget it. I don't want to know."

Geneticist Jonathan Rothberg, a pioneer of gene testing and sequencing and the founder of several biotech companies, has Huntington's disease running in his family. He tested negative, but his cousin did not—and committed suicide rather than face a disease that as a carrier she had a 100 percent chance of contracting as she grew older. Most DNA tests do not have this much predictive power, however; a "positive" for a mutation in most cases means an increase in risk, not that a person will actually get a disease.

Despite my nonchalance, I like to feel healthy, and I have always been mildly obsessed with exercising regularly. This has less to do with a sense of mortality than a realization that I feel out of sorts if I go a day or two without riding my bicycle or visiting a gym. As I grow older, I have given more than a passing thought to what I eat, having spent the last few years casually following a low-carb, high-protein diet that I have discovered during this investigation might one day kill me, given my genetic and physiological make-up. I have since discontinued this diet, a story I'll describe later in the book.

Susan Sontag, in *Illness as Metaphor*, wrote that illness is as much a part of life as good health is. "Illness is the night-side of life, a more onerous citizenship," she wrote. "Everyone who is born holds dual citizenship, in the kingdom of the well and in the kingdom of the sick. Although we all prefer to use only the good passport, sooner or later each of us is obliged, at least for a spell, to identify as ourselves citizens of that other place."

As your guide on a voyage into the day-side of Sontag's equation, the part of life that is healthy, I will explore secrets and clues, ultimately trying to see whether it is possible to predict when a shadow might fall on the kingdom of the well I've been lucky enough to inhabit for fifty years. Thus, the night-side of Sontag's dual kingdom will hang over the narrative and in many ways drive it. As Sontag says, no one has yet figured out how to deny the night-side of our citizenship, although at the end of this book I will discuss the possibility that science may be able to extend the daylight, perhaps for a very long time.

Checkup with my internist (the plan and three rules)

Before plunging into my tests, I visited my internist for a routine exam to establish a baseline of my health according to today's standard practices. On a bright day in June I met with Josh Adler, a fortyish general practitioner on staff at the University of California at San Francisco. He is also the medical director of its ambulatory care clinic, a building crowded with patients, white-coated physicians, nurses, and attendants pushing wheelchairs. Some patients look ill, others nervous, several healthy, and a few bored with waiting. Intravenous devices on stands, portable X-ray machines, defibrillators, and computers tucked here and there in hallways and on desks barely hint at the high-tech arsenal of cutting-edge technology available at this world-class medical center in the early twenty-first century.

Behind closed doors, deeper in the complex, are magnetic resonance imaging machines, catheterization labs, and high-resolution ultrasound scanners. But virtually all of this is reserved for the ill or those with current medical conditions. For the healthy, the procedure for getting a physical isn't markedly different from what my grandparents and perhaps their grandparents would have recognized: a small room smelling faintly of antiseptic with a blood pressure device and the stethoscope wrapped around a doctor's neck. One small difference is that Josh uses a digital thermometer to take my temperature, which is a normal 97.8 degrees Fahrenheit.

Josh's exam room is on an upper floor of a clinic built on a ridge with a dramatic view of Golden Gate Park and the Pacific Ocean. On a very clear day, the floor-to-ceiling windows give an impression of a building floating high above the treetops in the park. Far off in the distance to the north I can see the light gray cliffs of Bolinas, where a few weeks later I will go fishing with Josh Churchman.

Josh Adler says hello and leads me into the exam room. One anomaly in this physical is the extra time he gives me to chat about my project. Normally, in this age of managed care, he would have to be thorough but quick. I will be returning to him several times during the next year to show him my Experimental Man results and to get

his impression as an internist on the front lines of medicine. What did he think was useful in my tests? What had he been trained to deal with, and what were the possible dangers in any of the testing or in the results?

Josh is lean, with longish unkempt hair, clothes that look slightly frumpy, and glasses with a frame the size of those old "aviators" people wore in the 1970s. He exudes empathy and intelligence, which combine with his slightly rumpled look to make me, as a patient, feel as if I have a doctor who cares far more about me than about fashion and will be there for me even late at night if I need him. Josh smiles a lot and patiently waits for me to finish talking. Unlike those physicians who seem harassed and busy, he is calm and unhurried.

Josh has strong opinions about my project. Mostly, he thinks I am delving into technologies and tests that are still a work in progress and are not yet ready to be useful for healthy patients. Like many physicians, he hopes the project will be helpful as an investigation into the state of the technology, but he thinks I'll find little of value.

"I think the technology is exciting, but our ability right now to collect data far exceeds our ability to know what to do with such information. We are in a phase where we are able to collect data about human beings without any real sense of how to use this information in a way that would help a person or change his or her life. For instance, I'm trying to figure out how to guide my patients through certain genetic tests, which produce results that are sometimes hard to interpret."

"Do you have people asking you about genetic tests?"

"Not that many. I do have people asking me about genes for Alzheimer's, breast cancer, and colon cancer—I think there is some science behind these tests and a lot of media attention, so people ask about them. What people really want to know is that they're not going to get these diseases. But very few of these tests can predict that."

Take the BRCA tests for breast cancer, he says. Mutations of the BRCA1 and BRCA2 genes are carried by 5 to 10 percent of breast cancer patients. Patients with breast cancer in their families often take the tests, but having the gene does not mean a person will ever get breast cancer. Also, 90 to 95 percent of the people with the disease do not carry the gene. "Testing positive for the BRCA genes gives a person an increased risk of getting breast cancer," says Josh. "Trying to explain this to people is difficult, and to help them make choices about what to do. In many cases, if there is no physical sign yet of the disease, it is helpful because we can keep a close eye to see whether it develops."

Many other tests are next to useless, he says. "Most of them are association studies that give a person a slightly elevated risk factor for something like diabetes. But I can tell you that without a genetic test. A person's diet, age, and family history tell me more than a genetic test. What people want is a yes-no answer, and they will not get this with these tests."

Josh is wrapping a blood pressure cuff around my arm as he talks. He pumps it up with air and watches the metallic liquid in the device rise and fall.

"Your blood pressure is slightly elevated," he says. "But that's not unusual when people come to visit their doctors. On the other hand, you're reaching an age when blood pressure begins to go up in about twenty-five percent of people. We'll keep an eye on it."

"Is there any danger in taking these tests?" I ask him as he begins my ear exam.

"The genetic tests aren't dangerous," he says, "but the scanning could be. You have a small radiation risk with some of the CT scans. But the real danger involves taking the next step beyond the test. What happens with most of these tests is that we won't know what they mean, and there will be the possibility that there is something serious. You might come to me with a CT scan with a nodule on your liver. Could it be cancer? It's unlikely, but we don't know. The next step to find out for sure would be a biopsy, and that could be dangerous— there is a small chance of infection, bleeding, or accidently making a perforation, like poking a lung. So I see my job here as protecting you from getting hurt."

"To make sure you as the physician do no harm, as Hippocrates said?"

"Exactly."

He tells me that my ears look fine and checks my eyes and throat. He asks me to lie down on my back.

"Turn your head on the side," he says. "This is a very low-tech test to see if there is pressure in your jugular vein, which would indicate a problem with your heart." He listens for a moment.

"It's particularly important to contrast the tests you're taking with the more obvious and simple aspects of health care," he continues as I sit up. "You don't need a genetic test to tell you to eat healthy—though some people may need a test to convince them to do this. Don't smoke, and be sure to exercise, get a good night's sleep, eat plenty of vegetables. These are things that really do make a difference."

"And we don't need those fancy tests to tell us this," I comment. "Which makes me wonder if after all of these tests, I'll basically learn what I already know: that I should eat right, exercise, and sleep well."

"It wouldn't surprise me," he says.

We're both quiet again as he uses his—very cold!—stethoscope to listen to my insides. "Breathe deep and hold it," he says several times, doing the "hmmm" routine that doctors must have done to the mild frustration of patients since the invention of the stethoscope. About this time, Josh is called out of the room for a moment, and I'm left alone in my boxer shorts sitting in that tiny room. It is an odd moment that leaves me feeling abruptly vulnerable. I am in my underwear in a strange little room with no windows, the supplicant to the learned man in the white coat who at this moment has an enormous amount of power over me, to inform me whether he thinks I am sick. He is the augur in this room, the expert who can with a few words about an unexpected finding change my life and my conception of myself.

My confidence in my health and in running tests for this book is dipping, and I feel a pang of apprehension—an ever-so-slight wish not to be here, to take the risk that my personal vision of myself might be challenged.

Josh returns and apologizes for having to step out. He tells me that everything looks fine—so far. "We'll get the chemistry back in a few days, but I don't expect much there, either."

I let myself exhale and then jump back into reporter mode, remarking that the exam seemed remarkably low tech. Other than the stethoscope and the blood pressure device, much of the exam could have been done by the ancient Greeks.

"Perhaps, though I'd like to think we have learned a few things since then. It's more along the lines of an exam that developed in the past two or three hundred years, with much knowledge added since then, though the critical part of the exam is still the patient history—the conversation about how you feel, your family's history of illness, and so forth. This leads to about seventy-five percent of the diagnoses we make. The rest of it, including blood tests and the rest, is a lesser part of it."

"Do you think that will change with all of this new technology and knowledge?"

"It already has changed in certain arenas, in diagnosing things like prostate cancer or presymptomatic diabetes with various blood tests. Although in the history we try to identify who is at risk."

Josh had already taken most of my medical history and had come up with an unremarkable story that says I'm basically healthy. The only significant ailment I have had is a disk in my back that was herniated twelve years ago. They didn't operate, I tell Josh, but it took me six months to recover with physical therapy, and the back still bothers me now and then.

"So, Doc, what is my prognosis?" I ask when he is finished.

"Based on your physical and your family history, you are not at risk for any major disease that I can identify," he says. "Everything was normal in your exam, except your blood pressure—which we will watch, though I'm not overly concerned. Otherwise, the prognosis for you living a long and healthy life is quite good."

Josh orders a typical regimen of tests: white blood cell count, hematocrit count, and blood sugar. He suggests an EKG for my heart, which uses electrodes attached to my chest to check the electrical action of my beating heart, just to be sure the blood pressure isn't indicating something more serious than he thinks.

"At this point, you could order many more tests if you wanted to, right?" I ask. "Setting aside cost for a moment, why wouldn't you test me for anything that's not dangerous?"

"As I said, the only important clues in this exam would be from the family history," he says. "There also is no evidence that you are functionally declining. Beyond this, there is no way to know your risk, except as an average risk. I don't see a need to subject you to endless batteries of tests for no reason. It is costly, and we need to reserve them for people who really need them."

A few days later I get back my lab results, and everything is normal except my cholesterol. It's 209—slightly over the threshold of normal, which is 200 or less. Josh says not to worry. "We'll watch it," he says. "Cut back on meat and fatty foods."

"So I'm still healthy?" I ask, feeling that vulnerability creeping in again.

"You are not going to die today."

Fundamentally, I trust Josh Adler that I'm fine—a prognosis that fits in nicely with my core belief that I'm healthy. But I'm about to find out much more about myself than Josh can tell me—information that may give me some useful advance warning about Sontag's nighttime side of life but might also frighten or confuse me. For some people, knowing vast quantities of information, much of it incomplete and a work in progress, about risk factors and possible outcomes could plunge them into a kind of twilight between the kingdoms of the sick and the well.

The plan of my investigation and of this book is divided into four parts—genes, environment, brain, and body, with a short epilogue called "eternity" that will have a surprise result for me concerning my longevity and will assess technologies that may radically increase life span. Each of these sections will contain personal stories that could be your stories, too—my life's experience as a human organism and how the secrets inside my genes, cells, and organs have influenced my own and my family's lives. I'll weave in as much science as I think the reader can bear, while also pondering the usefulness and meaning of my results, the science, and what they might mean for society.

Section 1, about genes, will describe the current state of the science for hunting down genes and DNA markers that have been associated with diseases and other traits. These are variations from the norm that contribute to why some of us get sick from a specific disease and others don't. I will be tested for millions of genetic markers and hundreds of genes and will find out what mutations I have, along with my parents, my brother, and my daughter. I will also delve into genes that link me to my immediate past with my ancestors and to more ancient and primordial times, as the history inside my genome spirals backward in time to a DNA record of the earliest forms of life.

In section 2, I'll test levels of hundreds of manmade chemicals that I may have accumulated just from living on Earth—pesticides, plasticizers, flame retardants, heavy metals, and much more. I'll assess the impact of stress and the wear and tear on my body of living in the twenty-first century. As I did in my fish story, I'll delve into how these environmental factors interact with my genes and my body to create a profile of my defenses against potentially toxic input. I'll travel to places where I have lived and may have been exposed to certain chemicals, and I'll spend time trying to collect data as best I can on the environmental factors I have encountered.

In section 3, I'll peer inside my brain and have scientists whom I'm working with build a schematic of my brain's architecture. I'll put my head into magnetic resonance imaging machines—MRIs—logging more than twenty hours in the coffinlike tubes of these scanners, while neuroscientists run amok inside my noggin, testing me first for diseases such as Alzheimer's and then for how my brain responds to everything from fear and anxiety to hip-hop and Beethoven. I'll be tested for, among other things, my sleep patterns and cognition levels and how my brain makes decisions, takes risks, and even believes or doesn't believe in a supreme deity.

Section 4 will link together the previous three sections and will add results from tests, scans, and analyses of my body, including a full-body computed tomography (CT) scan and even a breakdown of the proteins in my blood. I asked several researchers to run my tests with sophisticated programs that model my heart to predict future disease, such as when I might get a heart attack. Assembling the pieces of the puzzle that is me—all of the tests—is an attempt to take a holistic look at a single person. Because most of the tests come out of disciplines that focus on one field, such as genetics or neuroscience, I didn't expect that much of this would fit together. But it did in a preliminary way, an example being the fish experiment, which incorporated genes and the environment and could have also tapped into my brain if I had wanted to look for neural damage from mercury. Although I'm willing to get virtually any test, this seemed unnecessary, given my minute exposure levels and my thankfully normal genetic profile for the GCL and GSTP1 genes.

I started with three rules for this project, following Josh Adler's advice. First, all tests would be done under the supervision or with the knowledge of a physician. Second, if my tests revealed any medical condition that needed to be acted on, I would follow through with appropriate therapies and interventions. And finally, my tests would be passive and cause me no physical harm. I conducted no self-experiments of the sort performed by Australian scientist Barry Marshall, who swallowed a broth of bacteria to prove that these critters cause ulcers, a stunt that proved his hypothesis, garnered him a Nobel Prize—and gave him ulcers, which were cured using antibiotics. Nor am I trying to emulate John Paul Stapp, who rode in a rocket sled traveling at the speed of sound and then abruptly stopped to test the impact of great force on the human body. The most dangerous test

1

GENES

Know then thyself, presume not
God to scan . . .

—ALEXANDER POPE

Not a genetic virgin

I lost my genetic virginity in 2001 on a typically bright, still afternoon in La Jolla, California, near San Diego. I could see the heat wilting a stand of palms outside the office where I was spending my last minutes of complete ignorance about my genes—those specific combinations of genetic markers inside my cells that might reveal a proclivity for a future disease or a behavioral flaw. For the forty-three years of my life up until that point, this information had been as hidden and secret for me as it had for nearly every human in history.

A few weeks earlier, I had given up five 9 milliliter vials of blood, less than 2 ounces, to a La Jolla start-up called Sequenom. Its scientists had teased out hundreds of my genetic markers—my As, Ts, Gs, and Cs—in a lab down the hall in this then-new, silver-metallic headquarters perched on a desert mesa close by the sea. The year 2001 was a time that today seems Jurassic, given the lightning-quick pace of genomics. Just a year earlier, President Bill Clinton had announced the completion of a rough draft of the human genome in a ceremony at the White House. The final draft would not be finished for two more years, in 2003.

I was working on a story for *Wired* magazine explaining this newfangled field called genomics, and I had come up with the idea of having myself tested and publicly revealing my results—something that had not yet been done by anyone. (Craig Venter had not yet revealed that the "anonymous" DNA his former company, Celera Genomics, had sequenced for the Human Genome Project was actually his own.) The idea had seemed silly at first, a gimmick. My job was to report on what scientists did, not to inject myself into the action. Yet I hoped that by having a real person take the tests, readers would connect with this highly abstract new science that involved such novel concepts to the public as nucleotides, DNA code, genes, and amino acids.

Sitting across a desk that day was Andi Braun, then Sequenom's chief medical officer. Tall and sinewy, with a long neck, glasses, and short gray hair, the then forty-six-year-old Braun was jovial, with a light German accent, as he called up my test results on his computer. I tried to maintain a steely, reportorial facade, but my heart raced just a bit, since the tests involved a range of frightening diseases that I might have a genetic risk factor for—or not.

Braun turned his monitor so that I could see it, and I read names of genes popping up on the screen: connexin 26, implicated in hearing loss; factor-V leiden, associated with blood clots; and alpha-1 antitrypsin deficiency, linked to lung and liver disease. Beside each gene was the location of a DNA marker that scientists had linked to a risk factor for these diseases: 13q11-q12, 1q23, 14q32.1. Braun explained that these are addresses on the human genome, the "PO box numbers of life." For instance, 1q23 is the address for a gene marker that when mutant can cause vessels to shrink and impede the flow of blood; it's on chromosome 1. (Humans have twenty-three paired chromosomes.) Thankfully, my result for this awful-sounding malady was negative.

"So, David, you will not get the varicose veins. That's good, *ja*?" said Braun.

Next up was the hemochromatosis gene. This causes one's blood to retain too much iron, which can damage the liver. As Braun explained it, somewhere in the past, an isolated human community lived in an area where food was low in iron. Those who developed a mutation that stored high levels of iron survived, and those who didn't became anemic and died, failing to reproduce. Now that most people get plenty of iron, however, hemochromatosis is a liability. The treatment? Regular bleeding.

"You tested negative for this mutation," said Braun. "You do not have to be bled."

I was also clean for cystic fibrosis and for a genetic marker connected to lung cancer.

Then came the bad news. A line of results on Braun's monitor showed up red and was marked "MT," for "mutant type," for a gene called ACE (for angiotensin-I converting enzyme). Many of my friends had long suspected that I was a mutant, but in this case it meant that my body makes an enzyme linked to high blood pressure. In plain English, I was a potential heart attack risk, if this marker was to be believed. Then a second red "MT" popped up on Braun's screen: another high blood pressure mutation. My other cardiac indicators were okay, which was good news, although I remember being surprised that I had any bum genes at all. I had been told that everyone has them, since we will all die of something, but it hadn't occurred to me that anything could be wrong with me. At the age of forty-three, I felt great—I still do at age fifty—and I came from a healthy enough family that I seldom thought about illness, disease, or death for myself.

In my *Wired* account of this scene in Braun's office, I wrote that I reacted to the bad news by wanting to find out everything I could about heart disease and those rascally ACE markers. The reality is that I didn't, in part because of my thickheadedness about believing I would never get sick or die, an attitude reinforced by what a Sequenom physician standing next to Braun told me that day in San Diego.

"These mutations are probably irrelevant for you," said Matthew McGinniss.

Braun agreed. "Given your family history, it's likely that you carry a gene that keeps these faulty ones from causing you trouble—DNA that we have not yet discovered."

I got more bad news that day: I don't have a marker called CCR5 that prevents me from acquiring HIV, should I indulge in unsafe sex; nor do I have one that seems to shield smokers from lung cancer.

"*Ja,* that's my favorite," said Braun, himself a smoker. "I wonder what Philip Morris would pay for that!"

At the time, Sequenom was exploring the idea of starting a business to test people for a string of genetic tests, something like what online genetics companies such as 23andMe and deCODEme are offering half a decade later. They even calculated a life score for me, based on genetic mutations that I have and don't have, and came up with a crude calculation of how these might affect my life span. My score was about 14 percent higher than the norm, indicating that I would live to be ninety-one years old, although, of course, this score was based on very preliminary studies that associated genetic markers with diseases. The score also failed to take into account, as McGinniss and Braun suggested, thousands of other genes and risk factors that might kill me either before or after the age of ninety-one. And then there was that piano that might fall on my head.

In 2004, I was tested again for different markers by another genomics company, deCode Genetics, based in Reykjavik, Iceland. Founded in 1996, deCode is a world leader in hunting down genes related to diseases that range from diabetes and heart disease to restless leg syndrome, discoveries that frequently land deCode on the front pages of newspapers around the world. Its approach has been to first search for genetic associations among the three hundred thousand people of Iceland—about 40 percent of the population (60 percent of the adult population) have given deCode their consent—and then to replicate these studies in larger populations such as North America. In 2007, the company launched deCODEme, an online genetic testing site that offers anyone who wants to pay $985 results on about twenty-nine genetic markers that are linked to diseases such as obesity and cancer.* DeCode is also developing drugs for stroke, heart attack, and Alzheimer's, although as of this writing, the company was facing financial difficulties and an uncertain future.

The company was founded by Kari Stefansson, a tall, charismatic Icelander with a white, pointed beard. He first made a name for himself as a neurologist at Harvard before returning home to start his company. Stefansson is brilliant and filled with an infectious passion; he also can be rude, loud, and demanding, a combination of attributes that he undoubtedly inherited from his Viking forebearers who sailed in longboats to Iceland and settled there in the ninth century.

I had shipped several vials of blood to Iceland, where Stefansson's geneticists cracked open my cells and extracted the DNA to test me for a new genetic marker the company had discovered that seemed to confer a higher than normal risk for stroke. On a gray, rainy summer day in Reykjavik, I arrived at deCode's headquarters on the edge of a lava field on this volcanic island to get my results, only to be invited by Stefansson to play one-on-one basketball at a Reykjavik gym.

"This is the time of day I get exercise. You need exercise. Listen to me, I am a doctor, and I know you need exercise," he said.

"Ah, okay," I stammered, not expecting this. I realized instantly as I sized up this man—he's 6 feet 5 inches, while I'm just over 6 feet and pathetic at basketball—that playing basketball with him would lead to my humiliation. As a journalist, I sometimes run into subjects who test my mettle, as if I need to prove my manhood to them. This might have been the case here—or possibly Stefansson just wanted a partner to play some hoops.

The game was predictably one-sided as we roared up and down a half-court, until at last I grabbed the ball and was in the air about to

---

* DeCode is continuously adding diseases and markers to its site.

make a basket. It was at this moment that Stefansson decided to give me a general impression of my genetic proclivities. He was behind me, pushing, as I was about to dunk the ball into the net, when he blurted out, "I have your DNA results."

"Yeah?" I said, suspended for a moment in the air, feeling that electric rush that said this ball was going to connect.

"You are genetically defective."

I hesitated for a split second, and he jumped up high, grabbed the ball, and raced down the court, dribbling and flashing me a maniacal Viking smile.

Back in Stefansson's office, he told me that he was just kidding on the court, although when he saw my test results, he was embarrassed to admit that I did indeed have a mutation that is associated with a higher incidence of stroke. He then explained to me what this meant, introducing me to the rapidly developing world of genetic testing, association studies, and risk factors as it had evolved even since the Sequenom experiment in 2001.

"We have established that you have a series of genetic markers that give you something like a two to seven times greater risk for developing a stroke than if you didn't. You have this entire haplotype"—a sequence of DNA that is passed down from one generation to another with little or no change—"so you probably have three times the risk. If this turns out to be the case in the American population, you are genetically predisposed to stroke."

I was surprised, responding that no one in my family had ever had a stroke, except my maternal grandmother, in her eighties.

"The only thing you have done is to inherit a predisposition," he said. "What does that mean, eventually? It means that if you stay in a certain environment, or if you are born in a certain environment, you will develop stroke. But you are not going to develop stroke, all right? You now know that you have three times the possibility of the average individual to develop stroke. So you have a strong incentive to take measures to prevent stroke. One of them is to make sure that you don't have high blood pressure; one of them is that you will not smoke. One of them is you will drink alcohol only moderately, because intake of large amounts of alcohol, binges, dramatically increase the probability that you will develop a stroke."

"But this genetic profile for stroke has not been tested for Americans. You've only tested Icelanders. Right?"

"Yes, before you can get too excited as an individual, you have to do a clinical trial in the population where you can use it, like in the American population," he said. "For some traits, ethnicity can be important. Some populations are at higher risk for some diseases."

"Those odds still make me want to go and have a drink."

"You cannot drink anymore."

That night, I met Stefansson for drinks at an Italian restaurant that served, among the usual pasta and veal, horsemeat, apparently an Icelandic specialty. After I drank enough red wine to give me a stroke for sure, Stefansson said goodnight and told me that all of that wine tonight would certainly kill me, that I would have a stroke by morning. He was kidding again, but as I walked to my hotel through the eerie lightness of the Icelandic "midnight sun," with the streets slick from dampness in the air and the distant volcanoes black and steaming, I wondered whether I should believe him.

I took a breath and decided that I felt great. I also reassured myself that I wasn't Icelandic, so the studies might not apply to me. Later, though, other genetic tests revealed that I have a rare DNA signature in my mitochondria that does connect me with Stefansson's Vikings. Mitochondria are structures in human cells that stay very stable over thousands of years and therefore can be used to trace a person's roots. This particular stretch of DNA, which I will describe in detail later, is not passed down in the usual manner as a mix of one's mother's and father's attributes. Rather, our mothers pass it to us in a matrilineal line. This makes sense for me, since my mother's family, through her mother, tracks back to Scotland, where Stefansson's Viking forebears grabbed many of the women they took with them to Iceland—including, perhaps, a long-ago ancestor of mine.

Predicting the future

Three years and another epoch or two later on the time line of genomic advances, I am back in Iceland to take more tests for the Experimental Man project. Within hours of landing, I find myself facing Kari Stefansson once again. This time he's holding not a basketball but a needle and a syringe.

Outside, fields of black volcanic rock stretch in all directions, and the sky boils with the same gray-white clouds I remember from my last trip. But I'm not thinking about the scenery as Kari Stefansson tightens a rubber strip around my lower-right bicep and searches for a vein. He and his team need a throbbing blue vessel to poke so that they can draw out enough blood to isolate a few white blood cells and to obtain yet another complete copy of my genome.

Tapping into my blood vessel is the first step for deCode's scientists to run my DNA through advanced machines that will eventually untangle more than one million genetic markers, or about .0001 percent of the nucleotides in my DNA. These markers are not complete genes, which are often composed of hundreds or thousands of nucleotides; they are individual letters within genes (and sometimes outside of genes) that appear in a single "base pair." DNA is made up of long, coiling strands of base pairs—written, for instance, as a pair of nucleotides that looks like this: AG or TC—that are attached to the double-helix superstructure of DNA like rungs that fit into a very long ladder. (Nucleotides are the individual As, Cs, Ts, and Gs in a genome.)

Most of the three billion base pairs in a human's DNA, some 99 percent, are identical with every other human's, but several million are not. That is, in one person, a specific letter in a certain base pair might be a G, and in another person, an A. These variations, called single nucleotide polymorphisms, or SNPs (pronounced "snips"), can increase one's chances of having blond hair or red hair, or of getting cancer or not. This is one form of a mutation, defined as any change, or divergence, in the nucleotide sequence of a gene. DeCode planned to scan my genome for about a million SNPs, which is one way of checking for large numbers of genetic differences without having to scan a person's entire three billion base pairs. Later, other companies and labs would scan me for millions more.

Like many physicians who think they can do almost anything, Kari Stefansson assumes that he can draw blood better than a trained nurse can. Earlier in the day, Stefansson jokingly dared me to let him draw my blood himself, even though he couldn't remember the last time he had drawn blood. I agreed, a decision I now regret, as a rather nervous-looking Viking geneticist waves a very sharp needle above a swelling vein. Thankfully, the "do no harm" side of "the doctor," as his close associates call him, kicks in, and he hands the syringe to a nearby phlebotomist. The nurse rolls her eyes at her boss and me. Taking the needle, she sticks it in my arm with such expertise that I barely feel it as she drains three vials of blood.

My gene tests are run on a flat chip called an "array," the size of a thick playing card: the Human Hap330. Made by San Diego–based Illumina, the array is one of several on the market that tags SNPs from an individual's DNA and identifies their location on that person's genome. Illumina's technology uses machines called "oligators" that synthesize millions of short, single-stranded fragments of DNA known as oligonucleotides, or oligos. Each oligo is attached to a glass bead that seeks out and latches onto a target sequence of a DNA sample. Lasers are used to identify a specific oligo, which corresponds to specific DNA sequences.

Later, Kari Stefansson will show me the fully loaded Illumina lab situated on deCode's ground floor, which will analyze my DNA. In a glass-encased room, robotic devices do most of the work, moving samples from station to station as machines douse my DNA with chemicals and run it through the array, the process monitored by computer screens with complicated, colorful readouts of numbers and other data. A few weeks later, I will visit Illumina in San Diego to be tested on an even more advanced array, which scans 1 million genetic markers. My DNA will also be parsed by other chips in this highly competitive industry, such as the Affymetrix Genome-Wide Human SNP Array 6.0, which tests for 1.6 million markers.

As the blood drains from my arm, I already know that few serious glitches will be found, since I have lived nearly fifty years in reasonably good health. Yet tucked inside my cells are more subtle mutations and combinations of mutations that are guaranteed to contribute to future illnesses and, eventually, to my demise. The question is: which mutations are there, and, more important, what could this information tell me about my future health?

It's the second part of this question that is the essence of genetic testing for healthy individuals. Predicting the future has been a core goal of genetics since Charles Darwin and others theorized that an inheritable substance passes traits from one generation to another: a substance in humans that can provide insight into who might go bald, grow tall and lean, or contract a dread disease. Reducing the risk of future maladies for groups of people and for individuals was also a major goal of the Human Genome Project and a key reason that

Congress allocated $2.7 billion for that effort. My own experiment will take a snapshot of my inner self, including what my DNA says about my present and, perhaps, about my future, keeping in mind that I will also be investigating how my DNA interacts with environmental input, and what impact my genes might have on my brain and other bodily systems in the present and beyond.

The quest to know what lies ahead has always been a part of human nature, an obsession dating back at least thirty thousand years, when Stone Age scientists fashioned the earliest calendars using scratches on bones and stones to keep track of the days until the next full moon and other events yet to come. This was one of the first uses of a then fledgling technology to make predictions about what had not yet happened. Later, ancient civilizations developed elaborate rituals to augur the future: cutting the entrails out of animals and tracking the movements of planets and stars to forecast the health of rulers and of families, clans, and kingdoms.

In modern times, experts from insurance actuaries to hedge fund managers crunch numbers to project the risk that a customer will die or to predict the behavior of stocks and markets. (In the fall of 2008, these predictions turned out to be dramatically wrong.) As a society, we use a range of predictions and forecasts to plan our days, weeks, months, and years, from reports on the weather the next day to projecting how much gasoline we can afford to drive across town.

Health outcomes, too, have become more predictable in the last century as statistics of mortality and morbidity and data on actual outcomes for diseases and therapies guide physicians in making decisions about their patients, both healthy and ill. When Josh Adler examines me, he can augur a great deal by feeling my liver through my skin and having me take a few simple tests, such as the one that measures my cholesterol level. He is assisted in interpreting these tests by the statistics of millions of patients who have also been poked and prodded. When Josh took my blood pressure and it was slightly high, he was using threshold numbers that are highly accurate predictors of when to worry—though, of course, there always is the possibility that a person is an exception to the norm and can tolerate higher blood pressure better than most, or worse. I ask him how often this happens.

"It happens," he says, "but it's rare."

Another common predictive number in health care is the odds given when patients are diagnosed with cancer and other terrible diseases.

"You have a twenty percent chance of this cancer going into remission," says a physician—or a 5 percent chance or a 70 percent chance. These numbers are not always precise and usually have more to do with a group's survival rate than with that of an individual with his or her own genetic makeup and circumstances. But these percentages offer a broad sense that a disease is either fatal most of the time or not, which is more information about our future health than previous generations ever had.

With all of our advanced numbers and statistics, however, most seemingly healthy people have little or no inkling about their medical futures. This very day I might feel a lump somewhere where it shouldn't be and discover it is a tumor that might kill me. I could have levels of glucose in my blood go critical next week and cause me to break out in a cold sweat and get dizzy, prompting a diagnosis of type 2 diabetes. But what about what might happen next year or in ten years? Can genetics, testing for levels of environmental toxins in my body, or scans of my brain allow me to predict my future?

In a broad sense, yes, although, for most people, the predictions do not involve drawing blood and having their DNA markers scanned. Much of what a person needs to know about a possible genetic and even environmental and behavioral future can be determined for free by checking his or her family history. One of the most accurate predictors of disease is whether your parents or grandparents have had diabetes, colon cancer, or schizophrenia. "Family history for many diseases is still the best indicator for determining risk factors," says Francis Collins, the leader of the international consortium that sequenced the human genome and, until recently, the director of the National Human Genome Research Institute. Collins advises people who want to have their genes tested to check in first with a genetic counselor to go over their family history.

Years ago, I did just that for my story in *Wired*, visiting Ann Walker, a genetic counselor and director of the graduate program for genetic counseling at the University of California at Irvine. Her job was to explain the whats and hows and the pros and cons of DNA testing to patients who faced hereditary diseases, expectant couples concerned with prenatal disorders, and anyone else contemplating a genetic evaluation. I was an exception to her usual queries at the time, being a healthy person who wanted to get my DNA tested. Mostly, Walker was dealing with patients who had some reason to believe they might have inherited a disease and wanted to find out what happens if they tested positive for the gene in question.

She started by asking about my grandparents and their brothers and sisters: what they suffered and died from and when. My Texas grandmother died at eighty-one after a series of strokes. My eighty-six-year-old Missouri grandma had breast cancer in her late fifties and, years later, ovarian cancer. Their men died younger, although, as I have said, each had a brother who lived well into his nineties. To the mix, Walker added my parents and their siblings, all of whom were healthy in their mid-seventies—and still are five years later. Then she asked about my generation—I visited Walker before we knew that my brother, Don, suffers from a rare genetic disorder I will describe in an upcoming section—and finally about my children. She looked up and smiled. "This is a pretty healthy group."

Normally, Walker said, she would send me home with a recommendation that no genetic tests were called for. But I was sitting across from her not because my parents carried some perilous SNP, but as a fit man looking for a health forecast. We are just beginning to train for this, she said, and told me the two general rules of genetic counseling prevalent at that time: no one should be screened unless there was an effective treatment or readily available counseling; and information from a screening should not bewilder people or create unnecessary trauma.

Even if no obvious disease runs through my family (except for my brother), the genes that I carry offer a wide range of predictive power. For instance, I have a series of SNPs in six different chromosomes that gives me a very high probability that I have blue eyes: a 94.17 percent chance, to be exact. One of the more telling markers is located on chromosome 15, near a gene called OCA2. Like many differences among people, this marker involves changes in the letters in a single base pair in the OCA2 gene. Like most SNP markers, the one associated with blue eyes has three possible letter combinations involving the two letters of this base pair—in this case, G and A, with G being the letter associated with the trait in question, blue eyes. (A is associated with having brown eyes.) The possible variations in this SNP are: GG (very high chance of having blue eyes); GA (chance of having brown eyes); and TT (chance of having brown eyes). About one-third of Caucasians in North America have blue eyes and are likely to carry the GG code. In Iceland, the number approaches 80 percent.

If you had never met me or seen a photograph, a glimpse at my results for this SNP—I am GG—and others associated with eye color would be an almost surefire predictor that when we did meet, you would be looking into baby blues. It's not 100 percent certain, however. I have a very small chance of having brown eyes—2.3 percent— or green, at 3.52 percent. But in genetics, upward of 94 percent is a very solid prediction.

Eye color, though, is a benign trait, with DNA markers that are easily verified by checking the hue of a person's irises. For most diseases, the evidence is less obvious, although there are some that are very certain if a person tests positive for a particular genetic marker or markers. Take spina bifida, a tragic disease that causes a baby's spine to be exposed and malformed at birth, a result of genetic glitches that can be detected by a prenatal test with great accuracy. Other genetic signatures will manifest with great certainty later in life. These include Huntington's disease, which can strike a carrier before the age of thirty, with some cases coming later, and which causes a grad- ual and lethal degeneration of nerves that control movement. The disease is not caused by an SNP, a single letter mutation, but by a sequence of three nucleotides, CAG, in chromosome 4 that repeats too many times. In a healthy person, the sequence can repeat up to thirty-nine times. For those who get Huntington's, as the repetitions approach or exceed thirty-six, a malformed protein is made that collects in the neurons, gradually causing them to malfunction. Victims face a steady degeneration of motor and cognitive function and eventually death at a young age. Testing positive for the Huntington's sequence has an almost 100 percent "penetrance": that is, almost 100 percent of people who carry the deleterious CAG repeats will get the disease.

"It is a death sentence," says Jonathan Rothberg, the geneticist who has seen the disease's devastating effects firsthand in his own family. He has had himself tested and, thankfully, came up negative.

If every genetic trait were this virulent, genetics would be a highly accurate crystal ball we could use to predict our physiological future. But the penetrance of Huntington's is very rare. For nearly all common diseases, SNPs and other DNA markers offer probabilities of possible futures, with many having a very low power to predict a specific disease in a specific individual. Millions of people who carry DNA markers associated with ailments will never get the diseases, while some people who don't carry the markers will.

To return to the eyes, I have been tested for several SNPs associated with age-related macular degeneration (AMD), a malady that usually afflicts people over fifty years old. It causes the degeneration of cells in the eye that are responsible for our sharpest vision ("dry" AMD) or causes new blood vessels to grow under the macula in the back of the eyes ("wet" AMD). Both cause a loss of vision, with wet acting more quickly than dry does. AMD is the leading cause of blindness in the United States for people over age fifty. Approximately one out of three Americans older than seventy-five will get the disease. There is currently no treatment for dry AMD; effective treatments for wet AMD have appeared only in the last few years.

My results for twelve different AMD markers all fell into the low- or average-risk categories, which is great news. For instance, my result is AA for a marker on a gene called CFH that is connected with AMD, which gives me an average-to-low risk factor. Those who carry a GG have a higher risk, and those who have AG also have a moderate-to-low risk.

This sort of prediction tells me very little about my prospects, how-ever, and points up several weaknesses in using SNP testing to foretell the future. The first is how SNPs are identified as being connected to disease. For all but a few SNPs, there is no direct functional link established—no mechanisms of cause and effect, although sometimes there are hypotheses. Instead, researchers use "association studies" that take a population of people who have a disease such as age-related macular degeneration and compare them with people who don't have the disease. Using gene chips made by companies such as Illumina and Affymetrix, gene hunters run scans of each person's genome or, more accurately, those one million or so gene markers covered by the chip that offer a kind of outline of a person's genome. They home in on SNPs that pop up in people who have AMD, which are narrowed down to those mutations—and the nucleotide letters associated with them (in this case a "G")—that show up in the most patients who have the disease. Data is then collected on how many people have and don't have the disease and carry each of the three possible variations on the base pair associated with AMD: GG, GA, and AA.

Carrying a high-risk marker does not mean you will get the dis-ease, though. Nor will everyone whose genes contain the muta-tion come down with the disease. "Just because we have identified a gene doesn't mean its function or its impact has been thoroughly understood or that having a gene has any real predictive value," says

Francis Collins. He believes, however, that the science of SNP discovery is moving so fast that the number of validated markers will increase from about fifty, when I talked to him in 2008, to perhaps five hundred by the end of 2009 and keep going up. Every few days, new SNPs linked to disease are discovered, for everything from stroke to lung cancer.

"A Moore's law for bioinformatics is under way," says Randy Scott, the CEO of Genomic Health, a genetic diagnostics company. Moore's law is the axiom that computing power doubles every two years. "Like mainframes to personal computers, this is an unstoppable force."

Even SNPs that are validated across large and diverse populations have another drawback, since only a few have been tested in clinical trials on real people in a hospital or a clinic to see whether SNPs really can predict whether a person will come down with a disease. "There is a missing link," says bioethicist Arthur Caplan of the University of Pennsylvania. "The technology right now allows researchers to do cool stuff, but it's not validated by clinical data. Large clinical studies are needed to get actual outcomes; the predictive value needs to be validated."

Randy Scott adds, "The notion that matching gene risks to lifestyle choice is in a good position to do very much of interest in a predictive sense seems to me to be a lot of heavy breathing but not much romance yet."

SNP chips also cover just a tiny fraction of the three-billion-plus base pairs of a human's genome or even the ten million or so SNPs and other types of mutations that are responsible for most of the differences in people. No ten-million SNP array chip is yet available, partly due to cost, although I'm aware of at least one company that is developing such a chip—and has tested me on it, though I do not yet have my results. Meanwhile, researchers have built maps of the genome that identify locations for many known SNPs and assign regions, or neighborhoods, where disease genes seem to cluster near certain SNPs. These regions are tagged by the mappers with an SNP that says, "There may be a trait here, but this chip can't tell us what exactly or where." Also, gene chips don't cover the 25 percent or so of traits that are determined not by SNPs but by insertions and deletions of nucleotides and repetitions of genetic codes called copy number variations. (The exact percantage is not known.)

To truly understand everything going on in a single person's DNA, to chisel out every last detail of your inner you, we would need to sequence an entire genome: all six-billion-plus nucleotides. So far, a handful of people have had their genomes sequenced, such as James Watson, the co-discoverer in 1953 of the double-helix shape of DNA, and also geneticist and Human Genome Project leader Craig

Venter. "Gene chips don't compare to whole genome sequences," says Venter. "What we need is more genome sequences, thousands of them, to make sense of genetics for individuals." No one disagrees with this, although the cost of sequencing an entire genome as of this writing is somewhere between $100,000 and $350,000. This is still a bargain compared to the $2.7 billion public price tag for the Human Genome Project, finished in 2003, but it is prohibitive for a large population study, even though advances in technology are pushing the price ever downward. Several projects are under way to eventually sequence dozens and even thousands of genomes for perhaps a few thousand dollars apiece. "When sequencing is as cheap as buying a Chevrolet, then we'll be able to make this science really work," says Watson.

In 2007, Venter published a sampling of findings from his sequencing in a study in the journal *PLoS* (*Public Library of Science*) and in a book, *A Life Decoded: My Genome: My Life*. He also posted the raw results on a Web site for anyone to peruse. Venter revealed deletion and insertion markers and copy number variations that SNP arrays cannot always pick up and also a number of markers that give him a higher than average risk for everything from Alzheimer's disease to drunkenness, although he told me that neither of these genes is a very powerful indicator of anyone's future. A few SNPs did get his attention, including some linked to heart attack.

"My father died of a heart attack when he was fifty-nine," Venter told me. Having just turned sixty when we talked, he had started taking low doses of statins: drugs that reduce cholesterol buildup that can lead to a heart attack. He is also eating more healthy foods and trying to drink less alcohol—despite, he said with a devilish grin, having markers that peg him as prone to imbibing a wee bit too much.

Ann Walker raised perhaps the most significant unknown about genetic testing when she asked, "How will people react to this information?" In part, that's what this book is about, although I recently saw data presented by genetic policy expert Kathy Hudson of Johns Hopkins suggesting that over 80 percent of people polled are okay with finding out their genetic results for diseases, even if there is no treatment or cure for that disease. In 2007, the NIH began the Multiplex Initiative project at Detroit's Henry Ford Health System to offer to volunteers DNA tests on fifteen genetic markers, those associated with type 2 diabetes, high blood pressure, coronary heart disease, high cholesterol, osteoporosis, lung cancer, colorectal cancer, and malignant melanoma. Seattle's Group Health Cooperative is also offering the tests. Participants are asked how they responded to the experience and whether the results prompted any lifestyle changes or calls to their physicians.

Despite these deficiencies, the prospects for these tests and their predictive power are still dazzling, even if many of them are not yet ready for prime time. "It's amazing to sit down and look at a genome," says Jonathan Rothberg, whose former company, 454 Life Sciences, sequenced James Watson's genome. His new company, RainDance Technologies, is one of three outfits that have promised to test all or most of my genome sometime soon, although the cost remains a problem. Since I left Iceland at the beginning of the book project, I have had millions of genetic markers tested and several genes sequenced. For now, this seems quite enough, given the challenge of deciphering and understanding the mass of data I have already accumulated.* I can't imagine trying to sort through *billions* of nucleotides, but I'm willing to try.

I'm doomed. Or not

Kari Stefansson is unhappy about having to deliver my latest results over the phone: those derived from the blood I gave at deCode headquarters. I am now in San Francisco, and he is in his Reykjavik office, simultaneously e-mailing me my results.

"You have a very bad result," he announces. At first, I'm sure he's kidding, as he did on the basketball court, but he's not. "It is a SNP for myocardial infarction: heart attack. It is a very serious SNP, one that has been validated by many studies. You are homozygous for this SNP on chromosome 9," meaning that I have two out of two high-risk nucleotides on this base pair associated with a mutation. "For this gene, you need to go on statins immediately!" he says sternly.

An e-mail containing my results pops up in my in-box, telling me the reference SNP number of this deleterious marker: rs10757278.†

---

* For my complete results, go to www.experimentalman.com.
† Note that "rs" refers to a "reference SNP" number. Each new SNP that is discovered is given an "rs" number by the National Center for Biotechnology Information (NCBI).

I open the attachment marked "Myocardial Infarction" and see on a simple table that the high-risk result is "G." This is bad news for me since I have two Gs, giving me a risk factor of 1.64—that's a 64 percent greater risk of having a heart attack compared to a person who has no Gs. Having just one risk-prone "G," called heterozygous, would give me a less elevated risk factor of 1.24 times the average. To put this in context with other risk factors for heart attack, having high cholesterol gives one a 2.0 risk factor: a doubling of the risk compared to people with a normal cholesterol level.

Scientists at deCode and elsewhere have found that this SNP confers an even higher risk factor for early-onset heart attack: a twofold increase over the average person, the same as for high cholesterol. I ask Stefansson how early-onset he is talking about, and he says, "Before fifty [years old]."

"So I've got about a month to worry," I say, having gotten this news about four weeks before my fiftieth birthday.

"Don't play games! This is serious!" Stefansson erupts in his gruffest voice. "Listen to me! I want you to call me when you have gone on statins."

I'm only a little jarred by this, in part because Stefansson often talks curtly for effect. I also have the same trump card for heart attack that I had when I received my genetic results for the *Wired* story several years earlier: my family history of limited heart disease. Stefansson agrees that family history does play a big role and that it is not factored into the 1.64 percent risk factor for the SNP. There could also be other genetic markers that counteract this one, he says, although so far, none have been discovered that have the predictive power of rs10757278.

"Is it possible that you have something that neutralizes this? Yes, of course," says Stefansson. "For fifty percent of the people with your variation, they will not have a heart attack."

"But why statins?" I ask.

"Because they are usually safe in low doses with few side effects, and because the only way you find out if you are in the fifty percent who will have a heart attack is to have a heart attack, which can kill you."

Stefansson explains that this SNP is situated within a cluster of markers on chromosome 9 that is linked to heart disease. This grouping of markers appears in a stretch of DNA about 190 base pairs long that geneticists rather opaquely call a "linkage disequilibrium," meaning that heart disease tends to be passed down from one generation to the next mostly intact, for some people, as a block of DNA that does not go through the usual random reshuffling of individual nucleotides passed down by one's mother and father.

Curiously, rs10757278 and the other related heart markers sit outside a functioning gene, in a wilderness of noncoding or "junk" DNA.* But there are two nearby genes called CDKN2A and CDKN2B that these variations in SNP markers may influence in some way. These genes are primarily associated with regulating the growth and aging of cells, processes that can be disrupted in cardiovascular cells by high levels of cholesterol, leading to atherosclerosis: the inflammation of arteries that can result in heart attack. No actual link has been made between rs10757278 and these two genes, however, and some scientists believe they may be connected to other genes or active sequences that have not yet been identified—or their presence may simply be an indicator for some other undiscovered function.

As statistics go, rs10757278 and other related markers on chromosome 9 are better understood and have more validity than many SNPs associated with common diseases. The relevance of the chromosome 9 cluster for heart attack has been confirmed in several studies in the United States and Europe that have tested tens of thousands of people. This is one reason that Stefansson wanted me to pay attention. Yet I still have doubts about the relevance of this SNP to my future health. I also have more gene markers to be tested before I am prepared to take this too seriously, let alone do something as rash as popping statins every day.

I don't have to wait long for the next wave of results to come back. The next findings also come from deCode but not from Kari Stefansson. They are delivered online through the company's then brand-new genetic-testing Web site. Called "deCODEme," this launch was part of a mini-wave of companies in late 2007 and early 2008 that brought DNA testing out of the lab, making it directly available to any healthy person who is willing to send in a spit sample or a cotton swab of his or her DNA (scraped from inside one's cheek)—and to plunk down, in the case of deCODEme, $985. DeCode, however, was kind enough to run me gratis, initially delivering results for twenty or so diseases and genetic attributes, which have been added to over time.

---

* Genes are stretches of DNA that provide coding for cells to make proteins. About 95 percent of human DNA is "junk," meaning that it is not used to make proteins, as far as scientists know.

I click on "myocardial infarction—heart attack" first, expecting the worst.

What I see surprises me.

My "score" is a mere .81 times normal risk factor, suggesting that I have a *lesser* risk of having a heart attack than most people. This score is quite different from what Stefansson reported on the phone a few weeks earlier. As I study the pages of consumer-friendly explanation and risk factors, I realize that the SNPs used to make this assessment do not include the telltale marker that Stefansson had sent me, the rs10757278. But it does include another SNP from that block of DNA on chromosome 9, designated as rs10116277. In the deCode study, this was one of several SNPs that have a significant correlation with heart attack and a similar predictive power. For most people, whatever variation they have for one of these markers, they have the same one on the other. So people who are GG and higher risk, like me, on one of these SNPs usually have GG on the other. But not me. I am an exception, with a GT on rs10116277. This gives me a lower risk factor for this marker. Even better for me, I came up with another variation on a second genetic marker that confers a very low risk of heart attack: a .87 times chance of getting the disease. (See the table on page 45.)

What the heck? I mutter to myself, staring at my computer screen and trying to figure out how I could have all at once a high, medium, and low risk of having a heart attack, depending on the SNP.

When I get this news, it is early in San Francisco, with the morning light faintly blue-gray over the Oakland hills, which I can see across the bay from my house. This is the reality of online genetics: unexpected news delivered on your own computer screen, when you're alone, early in the morning. I had already gotten a few minor jolts from my Sequenom data and other results, but this one is less a jolt than a matter of confusion at my contradictory results. How can this be?

My first reaction is to contact Kari Stefansson and ask him what happened. His team quickly responds and leads me deeper into the universe of nucleotides, association studies, statistics, and, as Mark Twain said, damn statistics, although hopefully not lies.

My first question is why Stefansson's original SNP—rs10757278—was not included on the deCODEme site, if it was so important. The team explains that this was a simple matter of logistics. Its consumer site uses the standard off-the-rack Illumina SNP array, which covers only about 10 percent of the actual markers that account for most human variation.

The SNP rs10757278 falls into the 90 percent of SNPs not covered by the array, but rs10116277—the SNP tested on its site—does appear on the Illumina chip.

Stefansson explains that for most people, this doesn't matter, since they will get the same variation for both SNPs. "But you are this strange anomaly," he says. "You are in a small group of people tested who came up with these different results for the two SNPs."

"But which result do I believe? Am I high risk or low risk?"

He says the original SNP, rs10757278, that he called me about has a stronger correlation with the disease, so I should probably pay more attention to that one. "But we don't yet know how to deal with people like you," that is, to determine the risk factor for people who do not have the usual pattern for the SNPs on chromosome 9.

"Does one cancel out the other, or does one override the other?" I ask.

"It is a very good question that we do not have the answer to yet."

Adding to the confusion are my heart attack results from other online genetic-testing sites. Not long after my deCODEme findings came to me via the Web, I log on to the beta version of Navigenics, another direct-to-consumer genetic testing site, based in Redwood Shores, California, south of San Francisco. Its site gives me high-risk factors for two SNPs that are similar to my original results reported by Stefansson. One of these is a third marker located on that same block of DNA in chromosome 9. I am high risk for this third marker on chromosome 9 and for another marker listed on the Navigenics site from a different gene. If Navigenics is correct, these two SNPs combined give me a 62 percent chance of having a heart attack: a higher-than-average risk that one day I will join the 865,000 Americans each year who have heart attacks. Nearly 158,000 of these people die, says the Navigenics Web site, a not too cheery thought.

Now I have three sources—deCode, deCODEme, and Navigenics—telling me that I have different possibilities for when, or whether, I might drop dead from a fatal heart attack. Soon after this, I add a fourth, 23andMe, an online genetic testing company financed in part by Google and cofounded by a longtime technology executive and consultant, Linda Avey, and by former venture capitalist Anne Wojcicki, the wife of Google cofounder Sergey Brin. The company 23andMe (the name refers to the twenty-three paired chromosomes in a human) provides me with results from one additional SNP test for heart attack, yet another marker that is supposed to be virtually interchangeable

The Author's Heart Attack Results from Three Online Companies

Gene	Marker (SNP)	Company	Author's Results	Risk Factor*
CDKN2A/ CDKN2B[†]	rs10757278	deCode	GG	**1.64**
CDKN2A/ CDKN2B[†]	rs10116277	deCODEme	GT	1
CELSR2/ PSEC1	rs599839	deCODEme	AG	0.86
CDKN2A/ CDKN2B[†]	rs2383207	23andMe	GG	1.22
MTHFD1L	rs6922269	23andMe	AA	~1.2
CDKN2A/ CDKN2B[†]	rs1333049	Navigenics	CC	**1.72**
MTHFD1L	rs6922269	Navigenics	AA	**1.53**

*High-risk factors (over 1.5) are in bold.
[†]The link between this gene and the genetic marker cited has not been verified.

with other SNPs on chromosome 9. My score for the 23andMe SNP is a modest 1.22 times risk factor.

This rounds out a slate of results from the three major online sites and from deCode, telling me that I am either doomed or not or something in between, findings that suggest a few kinks need to be worked out in these sites before genetic association studies will be ready to provide meaningful information to an individual. The table above shows my heart attack scores delivered by Kari Stefansson in his original call and from the three Web sites.

If this contradictory information isn't baffling enough, soon afterward, I hear again from the deCode people, who add yet another wrinkle to my quest to find out whether I have a higher or lower risk of keeling over with a massive myocardial infarction. It turns out that there is more than one statistical method of determining genetic risk factors and that the three sites use different methods that sometimes produce dissimilar results.

The method most commonly used by researchers is called an "odds ratio," which for genetics takes people with the same variation, such as a GG, and compares those in this group from deCode who have had the disease to those who don't. This seems straightforward enough, although now I am hearing about "relative risk," which takes the total number of people with a high-risk mutation (those who have and have not had a heart attack) and divides that number by the total population being tested.

Stefansson's people bring up these statistical differences when they send me a revised risk assessment for the SNP that Stefansson called me about in such a dither. Originally, he reported my variation as having an odds ratio of 1.64. Recalculating this as a relative risk, however, changes it to a 1.26: seemingly a more ho-hum threat, although Stefansson does not back down about his insistence that I take this seriously. Differences in calculating risk is one reason I got confusing results from the three sites, since deCODEme uses relative risk; 23andMe employs something it calls an "adjusted odds ratio"; and Navigenics uses an odds-ratio calculation that takes several pages of equations to explain.

"Genetics is in great need of standards that everyone agrees on," says David Agus, a cofounder of Navigenics and an oncologist at Cedars-Sinai Medical Center in Los Angeles. As I finish writing this book, the companies are meeting to see whether they can agree on guidelines. The U.S. Department of Health and Human Services, Congress, and several state health departments are also exploring the possibility of requiring standardizations in reporting results and assuring validity and accuracy.

Once again, I get Kari Stefansson on the phone to ask him which statistic (damn statistic?) to believe of the two methods he has provided me for rs10757278. Should I believe the odds ratio of 1.64 or the relative risk of 1.26? Stefansson says that his scientists and statisticians have decided that for an individual, the more meaningful statistic is the one that results from being compared to everyone in the population tested— the relative risk. Kari Kaplan, a genetic counselor at Navigenics, agrees when I ask her, although she says that since most SNP association studies report risks as odds ratios, that's what her company tends to use. "We want to be true to the primary studies," she says.

Soon afterward, I learn there are two more additional types of risk calculations. The first is "absolute risk," a number that tells us our chances of getting a disease before we die, based on the average risk that everyone in a population faces, plus or minus our own individual risk factors. For instance, according to Navigenics, I have a 49 percent chance of having a heart attack just because I'm white and male and live in North America or Europe. That's my baseline risk. Add personal risk factors, such as my SNP results, and my chances go up or down or stay the same. The second (and hopefully final) category is "lifetime risk," which tries to factor into my average risk numerous factors that pertain to me: multiple SNPs, diet, age, smoking or not, weight, and so forth. Ethnicity is also important, since many diseases and other traits occur with more or less frequency in different populations (blue eyes among Icelanders, sickle-cell anemia among Africans, cystic fibrosis among Caucasians).

The Web sites 23andMe, deCODEme, and Navigenics all provide lifetime risk numbers to their customers, although they focus their scores on the SNPs they test for rather than on a standard or agreed upon set of criteria. All of the sites factor in gender and, on some sites for some conditions, age and ethnicity. None ask for a medical history to determine factors such as weight and smoking, although they say they are planning to in the near future.

Below are my life scores for heart attack from the three sites, presented as a percentage chance that I will drop to the ground clutching my heart over my lifetime—which, of course, reflect the high, low, and medium results of my SNP profiles.

Lifetime Heart Attack Risk

deCODEme: 42% chance

Navigenics: 60% chance

23andMe: 29.9% chance[*]

If this seems perplexing and vague to you, you're not alone. "The public doesn't get probabilities of this sort very well," says family doctor Greg Feero of the National Institutes of Health. "But neither do most doctors, who don't understand the differences between odds ratios, relative risk, and absolute risk. I'm not sure that *I* do, entirely."

Well, yes, I think, now thoroughly confused by my scores. I take my results to experts whether I can make sense of them and whether there is anything I should take seriously.

I start with Harvard's David Altshuler, a leading geneticist and a skeptic about using association studies right now to determine an

---

[*] The company 23andMe provides a score based on the chance of a heart attack between the ages of forty-five and eighty-four; the others are lifetime risks for adults.

individual's medical future. He compares such knowledge to tarot cards and fortune-telling, suggesting that both have about the same validity and utility. "Your heart attack results are a great example of why this information is suspect," he says.

When I mention rs10757278 on chromosome 9, however, he has different advice. "This one has been validated in several studies, although we still don't know much about its function or even what gene it is associated with. But because the first evidence you would have that it is truly a risk for you is a heart attack, and many people die when they have a heart attack, it's worth taking it seriously. Also, you can prevent a heart attack from happening with highly effective interventions. So this one has some validity and utility."

In an e-mail, my internist, Josh Adler, is surprised at the contradictory information. He doesn't say "I told you so," but he indicates that there is nothing in my genetic results that would change his prognosis from my checkup at the start of the Experimental Man project. He does not think I need to take statins and suggests that I work on lowering my cholesterol by eating less bacon and using nonfat milk instead of half-and-half in my coffee.

Myocardial infarction is not the only disease result that is delivered by Kari Stefansson and others running tests on my DNA. Perhaps the most significant news as I scan my internal programming is the same conclusion that I came to during the *Wired* story tests in 2001: that I have very few nasty genetic markers, other than the somewhat mixed-up slate of heart disease markers. For healthy people who survive to my age, this is likely to be a common outcome: a huge genetic relief for some or, for people like me who are less concerned about such things, a genetic shrug. On page 49 is a table showing a sampling of a few results and risk factors out of thousands tested, which come from a variety of studies and sources—not only 23andMe, deCODEme, and Navigenics.

These scores suggest that I mostly have SNPs that place me at an average or below-average risk for these diseases, although I do have a moderate risk variant for two SNPs for asthma and a high risk for two SNPs associated with rheumatoid arthritis. My father's mother had severe arthritis late in life, but as a stoic Scottish English Midwesterner she never complained. I have no sign

Disease	Gene	Marker (SNP)	Author's Results	Risk Ratio*
The Author's Sample Genetic Results				
Asthma	ORMDL3	rs7216389	CC	0.69
	PLEKHA1/ARMS2/Htra1	rs932275	GG	0.68
	TNRC9/TOX3	rs3803662	TT	1.42
	CDKAL1	rs7756992	AG	1.21
Psoriasis	HLA-Cw6	rs10484554	CC	0.85
	IL12b	rs3212227	AA	1.13
	IL23r	rs11209026	GG	1.05
Rheumatoid arthritis		rs6679677	CC	1
		rs6457617	CT	**2.36**
		rs11203367	TT	2.1
	HLA-DRB1	rs660895	AA	0.42
	PTPN22	rs2476601	GG	0.89
	STAT4	rs7574865	GG	0.87
	TRAF1-C5	rs3761847	AA	0.78
	RA	rs2327832	AA	1.04

*High-risk factors are in bold.
For more results and details, go to www.experimentalman.com.

of arthritis—yet—but I can tell you that if I get it, I only hope that I can cope with it as well as my grandmother did.

Less defined than the potential predilections for disease are genes that affect behavior and emotion. For instance, I have the SNP variation I've mentioned in the DRD4 gene, which gives me a higher-than-average probability that I seek out novel situations. This is not a shock to anyone who knows me or to anyone reading this book, although this data comes from a small study and needs further testing. I lack a high-risk variation for alcohol cravings, but I do carry one that makes me at high risk for attention-deficit/hyperactivity disorder. These, too, come from small studies that need to be validated by other researchers, and may mean nothing significant for me.

I have slightly higher-than-average risk factors on some SNPs for diabetes type 2, prostate cancer, and colorectal cancer. I also have a SNP in the CILP gene that is associated with the formation of collagen in the lower back. Like a majority of people in North America, I have a variation that makes me susceptible to lower back pain, which perhaps explains my herniated disk in 1995. I always assumed that I popped my lumbar 5 disk after years of abuse riding bicycles and engaging in other sports and not paying attention to how I sit, lift, and carry objects, and because I generally exhibit type-A behavior, which is known to put stress on the lower back.

It's nice to think that a gene might have caused my hernia instead of all of that bashing and battering I inflicted on my back, but, at best, it's a combination of genes and self-abuse. I'd also like to think that if I had known about my predisposition for lower-back issues, I might have taken it easier. But the truth is that having this knowledge would not have changed my behavior. Even after I felt pain and several times was forced to spend a day or two in bed with a spasmodic back, I didn't change my behavior. It took two episodes of having my back swell up to compress my sciatic nerve in my left leg, causing it to go limp, to finally convince me that I was not immortal and needed to take my health more seriously. For months, I worked with physical therapists and fully recovered, although I still need to do stretches and watch how I treat my back, or the disk lets me have it with a sharp pain that says, stop what you are doing *now*!

A tale of two brothers

My mother had a photograph snapped (see page 51) when I was around six years old and my brother was five. She sometimes liked to dress her two boys the same until it became too embarrassing when we reached the ages of nine or ten. In this photo, we are wearing light-hooded green jackets, brown jeans, and cowboy boots and are standing outside in Kansas City, Missouri, the city where my father's family had lived for five generations. Don and I are towheads with buzz cuts, skinny kids with thin arms and delicate features. On that fall day, leaves are swirling in the wind and a chill is in the air.

My mother, who is now seventy-five years old, is in her early thirties in the photograph, ultra-slim and blond, a former model who looked like Marilyn Monroe. Even if someone who didn't know us happened on this photograph, they would identify us as sharing genes: a mother with her two sons.

Yet deep inside, in virtually every cell in our bodies, my brother, Don, and I have at least one subtle difference in basic programming given to us by our mother and father that has caused the two boys who dressed alike to diverge dramatically. One of us has enjoyed a healthy, full life; the other is disabled, a tragic dichotomy that was almost certainly caused by variations in the genetic letters that make him Don and me David, those more than six billion As, Ts, Gs, and Cs that constitute the DNA that makes us human, brothers, and the sons of Patricia, my mother. As we know, differences in even one letter in a gene—an A for my brother, a T for me—can mean having a disease or not, or having blue eyes brown. Other flaws can involve rogue copies

From left: Donald, the author, and Patricia Duncan, Kansas City, c. 1964.

of DNA or insertions or repeats where they shouldn't be and deletions where they should be.

Most common diseases and traits, and quite a few rare ones, follow a predictable pattern of inheritance that clearly runs in families. Red hair, for instance, runs in my father's family. His grandmother had bright red hair; so does his sister (my aunt) and her daughter (my cousin). My seventy-seven-year-old father's hair is now mostly white, but it used to be brown tinted with red, a blend that came out in a genetic test showing that he has a 63.87 percent proclivity for red hair, 22.85 percent for brown hair, and 13.28 percent for blond. Since my father's red-hair gene is located on his single X chromosome, and fathers pass down only their Y chromosome to their sons, he could not pass this on to my brother or to me. This is why Don and I, in our genetic results, each have a less than 1 percent chance of having red hair and a roughly 75–25 split between brown and blond coloring. Sure enough, I have no red hair, although Don does have an auburn tint, suggesting that he acquired his hair coloring from some other source in his genes.

My parents also delivered—inadvertently, of course—genetic coding that has bequeathed to Don a rare anomaly we suspect is osteogenesis imperfecta (OI), which is also called brittle bone disease. It is slowly causing his bones to become as frail in middle age as those of a very old man. This condition is so unusual that until recently, the family didn't realize that the source of his affliction is genetic. Women on both sides, including my mother, have suffered from severe osteoporosis late in life, but no male whom we know of in either family has had abnormally brittle bones, even in old age.

This created a deepening mystery about what exactly happened to Don and how this trait has appeared in an otherwise reasonably healthy family. Our search for answers illustrates how genetics can sometimes provide illuminating explanations about our inner selves and why we are who we are but also can lead us down twisting, confusing mazes that promise clarification but arrive at dead ends.

We didn't realize it at the time, but the first clues occurred as early as that chilly autumn in 1964, when I did what older siblings too often do: I was mean to my little brother. Pumping up and down on a seesaw in a park near our house, I suddenly leaped off when Don was suspended high in the air. Without my counterbalancing weight, he plunged downward and landed, to my horror, on his ankle, which snapped. He bellowed in pain, and my mother screamed. That night, I was chewed out by my father and later by the matriarch of my father's family, my grandmother.

We assumed it was a typical childhood break. But Don kept breaking bones. Not all of the time, but enough that in our family—which knew next to nothing about human genetics, like the rest of the world at the time—my brother was described as "accident prone." He broke wrists and ankles as a child; as a young adult, he had more serious breaks. The most severe was during the summer before his senior year in college when a sudden rainstorm made a street in Washington, D.C., ultra-slick, as he made a high-speed turn on a bicycle. Don lost control and slid hard into a concrete curb, snapping the top of his femur below the ball joint that fits into the pelvis. At age twenty-one, he had a four-inch metal pin screwed into his leg.

A gifted photographer, Don recovered from his fall and launched into a career in Maine, where he attended Bowdoin College. After joining me for most of a bicycle expedition around the world just after college—he broke two bones during the trip—he set up a photo studio and married his college girlfriend.* Walking down the aisle at his wedding, Don hobbled on crutches with a broken leg. This didn't keep him from one of his greatest loves: traveling deep into the back country of Maine to shoot large-format black-and-white photographs of waterfalls, streams, trees, and mountains, breathtaking shots that used many of the techniques developed by Ansel Adams and Paul Caponigro, Adams's protégé. Caponigro was a teacher at the famed Maine Photography Workshops where Don took courses after graduating from college. I remember Don setting out in a battered Volvo with the trunk adapted into a field darkroom and racks on the roof packed high with camping gear and his cameras, lenses, tripods, and other equipment. Each year, Don's work was shown in an ever-expanding list of galleries and museums in New England and beyond. Having two little girls barely slowed him down. He took them on his trips when he could, and he loved photographing them.

---

* This trip was the subject of my first book, *Pedaling to the Ends of the Earth*.

In his thirties, Don noticed that his feet got sore when he walked or stood too long. By then he was in almost constant pain from his hip injury. Around this time, Don's wry sense of humor began to wane. He became quieter and seemed blue, if not depressed at times.

"It started in May 1998," he recalls, "when I reached a certain level of pain, and it interfered with my walking." In 1999, at age thirty-nine, Don had a hip replacement, but the pain persisted, in part because the metal anchors for the artificial hip caused the attached bones to crack as they became more brittle over time. Then an administrator at Bowdoin College as well as a photographer, my brother became increasingly frustrated. He stopped taking photographs and there was a period when he was uncharacteristically irritable when he wasn't numbed by painkillers. Eventually, he had to stop working altogether. My parents, who retired in Maine to be near Don, his wife, and their two daughters, didn't know what to do. They had no idea that my brother's situation might be beyond his control, and we all wondered how someone so young and talented could basically shut down.

During this time, I lived first in Maryland and then in San Francisco and continued to have a full life with friends, family, and work. I was raising my children, traveling around the world on assignment, and bicycling whenever I had time. Then in my forties, I had broken only one bone in my life: my wrist, from a bad fall while running.

Don continued to break bones. In 2002, he slipped on the ice and broke his hip again. The chips of bone are still there, he says, adding to his pain. He went to a specialist to have his hip-bone density scanned. This test uses a special X-ray that takes images of a patient's bones in his hips and lower spine and then uses computers to calculate the density of the bone. His physician didn't think Don needed to take the test, saying that statistically, men his age don't have significant bone loss. But the results showed something different: that Don's bones were indeed thin and fragile. By then, I was thick into writing about genetics and imme-diately suspected that a gene or genes were responsible.

Don went through a period when he was angry about his malady, which was quite natural. These days, he doesn't dwell much on his bad genetic luck. "I miss the photography," he tells me with a sigh. "I miss being able to do normal things, like playing with my kids, without being afraid I'll hurt myself."

When I embarked on the Experimental Man project, I did so mostly as a journalist wanting to explain science. But I also wanted to conduct a search to see whether there was a genetic explanation for my brother's malady. This led me to Peter Byers, a physician and expert on the genetics of bones. His lab at the University of Washington in Seattle studies two genes—COL1A1 and COL1A2—that are responsible for making collagen. This is a protein that causes skin to be stretchy and strong and also makes bones sturdy and flexible. Byers says that collagen in bones is something like the steel beams that hold concrete together in walls and columns. "Bone is very brittle, like concrete," says Byers. "It needs collagen to hold it together and to make it bend." Mutations in COL1A1 and COL1A2 can cause collagen to be malformed and are a major culprit in osteogenesis imperfecta.

About one person in twelve thousand has this disease, which is caused by a mutation that interferes with the body's ability to correctly make type I collagen. The most severe cases cause affected individuals to die of multiple fractures in the womb. Others have only mild symptoms and can function relatively normally. Most people with this problem have a blue tint to the whites of their eyes; you can see it if you look hard at celebrities who have brittle bone disease, such as the late jazz pianist Michael Petrucciani and Julie Fernandez, who plays Brenda on the BBC's *The Office*.

Usually, this condition runs in families, which was part of the mystery with Don. Both sides of the family have had women suffering from osteoporosis in old age, including my mother, but not osteogenesis imperfecta, although symptoms of the two diseases can overlap. Both lead to a weakening and fragility of the bones and an abnormally high level of breaks. Is it possible that my mother actually has a mild case of osteogenesis? Or did this malady appear spontaneously in my brother—something called a de novo (new) gene mutation? This has been known to happen in sufferers of this disease, for reasons that aren't fully understood, says Peter Byers.

Last summer, Byers agreed to sequence the COL1A1 and COL1A2 genes for Don and for myself, and soon after we had blood drawn—Don in Maine, and me in San Francisco—and sent to the University of Washington to be analyzed.

On a typically damp, late autumn morning in Seattle, I visit Byers to get our results. We plan to conference in Don from Maine. On the phone before we met, Byers was a bit gruff, a busy researcher and practicing physician. "I cover everything from the nucleotide to the patient," he said—one of those researcher physicians who somehow manages to do both.

When I arrive, however, his brusqueness is gone, replaced by an excitement at the research challenge posed by Don's case. With thinning gray hair and rimless glasses, the sixty-four-year-old Byers is surrounded by heaps of papers and awards that, as often as not, he uses as paperweights. "I think we found something," he says, "but I'm nervous to show it to you, because we haven't checked it, and it could be wrong."

He first explains the test and what he's looking for. As with most other genetic tests taken from my blood, his lab separates out the white blood cells and extracts the DNA and runs it through a process called polymerase chain reaction (PCR), a technique that exponentially replicates DNA and other molecules millions of times so that researchers have enough DNA to work with. Before PCR was developed in 1983, geneticists were severely limited by the micro amounts of naturally occurring genetic material they could extract. COL1A1 is a lengthy gene located on chromosome 17, says Byers, more than eighteen thousand base pairs long. (Genes range from a few dozen to thousands of base pairs long.) Byers's lab first tested our specimens for seventeen gene markers within the two genes that give them big-picture clues about mutations that can cause osteogenesis imperfecta.

"We didn't sequence the whole gene," explains Byers. "Much of it is noncoding, a desert." They then "amplified" the target regions, sequencing dozens of base pairs to get a detailed map. It's like going from a resolution the size of a state or a county on Google Maps and zooming in to see finer details of streets, streams, and parks.

"We're looking for mutations in one or both of the two strands of your DNA and your brother's in the COL1A1 gene." He draws on a piece of paper a picture of the familiar double helix of DNA and beside it a bone, explaining that the matrix of a bone is a mesh of *triple*-helix shapes: three strands of collagen intertwined that make bones strong and flexible. He turns to his computer, and the results begin scrolling across the screen, displayed on a program called Mutation Surveyor 3.0. He shows me strings of nucleotide letters neatly cued up and a cursor line that zigzags, rising up and down above each base pair like the needle lines of a polygraph, except that on this graphic these represent the type of amino acid that the gene codes for, with the program looking for anomalies.

"These are Don's results," says Byers. "Yours was normal." He keeps scrolling. "There," he says, pointing at a sequence marked in red. "This is where the sequence went wonky—in here."

"What is it?" I ask.

"It's a deletion in one of the alleles"—in one of the two strings of DNA that each of us carries: one given to us by our mother and the other by our father. Again pulling out a pencil and paper—"This is so high tech," he says, laughing—Byers makes twenty-nine dots on the paper, saying this is the normal sequence for this section of the COL1A1 gene. He then makes a series of dots alongside it, getting halfway through and then stopping. "There's a section missing in one allele, but it's normal in the other. So the two alleles don't match up, and this one allele ends before the other one does."

"It's like lining up two lines of twenty-nine people and removing person number three in one of the lines," I say, catching on, "which realigns the rest of the second line so that there are only twenty-eight people, rather than the twenty-nine."

"Exactly. With DNA, this can cause the protein made by the sequence to be made wrong or not to be made at all. If we can confirm this, it would explain the bone loss."

He says that they have not seen this exact deletion before, although they are discovering that brittle bone patients have a wide variation in where and how the deletion occurs. In the 1,000 patients Byers has tested, his lab has found 335 different mutations, a lot of variations. "We think there are numerous types of mutations that can cause OI," he says. This is very different from many other diseases with a strong genetic component that have a single SNP variation that always occurs in the same location on a person's genome. "With sickle-cell anemia," he says, "you can take one SNP and test a whole population and say whether they carry the disease. With collagen you have to run the sequences, and even then there is no simple marker that says yes or no. Don could be the only one with this mutation, if that's what we're seeing."

Byers dials up Don in Maine and takes a brief medical history over the phone.

"How many fractures have you had?"

"Fifteen to twenty in my lifetime," says Don, though I wonder about this. I know of about at least a dozen, and he has had many other breaks, but I let it go.

Byers asks about Don's bone density, and he gives his scores: measurements taken by X-ray of the level of calcium and other minerals in his hips and spine. My brother tells Byers that he has tried standard drug therapies and has participated in testing experimental drugs, but nothing has helped his overall condition.

Byers explains what his lab did to test Don's DNA. "We extracted DNA from your blood and David's blood and looked for genes associated with collagen. We looked at David's DNA and didn't find anything unusual. We ran yours and found something else. We are rerunning part of this. If this is correct, then you're missing part of a gene, and this could cause your collagen to not form properly. This will take a week to confirm. If this is correct, it will explain why you have bone loss. It's heritable; it's a form of OI."

"I have been checked for OI, but not genetically," says Don. "I didn't go for the DNA test because it wouldn't impact my treatment. From my perspective, I'm much more interested in what to do than the 'why.'"

"This is why I got into this, to see if we can impact treatment," says Byers. "We're trying to understand the disease so we can develop drugs and treatments."

"I have a random question," says Don. "Does coffee impact bone loss?"

"We don't know. The only environmental factor we know about for sure is gravity. Astronauts lose density unless they exercise. Also bedridden people. We're designed as gravity-bound organisms, for our bones to bear weight."

After saying good-bye to Byers, on the way to the airport, I call Don, and he says he found the results interesting but not helpful in any practical way. "I'm looking for something that can treat me," he responds with a barely perceptible weariness.

I realize that for my little brother, this is merely one more session with a specialist that gets him no closer to what he really wants but dares not talk about anymore. He wants the pain to go away. He wants to roughhouse with his girls, to take a camera up a steep ridge and photograph up close the veins of a leaf turning red in the autumn, to slip on the ice and get back up with just a bruise.

Unfortunately, a few days later, what seemed like an answer to our family mystery is upended by an e-mail from Peter Byers that tells us that the anomaly he found in the first pass has failed to appear in subsequent samples. Don and I have different variations of SNPs in the gene, he reports, but not enough to account for the differences in our bones. He explains in an e-mail:

Dear David,

As you remember, we had some concerns about the sequence in one section from Donald. During our discussion I had said that I was a little uncomfortable presenting things that were not completed. As it turns out, to the best of what we can determine, the peculiar sequence is an artifact [a false-positive].

We repeated the amplification of his DNA another time and performed a series of additional amplifications and examinations of the original material that convinced us that the sequence contained this unusual artifact. I think there is a lesson there—we always confirm every mutation and at the time we were talking, we hadn't.

So, the end result at this point is that the sequences from both of you in both type I collagen genes are normal. From the SNPs we can conclude that you most likely received different copies of both genes from your parents. I have marked in red the positions at which you clearly have different sequences that could not be explained easily by recombination.

Best wishes,

Peter

"I'm a little embarrassed," Peter Byers tells me on the phone. "This happens, but not often." He says it was likely that a glitch occurred in the PCR. "PCR is notoriously finicky in cloning [copying] large molecules like the COL1A1 gene."

I ask him what else could be causing Don's malady. He says that there are other genes associated with bone loss, but none that provide an answer. "I suspect there are many genes involved," he tells me, "like diabetes. Maybe dozens or hundreds." There is also the possibility that Don is the only one who has ever had this disease: that he is a population of one for his version of brittle bone disease.

Byers had asked Don whether his daughters had a history of fractures. Don said no, and we hope this stays the case, although we all worry that the condition might have been passed down. This possibility is reason enough for us to continue to search for an explanation.

What went wrong with Don's internal coding remains a puzzle, although the experience has taught me to realize something that before I rather foolishly did not. As Mark Twain once quipped, "We do not deal much in facts when we are contemplating ourselves." Don's ailment has subtly altered the basis of my family's self-conception that we lead long, healthy lives. The sense that everyone in my family is a model of wellness has been part of my makeup for nearly fifty years and has shaped the visceral notion inside me that I may not be immortal, but mortality is not something I need to think about. Now I was being forced to reconsider this not only for my brother, but also for myself and for my children.

Sometimes I ponder why those two little boys in that long-ago photograph turned out so different. Why am I healthy and my little brother is not? I also wonder how my brother, and my family, would have reacted if there had been a genetic test available when he was born. Without a doubt my parents would have treated him differently if they had suspected he had this disease. As it was, our ignorance allowed Don to live the first thirty-five years of his life normally—riding bicycles in foreign lands, exploring the forests of Maine, and so on.

And what about Don's daughters; should they be tested? Do we want to know about a disease that has no cure and possibly alter their self-perception of how they will live in the future? These are questions confronting us all as science reveals the intimate secrets buried deep inside our genes, brain, and body.

My gene pool (mother, father, brother, and daughter)

In 1993, at the age of eighty-six, my grandmother woke up after an unexpected surgery that had interrupted her annual trip to Mazatlán in Mexico with Bill, her eighty-nine-year-old spouse. She had married him at age seventy, a tall, gregarious, and youthful man (for his age) from her church whose first wife, like my grandfather, had died a few years earlier. For several days, this remarkably healthy woman born near the turn of the twentieth century—she could remember when electricity, automobiles, and radio first came to her small hometown in Kansas—had been feeling ill. Her physician examined her and found that she had an aggressive cancer in her ovaries. Years earlier, a surgeon had removed them, but, apparently, tiny traces of the organ remained, enough for a few rogue cells to begin to replicate uncontrollably.

Even in her eighties, Grandma was at the center of our family, a youthful octogenarian always heading off to a meeting of the Rose Society or some other civic, church, or women's organization. She was a no-nonsense woman who could be stern but also laughed easily. She was always accessible to me, even to discuss matters I was reluctant to tell my parents about: girlfriends, as well as fears and anxieties about school and, later, work.

This was her second bout with cancer. Three decades earlier, when I was a small boy, she had been diagnosed with breast cancer. She had her breasts removed, although I don't remember her even being sick. Few people in the sixties talked about such things. The only reason I knew that anything had happened was that her arm sometimes swelled with fluid. They must have removed lymph nodes from under her arms. She beat the cancer and became a frequent speaker at events for survivors, something I found out much later.

The evening after Grandma's ovarian surgery, I got a phone call. I was living in Baltimore at the time and was planning to fly home to Kansas City the next morning to be there with her. The voice was weak but unmistakable. She said she felt tired and drowsy from the anesthesia, but that she had to tell me something important. I was the oldest grandchild, she said. "I will never forget the day you were born, it was such a wonderful thing." She had written a front-page story in the *Kansas City Star* about my birth and what it was like to be a new grandmother. (It must have been a slow news day.) "Word of First Grandchild Is Awaited with Eagerness" was the headline. The newspaper ran a photograph of her at age fifty-one, looking like she was thirty-five years old, with dark hair and confident-looking eyes, beautiful and strong.

"It's up to you to carry on the clan," she said faintly, "clan" being the Scottish term that my Scottish American grandfather—her first husband—had used to describe our family. I didn't entirely grasp what she was saying, but I promised her anyway. I told her that I would be seeing her the next day at the hospital. It was beyond my comprehension that this would be the last time she would speak to anyone in our family. Bill, who was there with her, told me later that just after the call, she went to sleep. She never woke up and, after slipping into a coma, died a few days later.

It's likely that my grandmother's cancers had a significant genetic component, since women who have both breast and ovarian cancer often have mutations on two genes called BRCA1 and BRCA2, pronounced braka-one and braka-two. The BRCA genes are "tumor suppressors" that normally regulate the growth of cells, making sure they divide properly and don't grow too quickly or otherwise veer out of control to grow into a tumor. BRCA2 also helps fix DNA that has been damaged by the sun, radiation, and other environmental impositions. Mutations in these genes account for about 5 to 10 percent of new breast cancer cases each year—about ten thousand patients out of a total of two hundred thousand—and about 15 percent of cases of ovarian cancer. Testing positive does not mean that one will get breast cancer, however. Female carriers have a 33 to 50 percent chance of coming down with breast cancer before the age of fifty, depending on the specific patterns of mutations, compared to a 2 percent chance without the gene. The odds increase with age, with carriers having a 56 to 87 percent chance of getting breast cancer before the age of seventy, compared to 7 percent among noncarriers.

When I was planning to write this book, the notion of breast cancer running in my family came up I mulled over the idea of testing the genes of my immediate family for dozens of diseases, including breast cancer. When I discussed with my parents and brother their being tested, we talked about the possibility of breast cancer running in the family, given my grandmother's history, although for males such as my father, brother, and me the risk would be very low. My mother's family has no history of having this disease. But what about my daughter, Danielle? Should she be tested? She is twenty years old, a freshman at St. Andrews in Scotland, and an avid kayaker. Petite like her mother, with blond hair and a love of animals and nature—she plans to be a marine biologist—Danielle was enthusiastic about wanting to participate in the Experimental Man project.

"Dad, this is my decision," she says during a video call from Scotland on Skype—the online video conferencing service that links people through their computers. She looks so young in the wavy video image, her long, blond hair framing her angular Duncan face with the pinched nose and thick eyebrows. "It's my genes, and whatever is there, I want to know."

"Yeah, but . . . ," I stammer, not wanting to say what is really on my mind: that she is still my little girl, and I want to protect her. But of course this is not possible with genetics. If she has a high-risk factor for a malady such as breast cancer, I could be genetically responsible. Yet there is nothing I can do about it. Watching her on my monitor, I remember when she was tiny and fragile but had the same defiant, willful expression that she is flashing at me now.

"But nothing," she says. "There is no reason not to test me. I'm going into biology, and I'm learning genetics."

"What if we find something?" I say lamely, thinking about my grandmother and her disease. "It might change you, especially since most of the tests I'm running have a limited usefulness. You might get a risk factor that is really small, which will scare you for no reason."

"The same thing could happen to you."

"I'm older."

"Tell me what I need to do."

I resist for a while longer, but the next time she comes home, Josh Adler okays a blood draw for her that is shipped to deCode in Iceland, where Kari Stefansson's lab processes her DNA on the Illumina HumanHap 330 array and, later, on Illumina's 1 million SNP chip.

By the time Danielle's blood was shipped to Reykjavik, I already knew some of my results for certain SNP variations related to breast cancer. To determine these, I first checked my results on the "me" sites, deCODEme and 23andMe, and found out for the genetic markers they tested that I am low-risk—a relief. But these sites offered only a few of the available tests for breast cancer.

For a more extensive repository of gene markers, I checked SNPedia, a Wiki-style Web site that carries reasonably user-friendly information on thousands of SNPs, including dozens associated with this disease. Anyone can contribute to the site, although its moderators require that entries conform to a standard template, and that information on SNPs, diseases, and other traits have appeared in peer-reviewed journals, even if the studies are small and have not been replicated or validated.

The first SNP associated with breast cancer that I checked on the site was a rare marker located on chromosome 17 that showed up in one out of twenty women tested in England (men were not tested). The risk factor among the three thousand women tested was 1.72 times normal for the GG and for the AG variations. I'm an AG, making me high risk. But this was not the worst news. This same study, conducted by the Institute of Cancer Research in London, also

analyzed the impact of twenty-five SNPs that appear in five different genes, including the BRCA genes, a multigene analysis that conferred on me an alarming risk factor of eight times normal for breast
cancer—if I were a woman.

I had to catch my breath, feeling a wave of fear about what I might
have passed on to my daughter.

A few weeks later, I get back the Illumina SNP results for my family,
and sure enough, my daughter and I share the same variation of the single SNP discovered in the Institute of Cancer Research study. My father
has it, too, possibly passed on from his mother. My brother is negative for
this SNP, although my mother has a different genetic marker on BRCA2
that gives her a higher-than-average risk, despite having no breast cancer
in her family. I am unable to test my daughter on the 25-SNP package, since many of the SNPs for that study are not part of the standard
Illumina chip that was used to test her, although her positive result for
the single SNP and for another one on the BRCA2 gene causes me to
look into this further.

I find myself going through what thousands of people face who
have a possibility of carrying a gene for a dread disease or, like me,
have family members who have an increased risk of becoming
afflicted. Except that having one or two SNPs is hardly conclusive.
But it is enough to make me want to take a test offered by Utah-based
Myriad Genetics—a patented and expensive analysis that sequences
long stretches of the BRCA1 and BRCA2 genes and is considered to
be one of the most valid tests in genomics.

Myriad's breast cancer test was among the first serious genetic tests
commercialized for a common disease. It dates back to the early 1990s,
when scientists at UC Berkeley pinpointed the location of BRCA1 in
the human genome, and researchers at the University of Utah success-
fully cloned (copied) it. In 1994, Myriad, the University of Utah, and
the National Institutes of Health filed a patent on the test; in 1996, the
company began to offer it. Since then, over one million people, mostly
women, have taken progressively more detailed versions of the test
and another one for BRCA2, which was characterized in 1995.

Deciding to take the Myriad test gives me an opportunity to work
with a company founded by entrepreneur Ryan Phelan, an early devel-
oper of Internet health sites who also founded in 2003 one of the first
online genetic-testing companies. Called DNA Direct, it differs from
23andMe and most of the rest by offering serious medical diagnostic

tests, including Myriad's breast cancer assessment. Phelan lives on the San Francisco Bay in a renovated tugboat (circa 1912) with her husband, Stewart Brand, the cofounder of the Long Now Foundation, a futurist organization, and the publisher in the 1970s of the legendary *Whole Earth Catalog*. DNA Direct's office is also on the bay, in downtown San Francisco near the Ferry Building. One morning I visit Phelan and Trish Brown, the company's vice president of clinical affairs, to ask about taking the test for BRCA1 and BRCA2 and about waiving the $3,456 fee, most of which goes to Myriad, to get both genes fully sequenced. (A less expensive and less precise version is available from Myriad through DNA Direct for $620.)

They agree, but only after Trish Brown makes sure that I have a good reason to take the test. DNA Direct does not turn anyone away, but it discourages people from taking the test without having diseases running in their families or some other strong medical need. Before heading to DNA Direct's office, I check its Web site to see whether the BRCA tests are indicated for me. According to the site, people with the following history in themselves or their family should take the test:

- Breast cancer before age fifty
- Ovarian cancer at any age
- Breast cancer in both breasts
- Breast cancer and ovarian cancer diagnosed in the same person
- Male breast cancer
- Ashkenazi Jewish ancestry

My grandmother's history of having both ovarian and breast cancer qualifies me and my daughter for at least two of the criteria, says Brown, although my daughter is three generations removed. I ask Brown whether she thinks we should be tested, and she suggests that I take the test first. If I come out positive, then it would make sense to test Danielle, too. "Most people who take the test get back a negative," says Brown, who provides personal counseling to me throughout the testing process. Phelan adds that a negative for this gene does not mean that a person will not get breast cancer.

Soon after my meeting on the bay, a white box with blue lettering arrives at my door: the Myriad BRCA tests. I take the box to a blood-drawing center, which overnights a small tube to Myriad in Utah—and wait.

Weeks later, Brown calls to say that she heard from Myriad. Her voice is low and quiet and when she pauses, I have a momentary scare—could it be positive? No, she says, "it's negative." A few minutes later, she faxes me the report from Myriad that says in big letters: "NO MUTATION DETECTED."

In an e-mail connecting me to DNA Direct's Web site, I get a formal report from the company with several pages of information and a letter prepared for sending to my doctor, should I want to inform him of my test results. An excerpt from the report, which in total runs several pages, is below.

BRCA Sequencing Report for David

1 Your Test Results

You have been tested for an inherited condition known as Hereditary Breast and Ovarian Cancer, or HBOC. The test you took, which is called BRCA 1 and 2 sequencing, looked for gene changes in the BRCA 1 and 2 genes associated with hereditary breast and ovarian cancer. This testing was recommended because your paternal grandmother had both breast and ovarian cancer, and you had several positive BRCA SNPs (single nucleotide polymorphisms) on a genome wide array. The concepts of a SNP and a genome wide array are not reviewed in detail in this report, but if additional information is needed, DNA Direct would be happy to provide it.

Your test was negative. You *do not* have any known disease causing mutations in your BRCA1 and BRCA2 genes.

You do have several polymorphisms, reported by the lab via a special request. These polymorphisms are considered by the lab to be benign (non-disease causing) changes in the BRCA1 and BRCA2 genes. They are not associated with an increased risk of hereditary breast and ovarian cancer, which is why they were not reported in the typical lab report. These polymorphisms include:

  BRCA1 Q356R
  BRCA2 N372H (rs144848)
  BRCA2 K1132K (rs1801406)
  BRCA2 S241S
  BRCA2 IVS16-14c>T

An excerpt from the author's test results for BRCA1 and BRCA2, genes associated with breast cancer, prepared by DNA Direct.

As usual, my daughter is unfazed when I tell her about the negative results over Skype, her image wavering in the blurry reception. She has on a knit cap, and her blond hair is spilling out the bottom. She has just come back from kayaking with friends, and she looks both very grown-up in her college dorm room and also like the little girl who used to fall asleep on my lap in the car.

"See, you didn't have to worry," she says.

But she has no context to understand what a positive result might have meant. She has not known illness or death in a close member of our family. She was age five when my grandmother died, and this California-Scotland daughter does not see her uncle Don much in Maine. For me, the fact that she still has a risk of getting this disease from other SNP results will become a small bit of knowledge I tuck away to ponder in quiet moments.

Beyond the possibilities raised by my grandmother's medical history, I haven't expected any serious genetic issues for common diseases to come up for my family. Once again I am probably kidding myself, since it's difficult to find a family that doesn't have something unpleasant implanted in its DNA—a reality that unexpectedly becomes apparent on the day that deCODEme activates the results of my parents, brother, and daughter on its Web site, and I call up my father in Maine to go over his report.

I have not yet looked it over, but he has. I ask whether he has seen anything of note.

"No, there's really not much here," he says.

"Let's take a look," I say, clicking on the first disease listed on my dad's deCODEme Web page: age-related macular degeneration. Like me, my dad has a below-average risk factor on two SNPs and a very low 1.1 percent (out of 100 percent) lifetime risk of genetically acquiring this disease.

"It's good to know that I won't be blind when I'm old," he says, chuckling.

"There's still time," I say, teasing my ridiculously healthy father, who at age seventy-seven wears glasses only when driving and watching television.

Next on the alphabetical list of ailments is one of the most dreaded of genetic tests: early-onset Alzheimer's disease, the degenerative brain disorder that slowly robs millions of older victims of their memories and eventually their lives. This marker, located on the APOE gene, is the only one that pioneering geneticist James Watson did not want to know when he became one of the first people ever to have his entire genome sequenced in 2007.

"I have Alzheimer's in my family, and I just didn't want the aggravation of knowing," the eighty-year-old Watson told me, since there is currently no cure or effective treatment. That makes administering this genetic test controversial for some geneticists and bioethicists who believe genetic tests should not be given for diseases for which a person might get a positive for a disease that has no treatment.

"Tests should be medically actionable," says David Altshuler of Harvard. Others believe that testing positive or negative for Alzheimer's when it runs in families can provide some relief for those who are negative and can allow people who are positive to prepare for the possibility that they will get the disease.

I already tested negative for Alzheimer's, and since we have no dementia in the family, I expect the same for everyone else. When I click on my father's results, however, I am astonished to see that he has the medium-risk variant for Alzheimer's. He is an AG, which confers a 1.74 times normal risk for this malady—a medium risk factor. Having the double-trouble version of this SNP—a GG—is rare but confers a close to ten times the average risk. Most people who have a GG get the disease. Fortunately, my dad does not have this variation.

"Dad, I hate to tell you this," I say haltingly, "but you have a marker for Alzheimer's."

"I saw that," he says. "But I'm seventy-seven years old, and I feel fine, so who cares?"

Typical of my family, I think, to dismiss anything medical, although he had a point. At his age, the risk factor conferred by this variation didn't mean much. But what about the rest of the family?

Results for my mother and my daughter are the same for Alzheimer's as mine: an average risk. My brother, however, turns out to have the same medium-risk variant as my father. We both had a one-in-four chance of getting my father's one G and a 75 percent chance of getting two As, since our parents collectively have one G and three As.

Uh-oh, I think. My brother hardly needs to hear that he has had more genetic bad luck. I consider going the Watson route and staying quiet about these results, although, of course, my brother has access to his scores on the deCODEme site and either hasn't checked them or doesn't realize what they mean.

I take my concern to Kari Kaplan, a genetic counselor at Navigenics. She advises me that having my brother's version of the APOE gene is not a huge risk. "It's quite modest," she says. "The homozygote [GG] is the one to worry about." So I call my brother in Maine to tell him the news, very much aware that it is odd to be delivering this information as his brother. Over the phone I hear the audio version of a shrug.

"I'm not worried," he says, and I believe him.

I could go on with family results, but I'll write about just one more: the heart attack SNP that Kari Stefansson got worked up about and Harvard's David Altshuler warned me to take seriously. As I have reported, I got the high-risk version of this trait. But what about the rest of my family? The answer: no one else got the high-risk version of this SNP—as the diagram below indicates.

But my mother and father do have the raw ingredients to provide an offspring (me) with this undesirable variant. This gave Don and me each a 25 percent chance of getting either the highest-risk factor

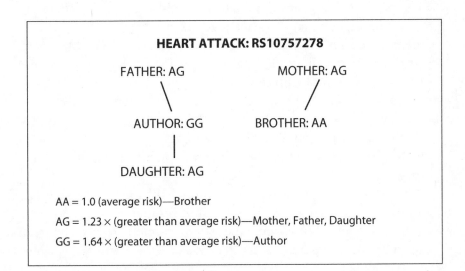

Diagram of how genetic variations are passed from parents to offspring.

or the lowest and a 50 percent chance of getting the medium-risk factor. This time, my brother caught a break and got the average-risk AA variant. My daughter inherited a medium risk, getting her higher-risk G from me and the lower-risk A from her mother.

When I tell my father that he and mom have given me the GG, he replies, "Sorry," sounding slightly amused, as if saying, "Well, what am I supposed to say? It wasn't my fault!"

Other highlights of my family's three-generation study of genetic markers are published on the Experimental Man Web site, drawn from the mind-numbing heaps of nucleotides tested: more than five million total DNA markers among the five of us. And this is just a tiny fraction of each of our entire genomes, which contains such a crush of data that at the moment it's difficult to make sense of it for any one person or for a family. Much of this information remains contradictory, like my heart attack data, which illustrates that using individual SNPs to predict the future course of a disease for a human being is problematic at present. Many more factors need to be factored in: age, diet, family history, and environmental criteria such as stress and exposure to toxins. This Age of Genetic Confusion will gradually fade away as software engineers design better programs to crunch through and analyze the data, and as clinicians validate the information on real patients.

I'm reminded of geneticist Jonathan Rothberg's comment about how fascinating it was to take the first look at James Watson's whole sequence. But I'm trying to imagine how he could possibly get a sense of all three billion base pairs or even the forty million lines of base pairs that make up the encoding part of our genes. Printed out on a spreadsheet, these encoding base pairs would need a wall 108 miles long to pin it on. My results for several thousand genetic markers on a chart would run about fifty feet if printed out and hung on a wall. (Following is a table with a small sampling of some of the results.)

After all of this testing, what my family has learned is something we already knew: that we should each live a normal life span and perhaps longer. This is an outcome that my grandmother, who charged me to make sure the clan carries on, was at least partly responsible for, as were generations of ancestors on both sides of the family who contributed to our DNA.

Three-Generation Study: The Author's Family (Selected Results)

Disease/Gene	Marker	Author	Risk	Father	Risk	Mother	Risk	Brother	Risk	Daughter	Risk
Arthritis											
APOE	rs440638	AA	0.51	AG	1.74	AA	0.51	AG	1.74	AA	0.51
PTPN22	rs2476601	GG	0.89	GG	0.89	GG	0.89	GG	0.89	GG	0.89
STAT4	rs7574865	GG	0.87	GG	0.87	GT	1.15	GG	0.87	GG	0.87
TRAF1-C5	rs3761847	AA	0.78	AG	1.03	AA	0.78	AA	0.78	AG	1.03
RA	rs2327832	AA	1.04	AA	1.04	AA	1.04	AA	1.04	AA	0.8
Colorectal cancer											
SMAD7	rs6983267	GT	0.99	TT	0.82	GT	0.99	GT	0.99	GG	1.2
Celiac disease											
HLA-DQA1	rs2187668	GG	0.3	GG	0.3	GG	0.3	GG	0.3	GG	0.3
IL21	rs6822844	TT	0.46	GT	0.73	GT	0.73	GT	0.73	GT	0.73
Novelty seeking											
DRD3	rs6280	CT	Med	TT	High	CT	Med	CT	Med	CT	Med
Psoriasis											
HLA-Cw6	rs1048454	CC	0.85	CT	2.67	CC	0.58	CT	2.67	CC	0.58
IL12b	rs3212227	AA	1.13	AA	1.13	AA	1.13	AA	1.13	AA	1.13
IL23r	rs11209026	GG	1.06	GG	1.06	GG	1.06	GG	1.06	GG	1.06
RA	rs13192841	GG	1.04	GG	1	GG	1	GG	1	AG	1

Rollo the Viking and me

Years ago in France, my mother met a short man with round cheeks and wrinkles around his eyes named Michel DuBosc, the French version of DuBose, her maiden name. The retired owner of a linen factory in Normandy, the seventy-something Michel has more than a passing interest in genealogy and can directly trace his family back to the early eighteenth century and, perhaps, indirectly back to Vikings who accompanied Rollo, the Norse conqueror, whose warriors swept into northwestern France in the ninth century. This was during the great era of conquest by the Vikings and the raids that took them in their sleek longboats as far away as Iceland, Greenland, and Vinland (Canada) to the west; to the Volga River and Moscow in the east; to Italy, Sicily, and Constantinople in the south; and to the eastern Mediterranean Sea.

When my mother met Michel, she had recently become intrigued by her DuBose roots, which ran back to Huguenots who had fled from Normandy in the seventeenth century to become landowners and gentry in Charlestown, South Carolina. As she later discovered, in France her family runs back further still to a line that includes a knight who accompanied William the Conqueror to the Battle of Hastings; a nobleman who fought in the First Crusade in the capture of Jerusalem; and a chancellor to King Charles VI (the Mad) of France in the late fourteenth century. This chancellor was named Nicolas DuBosc; he was also the bishop of Bayeux and a diplomat who negotiated a major peace treaty with the English in 1681. A painting of the chancellor-bishop in the Cathedral at Bayeux looks eerily like my mother's brother, Robert DuBose—or so I have heard from my mother and father. I have seen only blurry photographs of the painting.

"Could it be?" my mother asked after she met Michel. "Could he be a long-lost cousin?"

As genealogies go, my mother hit the jackpot with her ancient lineage, although once her family reached America, things did not always go well for her direct line. In a kind of reverse American-dream story, the family's time in the colonies began with Isaac DuBosc, a wealthy Huguenot who fled France when the Catholic king Louis

XIV cracked down on Protestants in the 1680s. Anglicizing his name to DuBose, Isaac bought up plantations and other property near Charleston and quickly became a prominent citizen of a town that still considers DuBose to be a venerated name. My mother's ancestral lineage then moved west over the next few generations to Alabama and to Camden, Arkansas, where the most recent generations live in rather humble circumstances in a slightly rundown Victorian house built by Isaac's great-great-great-grandson, Walter Winston DuBose, my mother's grandfather, who by all accounts was a genial man who did very little other than fish and play dominos.

Until we could run a genetic test, we had no real way of knowing whether Michel was related to my mother, since the two families do not directly link up in genealogical tables. Merely sharing a surname is not evidence of a common ancestor, since most people living in Europe centuries ago had no surnames until the sixteenth or seventeenth century, when many commoners took on the name of a local lord or whoever was the dominant aristocrat. Yet it was possible that my mother and Michel shared the DNA of a long-dead DuBosc ancestor, given the sheer numbers of descendants that anyone living six or seven centuries ago would have produced. For instance, Nicolas DuBosc, the chancellor, now has at least 130,000 living descendants, the product of seventeen generations of his progeny living, breeding, and dying since the fifteenth century.

This number is based on each descendant averaging two children, who themselves each had two children, and so forth, in an exponential growth pattern. By generation sixteen, my mother's father's generation, the number of offspring hailing from Nicolas would be 65,536; by generation seventeen, my mother's generation, it would be 131,072. Nicolas also had three brothers, who included the DuBosc my mother's family is directly descended from, Queffin du Bosc, the bailiff of Rouen in the mid-fifteenth century. All four brothers would collectively have close to a half-million descendants alive today, some retaining the DuBosc name. Add in the other members of this already large family, even in the days of Nicolas and Queffin, and legitimate descendants of the DuBoscs could number more than one million people. Two of these could be my mother and Michel.

In the summer of 2007, I met my parents in Paris, where they had organized a "family" luncheon with Michel and his family in a flat near the Eiffel Tower, the home of Michel's brother Yves. In my bag I carried genetic-testing kits with buccal swabs, long-handled cotton swabs that one uses to scrape against the inside of the cheek to snatch a few cells that can then be busted open to extract DNA. I had asked Michel whether he minded giving me a sample, which we would test against my mother's family's DNA to see whether there was a match suggesting an ancestral link.

The DNA region being tested is on the Y chromosome, the male sex chromosome that has proved useful to track ancestry because, unlike other chromosomes, it does not mix DNA provided by the mother and the father. From one generation to the next, it tends to stay stable, mutating very slowly. This allows geneticists to compare Y-DNA between two or more men to see how closely certain regions of the chromosome match up. Two people who are closely related, such as my father and myself, should have a near-identical Y-gene sequence, with no mutations or, at most, one mutation. Go back a few generations, however, and the DNA of two men living today descended from, say, the same great-great grandfather will be identical except for mutations that have occurred in the five generations between great-great-grandpappy and the two current-day fellows. The close match between them suggests that they share a recent common ancestor. Ancestral DNA companies and researchers can now determine (in genealogy-speak) the most likely estimate (MLE) for the time (T) when the most recent common ancestor (MRCA) lived.

The DuBosc flat in Paris was like an Edith Piaf song without the wartime sadness—at least, it was a space where I imagined Piaf would have felt at home. The furniture was mid-twentieth century, like the kind I remember in my grandmother's house, but with more of the ornamental flourishes that the French traditionally love: carvings on the dark wood of table legs and the backs of overstuffed chairs. The rooms were dark because in the rising heat of early summer, the shutters were partially closed, the Parisian answer to air conditioning. The meal was long, with several courses heavy on game and fresh fruit and cheeses—what we would call "organic" in California, and what they would call "normal" in France. In that country, food, I was once told, is an art rather than fuel, which is how it is too often perceived in the States. Each course had a different wine, from dry white to a sweet dessert wine. We had long, drawn-out conversations that were challenged by the language barrier. The DuBoscs spoke some English, and I speak a whisper of French, but we got by.

Toward the end of the luncheon, I pulled out the buccal swabs. Michel and his brother, Yves, dutifully scraped cells from their inner cheeks, and we tucked the swabs into their protective envelopes. The DuBoscs were amused by the process, although I reiterated to them and to my mother that it was possible there would be no match. They assured me that whatever the outcome, they enjoyed having new friends.

Walking out of the flat into the bright sunshine, I asked my mother, "Are you sure this is okay?"

She said, yes, although I sensed that she would be very disappointed if the match came up negative.

After lunch, Michel and his wife, a retired child psychologist, drove me to their home in Normandy, back to the DuBoscs' ancestral lands. They lived in a modernized sixteenth-century farmhouse near the small village where Michel had been born, about thirty miles southeast of Rouen on the Seine River. After leaving Paris in the DuBoscs' Mercedes, we passed through the rolling hills, farmland, and occasional tall hedges of northwest France, which were green and fresh as spring gave way to summer. Michel stopped several times at overlooks where we could see the bluffs of the river and ancient towns with castles, monasteries, and cathedrals, some in ruins, in the distance. Soon we were motoring through the heartland of the DuBoscs' ancestral lands; we arrived at Michel's farmhouse late in the day. His wife prepared a simple meal of chicken and apple cider from fruit grown on the property, and Michel and I sat down to talk.

He confided in me that he was pretty sure we were not related, and that his family was most likely descended from locals or tradesmen who took on the DuBosc surname when last names became fashionable two or three centuries ago. "We look like Celts," he said, referring to the most ancient stock of people in Western Europe. Traditionally, they are a smaller, darker-haired people than those descended from Rollo's Vikings and other blond, blue-eyed Northerners.

I pointed out that my mother's family has a mix of tall and short people: her father was only slightly taller than Michel, although his son, my mother's brother, stands tall at 6 feet 2 inches, and I'm just over 6 feet. "The Celts and Northerners must have been mixing in France for centuries," I say. Yet Michel has been unable to link his line to the main line of the DuBosc family, though he would love it if he could. He first got interested in genealogy in the early 1950s, when he read about Nicolas DuBosc, the chancellor, in a book he found: *La Généalogie de Chanceleiros de France.* He went on a search to trace his family but came to a dead end because most of the oldest records of the DuBosc family had been destroyed when the Allies bombarded German-held Rouen in World War II. As we sipped more wine in front of a big fire in the DuBosc farmhouse, Michel worried that my mother would be upset if the genetic results turned out to be negative.

When I returned home, my mother insisted that I carry on with the test. However, because it involved the Y chromosome, her buccal cells as a woman would not work. So I mailed a kit to her brother, my uncle Bob, a retired U.S. State Department officer and a former ambassador to Barbados, who was living in Harpers Ferry, West Virginia. He is the one who is said to resemble the portrait of Nicolas DuBosc.

Now armed with buccal swabs from both sides of the Atlantic, I mailed them to Family Tree DNA, one of the leading ancestral DNA-testing companies on the Web. Founded in 2000 by Houston entrepreneur Bennett Greenspan, the company has already tested 123,591 Y chromosomes as of spring 2008, according to its Web site, including those of more than 83,000 unique surnames. Greenspan's site partners with thousands of "surname projects": groups of people with the same or related last names who are linked to one another genetically by Family Tree's analysis. One surname project is DuBose, which includes DuBois, DuBoise, DuBoice, DuBosc, and other variants.* The DuBose project has its own Web site, with genealogies of people linked through their Y chromosomes. Once we got back the results from Bob and Michel, we would compare their DNA to that of people who were already on the DuBose site.

Bennett Greenspan became interested in ancestral DNA when he got stumped during his own genealogical search. He had found through conventional records the details of where one of his great-grandparents came from in the Crimea, near the Black Sea. He had found a man living in Argentina who also believed he was descended from these Crimean forebearers. But how could they be sure? Greenspan had read about a University of Arizona geneticist, Michael Hammer, who was a leading expert on the Y chromosome.

In 1997, Hammer and a team of colleagues published a study in *Nature* that used Y-chromosome markers to show that a large proportion

---

* Some of these names can also come from different lineages. The DuBois family, for instance, has a long line in France, although occasionally people from the DuBosc line called themselves or were called DuBois.

of Jews who traditionally belong to a clan called the Kohanim, which historically traces to Aaron, the brother of Moses, do in fact share a common male ancestor. They are linked by a Y-chromosomal pattern called the "Cohen Modal Haplotype," which consists of sequences of DNA that are passed down from one generation to another with little or no change. In another study, Hammer's team found that while Jews share genetic links with non-Jews from the Middle East (such as Palestinians, Syrians, and Lebanese), Jewish communities have generally not intermixed with non-Jewish populations, despite the complex history of Jewish migration out of Palestine in various diasporas. (The last major one occurred in 70 C.E., when the Roman emperor Vespasian dispersed the Jews after an abortive rebellion.) If Cohens had mixed with others in the many places where they settled after the final diaspora, then their descendants today would not share the same genetic signature in their Y chromosome.

Greenspan contacted Hammer, who helped him determine that the man in Argentina did in fact share a recent common ancestor from near-central Asia—a process that gave him the idea to start Family Tree. Greenspan's company also coordinates testing for *National Geographic*'s Genographic Project to use Y and mitochondrial DNA patterns—those stretches of DNA that are outside of a cell's nucleus and are passed down by one's mother—in people around the world to create a time line backward tracing how humans left Africa and spread around the world.

Several weeks after mailing in the swabs from Michel and Uncle Bob, I got an e-mail from Family Tree that contained the results. Thinking about my mother, I clicked on and took a peek.

"Uh-oh," was my first response: the two panels of DNA markers were not a close match. As I could see, the two samples differ on four out of twelve "loci," although not by much. "These men are not closely related," Greenspan would later tell me. Yet they were not unrelated; it's just that their link was centuries ago. I sent a note to my mother, trying to make the best of the results, and she interpreted what I wrote as great news in a letter to our French "cousins":

Dear Michel et Yves,

I just received the report on the DNA tests that David took in Paris and with my brother Robert (Bob) and we are related, that is, we have a 90% chance that we have a common ancestor or

relative in the last 23 generations (575 or so years ago) and a 50%
chance of a common relative in the last 175 years (1826). This
means that it is almost certain that we are truly cousins, even
if very far back (which I always suspected as DuBosc is a very
large family) and we spring from the same blood lines.

This is exciting!

Much love,

Pat

My mother was perhaps a smidge optimistic, although I was surprised
about the 90 percent match in the last five hundred years, given what
Michel had said about his suspected Celtic bloodline. "This is the
power, and the fun, of ancestral DNA testing," said Greenspan.

The rest of Bob's results from Family Tree traced his ancestry back
much further than even Nicolas DuBosc and Rollo the Viking. Each
human male belongs to a human haplotype that shows a pattern backward all the way to Africa, the mother continent, from which the first
bands of modern humans departed about fifty thousand years ago. Uncle
Bob turns out to be a member of the R1b1c haplotype, which traces
its lineage back to one man, a hunter-gatherer who is thought to have
arrived in Europe about thirty-five thousand years ago. His offspring
eventually settled in Italy. They then moved north into France, where
they painted the magnificent cave paintings along the Dordogne River
Valley at Lascaux and Altamira before retreating into Northwestern
Spain (Basque country) during the last ice age, ten to fifteen thousand
years ago. As the ice retreated, they moved into other parts of uninhab-
ited Europe, mostly in the west and the north. About 60 to 70 percent of
all males in the British Isles and Spain are R1b1c, and it is the prevalent
haplotype group for Western Europe but decreases among present-day
humans as one moves east.

On my father's side, I had my own Y chromosome run, testing
more than the twelve loci that were run on Michel and Bob. A sixty-
seven-marker panel offered by Family Tree is a high-resolution test
that gives a more accurate comparison between me and a potential
long-lost cousin to see whether we share a common ancestor and when
this forebear may have lived. The first results I looked at indicated that
I'm an R1b1c, like Uncle Bob, which tracks my father's family
on the same progression into Europe as my mother's line. Both sides of

the family moved out of Africa fifty thousand years ago and first moved northeast into Mesopotamia and then farther north to the Caucasus, finally turning west into Europe before the last ice age. This progression was verified by deCODEme, which runs tests on ancestry as well as disease. DeCODEme also compared my DNA to studies that break down all humans into fifty basic ethnic and geographically based haplotypes. Researchers are working on refining these designations to several hundred classifications.

My Y chromosome compared most closely to the "Orcadian" haplotype, which is associated with people who lived in Scotland in ancient times. My second-closest match is French, my third is Basque, and so forth, in a progression that matches in reverse the movements of the R1b1c group as its members slowly migrated north, then west. Here are my top six matches to ethnic haplotypes:

1. Orcadian (Scottish)
2. French
3. Basque
4. Tuscan
5. Italian
6. Russian

I am furthest genetically from the San people in southern Africa, but all humans, whether they are Orcadian or San, are still remarkably similar, perhaps less than .1 percent different, although recent studies suggest we may be as different as .5 percent. Here are the groups that are most distant from me:

45. Mandinka
46. Yoruba
47. Bantu South Africa
48. Biaka Pygmy
49. Mbuti Pygmy
50. San (Bushmen)

Unlike the DuBose side of my family, on the Duncan side I had no suspected cousins who were descended from long-ago ancestors: no Michels, or Yves. But I knew that there are many thousands of descendants out there with whom I would share common Duncan ancestors, if only I knew who they were. Without genetics, I would never know. With Family Tree, however, I was able to run a Y-chromosome search to see whether anyone I match with had signed up to have his Y chromosomes tested. Through Family Tree I also was able to link up with the "Clan Donnachaidh Project"—which is looking for genetic matches among Duncans—Donnachaidh being the Gaelic name for the clan that includes people named Duncan.

Before sending in my DNA, I already knew quite a bit about my Duncan family history. My knowledge started with my grandfather Duncan telling stories to me as a boy. A reverend in the Congregational Church, an architect, and a civic leader in Kansas City, Granddad claimed that we were descendants of King Duncan of Scotland, who was murdered by Macbeth (or by men loyal to this treacherous duke) in 1040, a story recounted with loose accuracy in the Shakespeare tragedy. As I grew older, I had my suspicions about this lineage, although my grandfather underscored our Scottishness by joining the St. Andrews Society of Kansas City, the local chapter of this national organization of Scottish Americans. One year he traveled to Edinburgh to purchase a kilt made out of the ancient tartan of the Duncan clan. As a boy, he took me to compete in races in the St. Andrews Society's annual "Highland Games," and on the birthday of Scottish poet Robert Burns I was forced to eat the national Scottish dish, haggis, made from various innards of sheep cooked in a sheep's intestine. I tried to learn the bagpipes around the age of ten but hyperventilated when I attempted to blow enough air into the bag that one squeezes to make the sounds. When my dizziness passed, my mother suggested to my grandfather that I stick with piano lessons.

My grandfather, like his father, was also a Mason. Masons are members of an organization that traces its history back to Europe in the Middle Ages. "My father loved the pomp and pageantry of these organizations," recalls my father. "He was sure that the link with the Masons included our ancestors in Scotland." By the late sixties, Granddad was a master of Kansas City Lodge 220 and the statewide grand chaplain of the Grand Lodge of Ancient, Free, and Accepted Masons of Missouri. The only other detail about our family history that I learned as a child was that the Duncan who originally came to Kansas City was named Nathaniel Ewing Duncan. My grandfather's grandfather, Nathaniel, had died before Granddad was born.

This was the extent of my knowledge when Granddad died in 1972, and for years afterward, until one day I was researching an unrelated topic at the Library of Congress in Washington, D.C. While waiting for some books to be delivered, I was killing time and I wandered into the library's busy genealogy room. It was filled with people painstakingly searching for tidbits of information that might reveal the identity and perhaps more about a grandfather or a great-grandfather: a mention in a census or on the manifest of a ship or in the arrival logs at Ellis Island. I strolled over to the rows of books on shelves, looked under "D," and took down a slim volume titled *The Story of Thomas Duncan and His Six Sons*, published in 1928 by Katherine Duncan Smith, then ninety-four years old. She was a prominent citizen of Birmingham, Alabama, but had been born in Pennsylvania in 1844. The book was filled with genealogical family trees and descriptions of various Duncans and had an index of names. I checked, and there was a Nathaniel Ewing Duncan listed on page 24. I turned to this page, and there he was, Nathaniel Ewing Duncan of Kansas City, listed as one of eight offspring of John Kennedy Duncan (b. 1803) and Anna Woodbridge Oliphant of Cumberland County, Pennsylvania. John was a cousin of the Alabama woman who wrote the book, and Nathaniel was John's third son, born in 1835. I flipped forward in the book to the early generations, and it became apparent to me that the Duncan clan originally came from Shippensburg, Pennsylvania, which they had helped settle in the eighteenth century.

In about five minutes, I had found what it often takes a lifetime to assemble: a whole family history starting with Thomas Duncan, who probably left Scotland for Pennsylvania in the mid-eighteenth century, where he was "a first settler, a brave pioneer of Christian civilization in a new country," according to Katherine Duncan Smith's book. "His father in Scotland was believed to be William Duncan," she wrote, a reverend from Perth who had a ministry near Glasgow. This William was "martyred in the time of Charles II," although it appears that he was an Anglican and belonged to the Church of England. This would not have made him a popular man during a time when the English were the overlords of Scotland, and the Scots were constantly fomenting rebellions and staging riots, often directed at the Church of England.

Once again, my family hit a genealogical bonanza with this history, which includes judges, doctors, large landowners, military officers, and a commodore of the U.S. Navy, John Duncan Elliott. Elliott was a hero in the War of 1812, when he helped rout British ships coming down from Canada on Lake Erie. He later became a notorious figure as the commodore of the U.S. Navy under President Andrew Jackson. He was a Jackson partisan who once tried to replace the maiden's-head ornament on the bow of his flagship, the USS *Constitution*, with a carved visage of his beloved president. He also once loaded Old Ironsides with wine and women to celebrate a victory over pirates during the 1820s. Nathaniel, a second cousin to the commodore, had a far less distinguished life. The only notable moment for him was becoming a first lieutenant adjunct during the Civil War and serving in a regiment from Iowa, where his family had moved sometime before the start of hostilities. Nathaniel was wounded and captured by the Confederates at the Battle of Shiloh on April 6, 1862, and was returned to his unit after a prisoner exchange in November 1862.

Adding to this trove of stories was a box that I discovered one day in my grandmother Duncan's garage—the grandmother who survived breast cancer. My grandfather had died years earlier, and she had no idea where the box had come from. It contained letters, documents, a statement from the U.S. Army announcing Nathaniel's 1897 death in the Soldier's Home in Fort Leavenworth, Kansas, near Kansas City, and buttons from an old army uniform. One of the most intriguing documents concerned his older brother, Ashbel Duncan. Folded up in an old leather packet was a certificate printed on thick paper and decorated with a blue ribbon. At the top are the all-seeing Masonic eye and the words: "Master Masons Diploma." In faded ink, the paper announced that Ashbel F. Duncan had become a Master of the lodge of Fayette County, Pennsylvania, based in Uniontown. The paper was signed by several members of the lodge and dated March 14, 1864—six months before Ashbel, a captain in the Union Army, died leading a successful charge against the Confederates near Richmond, Virginia.

Another document in the box was a letter written by Nathaniel's younger brother, Fidelio. He was a private in the Union Army writing from Harpers Ferry, West Virginia, just a few weeks after Ashbel's death. The missive is penned in beautiful script on the eve of what was apparently his first battle. It is addressed to his and Nathaniel's father (my great-great-great-grandfather), John Kennedy Duncan. Here is an excerpt:

Dear Pa,

I wrote to you yesterday, but as I leave for the front today, thought I would let you know of my departure. I am about to engage in the realities of war. I will be thrown among enemies but can look back with pleasure at the dear ones at home. . . .

Perhaps I may never be permitted to see you all again but I leave me in God's hands. He will do all for the best. If it is my lot to be taken in this war may I die as my dear Brother [Captain Ashbel Duncan] did doing my duty facing the foe with the Star Spangled Banner floating proudly over me. . . .

I am ordered to get in line to start. Good bye to all. Pray for me. Much love to all from your most affectionate son,

F. H. Duncan

The day after writing the letter, Fidelio was killed in battle. Nathaniel survived his wounds and battled on until the end of the war. He moved back to Iowa, where he divorced his first wife and married a much younger woman named Sarah Ray. After moving to Kansas City sometime after 1869, he was plagued by his war injuries and did little else that has been recorded, other than sire my great-grandfather Harry and two more children.

Less than twenty-four hours after I requested that Family Tree look for Y-chromosome matches, I got an e-mail from a woman in New York City named Kathy Duncan Crawley announcing that we were related. Since Kathy lacks a Y chromosome, the actual DNA link was from her forty-five-year-old brother, Daniel Earl Duncan, of Susquehanna County, Pennsylvania. I had a near-perfect match with Daniel, who is a mere one marker removed, giving us a 78.37 percent chance of having a common ancestor within the last six generations. In fact, we do share a common predecessor five generations back: Nathaniel's grandfather, Samuel Duncan. Kathy Crawley's ancestor was Nathaniel's uncle, also Samuel, who Kathy told me had moved with Nathaniel's father, John Kennedy Duncan, to Iowa, probably in the 1850s. Samuel later moved to South Dakota, where Kathy was born a few generations later, in 1954. When Kathy was a girl, her family moved to San Diego, where her policeman father was a homicide detective who died at age forty-two of an aneurysm after a motorcycle accident. Kathy's widowed mother then moved the family to her hometown of New York City.

On a rainy-cold afternoon in Manhattan, I met Kathy and her daughter, Kaitlyn, at a café in midtown, where I was surprised to meet a "Duncan" with dark red hair, hazel eyes, and a Mediterranean coloring to her skin.

"That's from my mother's side," said Kathy, "she's one hundred percent Italian."

So we naturally ordered cappuccinos and began to talk about the Duncan family. Kathy had done a great deal of research.

"I'm mildly obsessed," she said.

"More than mildly," said Kaitlyn, rolling her eyes.

The big family mystery, Kathy said, is the Scottish connection. Was the "martyred" Reverend William Duncan in Glasgow really the father of our ancestor who migrated here?

"Have you found a genetic link yet in Scotland to confirm we are indeed from Scotland?" I asked.

No, she said, most of the people tested on Family Tree were from the United States.

"Hmm," I said. "I think I can at least place us in Scotland with some testing I did a few years back."

This was during a visit to the United Kingdom for my original *Wired* article testing my DNA, when I visited geneticist Bryan Sykes at Oxford University. He's a professor of human genetics at the university and the author of the best-selling *The Seven Daughters of Eve* and other books about tracing ancestry through DNA. He had agreed to test my DNA through his then new company, Oxford Ancestors. Sykes had first made headlines in 1994 when he used DNA to directly link a five-thousand-year-old body discovered frozen and intact in an Austrian glacier to a twentieth-century Dorset woman named Marie Moseley. This stunning genetic connection between housewife and hunter-gatherer launched Sykes's career as a globe-trotting genetic gumshoe. In 1995, he confirmed that people purported to be Anastasia, the daughter of the last czar of Russia, Nicholas II, were impostors by comparing DNA of their remains to that of the czar's living relatives, including Britain's Prince Philip. Sykes also disproved explorer Thor Heyerdahl's *Kon-Tiki* theory by tracing Polynesian genes to Asia and not to the Americas.

Before heading to England, I had e-mailed a swab of my cheek cells to Sykes, a linebacker-size fifty-four-year-old with a baby face and an impish smile. In his lab near Oxford University, he delivered my Y-chromosome results, which he had run against a database of ten thousand other men's Ys to see which profile was closest to mine and whether any of them share a common ancestor with me.

After entering my Y DNA code into his laptop, Sykes looked intrigued, then surprised, and suddenly moved to the edge of his seat. Excited, he reported that the closest match was, incredibly, him: Bryan Sykes! "This has never happened," he said, telling me that I am a mere one mutation removed from him on the twelve-marker test and two from the average profile of a Sykes. He had not collected DNA from many other Duncans, he said, although it appeared that sometime in the last four hundred years or so, one of his Sykes ancestors, who lived just south of the Scottish border in Yorkshire, must have ventured over the border and had a child with a Duncan, or vice versa. "That makes us not-so-distant cousins," he said. We checked a map of Britain on his wall, and sure enough, the Sykes family's homeland of Yorkshire is less than two hundred miles south of Perth, the heartland of the clan and the original home of the Reverend William Duncan. I was also closely matched with other Duncans in Sykes's database who lived in Duffus, Grampian, and Peterhead, which confirmed that we have some genetic connection in Scotland in the last three or four hundred years, which Kathy Crawley was very happy to hear.

Sykes then tested me using a second method to trace human ancestry through genetics. It involves that band of DNA located in the mitochondria—the structure floating around in every cell that acts as a power factory for the cell, converting fuel (glucose) into energy. Eons ago, when life was young on Earth, mitochondria were parasites, or perhaps symbionts, that merged with early versions of cells and stayed, becoming an integral part of life, with their own DNA arranged in a circle—called mtDNA—separate from the double helix in the nucleus. Like the Y chromosome, mtDNA does not recombine when mom and dad mix up genes. Passed down through our mothers, it stays relatively stable over long stretches of time, with occasional mutations providing the same sort of timeline markers as the Y, except that it provides a matrilineal, rather than a patrilineal, history.

According to Sykes, deCODEme, 23andMe, and others, I belong to the mitogroup designated as H, part of the super-mitogroup R. H-ers like me and my mother can trace our lines back to one woman who lived thirty thousand years ago, possibly in the Middle East. Her offspring soon moved into Europe, following the boys (my Y forebearers)—or perhaps it was the other way around—south into lower France and Spain during the last ice age, then north and west into France and Britain. Almost half of all modern Europeans are descended from this woman, whom Bryan Sykes gave the name of "Helena," one of the seven daughters of Eve in his book. Some famous H people (Helena's children) include, according to deCODEme, Marie Antoinette, Prince Philip, Susan Sarandon, Warren Buffett, and the Apostle Luke: a troupe of ancestors that covers pretty much the spectrum of possibilities, from a tycoon to a dilettante, a snob, a saint, and a liberal actress.

Sykes had broken the news to me about Helena over dessert in Oxford, handing me a colorful certificate, signed by him, that heralds my many-times-great-grandma. He told me that she lived twenty thousand years ago in the Dordogne Valley of France. More interesting is the string of genetic letters from my mtDNA readout that indicates that like most Western Europeans, I'm predominantly Celtic, which makes sense. But other bits of code in Sykes's test reveal traces of DNA that make me Asian and about 3 percent African, which I already knew, but he also suggested that I have a smidgen of Native American and even Southeast Asian.

I told Sykes that the American Indian DNA didn't surprise me, but Southeast Asian? Where did that come from? None of my ancestral groupings for my Y or my mtDNA people ever got close to Southeast Asia. Sykes laughed. "We are all mutts. There is no ethnic purity. Somewhere over the years, one of the thousands of ancestors who contributed to your DNA had a child with someone from Southeast Asia."

"This is not serious genetics," Sykes added, "but people like to know their roots. It makes genetics less scary and shows us that, through our genes, we are all very closely related."

As I finished that journey into my ancestral past, I was aware that my DNA carries more information about my origins and lineage than simply what concerns the last fifty thousand years or so since *Homo sapiens* left Africa. The DNA that makes me human is a tiny part of the sequences that run up and down my double helix. What I wanted to do next was push back much further along the timeline of my nucleotides, long before humans and even most mammals. This is possible because DNA is an extraordinarily well-conserved molecule that keeps adding new codes to its original template as evolution works its way down the ages, retaining much of what came before it in the 3.5 billion years since life first appeared on Earth. Most of this conserved DNA is dormant and no longer functions; other bits are crucial for virtually every organism to function, including those creatures that have not roamed the Earth for eons—but remain a part of us.

My dinosaur DNA

I'm in Bozeman, Montana, to find out what ancient DNA might be preserved inside of me. Specifically, I'm checking to see whether I share a genetic sequence with a *Tyrannosaurus rex* that died sixty-eight million years ago in what is now Hell Creek, an isolated canyon hundreds of miles to the north, near the Canadian border. I'm running my fingertips over a stump of the *T. rex*'s cool, hard femur bone, or what's left of it after scientists sliced up and pulverized most of it in search of specks of soft tissue that miraculously held a fragment of the osteocyte cells that make collagen. This should have decayed eons ago. Instead, paleontologists discovered microspecks of a protein with an apparent similarity to type I collagen, a material found in creatures with bones—including humans. Type I collagen is made according to instructions from the COL1A1 gene that Peter Byers at the University of Washington in Seattle investigated for my brother, Don.

Paleontologist Jack Horner is holding the truncated femur in a basement storage room filled with old bones wrapped in heavy plastic. It's part of a subterranean research complex under Bozeman's Museum of the Rockies that also includes rooms for cleaning and sorting bones, analyzing them, and preparing them for exhibits. "Incredibly, we found an intact fragment of a protein in this femur," says Horner. Around us, a platoon of scientists, technicians, and graduate students from the University of Montana work at long tables on dinosaur bones as a fine coating of dust hangs in the air, a mix of plaster, dirt, and ground ancient bone. One room houses an X-ray machine and electron microscope; another one a CAT scan used by Horner's team to discover intact dinosaur embryos inside eggs millions of years old. This is a place where grown-ups get to play with dinosaur bones, says a grinning Horner.

In the 1970s and 1980s, Jack Horner made discoveries that helped redefine our knowledge about the terrible lizards that once roamed the planet. Nearly all of the thousands of dinos he has found come from Montana, where the rugged topography of rock and soil has been worn down by the elements to occasionally reveal a fossil. It's dry and desolate here now and cold much of the year, but during the dinosaur era, this region had a climate more like Louisiana's, Horner tells me, with lush foliage, swamps, and warm temperatures year-round.

As a scientist, Horner is most famous for discovering with his colleague Bob Makela that some dinosaurs were sociable, built nests, and nurtured their offspring. They discovered several nests in western Montana where a trove of some two hundred *Maiasaura*—this name, given by Horner and Makela, means "good mother lizard"—had been discovered, from embryos in eggs to adults. *Maiasaura* were large, duck-billed dinos that lived seventy-six million years ago, gathering in herds that could number in the thousands. Horner has had two dinosaurs named after him—*Achelousaurus horneri* and *Anasazisaurus horneri*. Horner has also studied the growth and development of dinosaur species from babies to adults, showing in some cases that specimens that were thought to be different species are actually the same species at different ages. In the museum, he shows me a dozen triceratops skulls, from a youngster dino with a small head, hood, and horns to an older triceratops with a massive head and hood the size of a cow. In 2000–2001, Horner's teams discovered several remarkably well-preserved *T. rex* specimens much larger than any previously found, including a monster they dubbed the Custer *T. rex*, which weighed well over ten tons, more than a semitruck.

The discoveries of this gentle man with the wild beard and hair were picked up by Michael Crichton and incorporated into the character of Alan Grant, a fictional paleontologist working in Montana in the novel *Jurassic Park*. When Steven Spielberg made the original movie and then *Jurassic Park II*, Horner was hired as an adviser. "They even gave me a chair with my name on it," he says. For *Jurassic III*, directed by Joe Johnston, Horner was even more closely involved.

"Steven and Joe have very different styles," says Horner. "Steven— well, he's Steven, and you don't just hang out with him. With Joe, I saw the editing process, everything." Horner did his best to keep the depiction of the dinosaurs up-to-date with the latest science during the course of the three movies, which included making *T. rex* in *Jurassic*

III a scavenger. "We now think that they were not predators," he said, which is unfortunate for the lawyer and other characters in the first movie who got munched by a *T. rex*.

Horner is shy and sometimes lets his sentences trail off. With his beard, he looks like a Santa Claus in sneakers and his staff clearly adores him, from the woman who works in the museum gift shop to colleagues in his lab. Horner has a touch for fund-raising and despite his shyness is at ease with Hollywood stars and billionaires. George Lucas is a donor; so is Silicon Valley financial titan Tom Siebel, the namesake of the museum's Siebel Dinosaur Complex. Susan Brewer, the former wife of actor Peter Fonda and the mother of actress Bridget Fonda, is a close colleague in the lab. As I walk past her bench with Horner, she looks up from some bones she is studying and says hello.

Horner loved working on the *Jurassic Park* films, he says, but he disagrees with the central scientific conceit of both the book and the movie: that dinosaurs could be re-created by using their DNA, which in the story was preserved in the bellies of Jurassic-era mosquitoes that had fed on their blood and were then trapped in sticky sap that turned to amber. "It's highly unlikely that could happen," says Horner. "Too many things would have to go right to preserve an entire genome of an animal in a mosquito. The DNA would break down almost immediately because of digestion in the mosquito, for one thing." But Horner, with a mischievous smile, says it could be possible to re-create dinosaurs using a different method.

As Horner explains, dinosaurs are still with us. "They're now called birds," he says. As pretty much any nine-year-old who is fascinated with terrible lizards can also tell you, modern avians share bone structures and other features—including, in some species, feathers—with dinosaurs. This strongly suggests that dinosaurs, such as the ferocious *T. rex*, evolved into modern birds. The short protein fragments that were recovered from the *T. rex* bone found in Hell Creek in one experiment matched most closely to the collagen in a chicken. (This analysis was organized by a former student of Horner's, Mary Schweitzer of North Carolina State University, and was published in *Science* magazine in 2007.)

Horner told me that chickens retain dormant sequences of DNA that would, if activated by bioengineering, cause a chicken to grow teeth and a dino tail and to sprout little *T. rex* arms instead of wings.

"We don't know how to flip these genetic switches," says Horner, "but we might one day find out. The question is: should we?"

The *T. rex*–chicken connection has led the whimsical Horner to propose that a real Jurassic Park could feature a "chickenosaurus" as an attraction. This might not equal the punch of a real *T. rex* running amok, eating attorneys and threatening cute children, or packs of viciously intelligent velociraptors scrambling between the legs of majestic brontosauruses, but those sharp little chicken teeth might hurt like heck if they grabbed onto a finger.

Strange as the *T. rex*–chicken story is, the 68 million years separating these two animals covers a mere 2 percent of the total time that life has been present on Earth. By the time the *T. rex* roamed the swampy bayous of Jurassic Montana 68 million years ago, the basics of genetics and cells had already been established for billions of years: genes with four nucleotides and proteins made from twenty amino acids. This system is present even in the oldest fossilized creatures ever discovered, single-cell organisms that lived in the Archean Age, from 3.8 or 4 billion to 2.5 billion years ago.

Some of the most ancient DNA inside me and you is our mitochondrial DNA. Scientists think that the original mitochondria were simple microoganisms that did not have a nucleus to house their DNA. Instead, they had DNA free floating, arranged, as I described earlier, in a circle, rather than a double helix. Roughly 1.7 to 2 billion years ago, more complex cells that had developed a nucleus absorbed these microbes, and the two creatures developed their symbiotic relationship. Other very old genes have to do with basic protein functions that exist in nearly all organisms, such as converting glucose into energy and maintaining cell membranes.

Since this book is not a lesson on the details of evolution, suffice it to say that as the years passed, organisms on Earth evolved from one cell each to the first multicellular creatures, such as sponges and algae, and onward to the divergence between plants and animals—though humans retain in their genomes elements of each stage of evolution. For instance, we share about 20 percent of our genes with plants such as rice.

Jumping ahead a few hundred million years from the plant-animal divergence, today you and I share about 75 percent of our genes with the tiger puffer fish. This animal has had such a singular success at surviving that it still lives today, having stayed largely the same for 450 million years. This is when the evolutionary split occurred that left one branch of this evolutionary line to remain puffer fish and another to eventually become humans and thousands of other species. The "shared" DNA between humans and puffer fish ranges from the basic genes of life to those that have only a loose similarity, having changed in terms of size, location, and in many cases function since our ancestors split apart.

Leaping ahead again, we share nearly all of our genes with mice, which humans diverged from about seventy-five million years ago, getting us up close to the time of the *T. rex*. The details of human and mice genes often differ, however, which accounts for the rather glaring discrepancies between us and our tiny cousins, such as size and life span. If we jump ahead in time past the *T. rex* for a moment, the organism that humans are closest to genetically is the chimpanzee, from which we diverged only five million years ago. We share some 98.7 percent of our total DNA with chimps.

When Horner and his crew discovered the intact protein fragment inside the *T. rex* femur, they compared the sequence to several modern animals and found that the closest match was, sure enough, a chicken. "It's a nearly identical match," says Horner.

To find out whether the *T. rex*–chicken sequence has also been conserved inside a human like me, I asked the biologist and entrepreneur Nathaniel David to compare the *T. rex* protein fragments of what appears to be the type I collagen protein to a public database containing the human version of this protein. David is the chief science officer of Kythera Biopharmaceuticals, a Los Angeles–area biotech company that is developing drugs based on, among other things, manipulating collagen.

"The recovered dinosaur-protein fragments were short (less than 20 amino acids in length)," he e-mailed me, "while the human-collagen protein is over 1,000 amino acids in length. But those we had were either perfect matches or different at only one amino acid position." Following is a comparison of one of the sixty-eight-million-year-old *T. rex* fragments to the human collagen sequence; each letter represents an amino acid:

T. rex:	GATGAPGIAGAPGFPGAR
Human:	GAPGAPGIAGAPGFPGAR

The only difference is a single amino acid, T (in bold), meaning that the *T. rex*–derived fragment contained a threonine amino acid at that position, while humans have a proline.

As David said, however, these eighteen amino acids tell an incomplete story about the evolutionary similarity of human and *T. rex* type I collagen. "We would need more of the protein to compare to the human to see how this protein has changed in evolution," wrote David, "or, even better, the DNA. But DNA isn't chemically stable enough for a *Jurassic Park*–style comparison." The experiment did prove, though, that there is at least a little bit of *T. rex* preserved in me and you, despite the vast gulf in time.

Back in Horner's lab, he's showing me a slice of dinosaur bone under a microscope—I marvel at the beautiful brown, yellow, and black patterns that look a bit like thin slices of petrified wood. A poster of *The Lost World* is on the wall above us. He explains in an excited, boyish voice how you can estimate the age of an animal through the scope. He points out holes where blood vessels once ran and green splotches that he says are bits of collagen. He also shows me a paperweight-size bone fragment that he discovered as a boy near his home in Shelby, Montana, which helped inspire the young Horner to devote his life to dinosaurs.

I ask him how he has come up with such unique departures from accepted knowledge about dinosaurs and evolution. He looks up and smiles.

"If you have preconceived ideas about science, you might as well do something else."

You show me yours, I'll show you mine

Being the gatekeeper for my family's DNA for this project has put me in the peculiar position of delving into other people's secrets, even if these people are my close relatives. How strange, I thought, that I have four files in my computer, each containing more than a million genetic markers for my mother, father, brother, and daughter.

In the past, sharing deeply personal information about everything from potential addictions to a proclivity for prostate cancer would not have been a topic of general discussion in my family—or in most others, I would imagine. I suspect this will change as the personalized medicine revolution unfolds, but for now, what lies hidden in our DNA is not a topic most families are likely to chat about over dinner, much less among friends and acquaintances. Yet I wonder— as we await results from various studies on how early adopters of genetic testing are reacting—how the small but growing number of testees outside of my family circle might be reacting to their results. Were other people more inclined to worry about their health after taking the tests and sweating over even the smallest of high risk factors?

Are some early adopters actually seeking out others to compare their results with? 23andMe claims it has the genetically curious interacting via a share feature on its Web site, although I'm not sure how many people are actually participating in this intriguing experiment. While we wait to find out if personal DNA will soon appear on Facebook alongside one's age and relationship status, I decide to share mine with a friend who is not in my family but is a paragon of early adopters: *Wired* maverick editor Kevin Kelly.

I visit him on a sunny day in the usually fog-bound town of Pacifica, California, down the coast from San Francisco, where he is soon showing me his, and I am showing him mine. He was recently tested on 23andMe and will later be tested on deCODEme, and he has no problem sharing his risk factors which are spread out on two giant flat-screen monitors in an office filled with computers and a wall of books two stories high. Kevin has cherubic cheeks and a rustic-looking beard sans mustache that runs from ear to ear, giving him the look of a Mennonite farmer. A longtime proselytizer for technology and for using technology to enhance the self and to define and redefine groups, Kevin is a beloved figure in Silicon Valley as a tech guru and a thoughtful optimist about human possibilities.

He is typically gung-ho about genetic testing, although he admits that morning in mid-2008 to being underwhelmed so far by the seventy or so traits then available on these direct-to-consumer sites.[*] Unlike me, who got most of these tests gratis, Kevin is an actual customer. He paid nearly $1,000 each for 23andMe and deCODEme.[†] "I'm not

---

[*] The sites are adding new genetic markers all the time; these numbers are current as of mid-2008.

[†] In September 2008, 23andMe lowered its price to $399.

learning that much about myself—yet," he says with a shrug, but he believes this will change. "It's like the first personal computers or fax machines. They were clunky and there wasn't much you could do with them." They were also very expensive for what you got. "These sites cost too much," he says, "but the first PCs cost, like, five thousand dollars, so this is to be expected. I assume it will get much cheaper."

Kevin asks whether I have found anything serious, and I pull up my contradictory heart attack results.

"Well, that's a bit confusing," he says. We pull up his heart attack results on 23andMe, and he shares with me the higher-risk variant of a marker in that cluster of heart-attack indicators on chromosome 9. We both came out GG, which gives us a 1.23 greater risk of heart attack than normal—not a huge risk factor, although I'm glad we both ordered healthy sandwiches for lunch.

We keep checking. Like me, Kevin is average or below average for most diseases, with the occasional slight bump up in risk factors. (See the table below.)

The main exception is a disease that he says runs in his family: glaucoma. I come out CC on this, baseline normal, but he comes out

DNA Disease Risk Factors: Comparison of the Author and Kevin Kelly (KK)						
Trait	Gene Marker	Risk	Author	Risk	KK	Risk
Age-related macular degeneration	rs1061147	A	CC	0.34	CC	0.34
Colorectal cancer*	rs6983267	T	GT	1.03	GT	1.03
Exfoliation glaucoma*	rs2165241	T	CC	Baseline	TT	7.2
Heart attack*	rs2383207	G	GG	1.23	GG	1.23
Obesity*	rs3751812	T	GG	0.8	TT	1.49
Restless leg syndrome*	rs3923809	G	AG	0.74	AG	0.74
Type II diabetes*	rs7903146	T	CC	0.82	CT	1.15

All traits are from 23andMe.com.
* Also on deCODEme.com.

with a TT, which confers a very-high-risk factor, about seven times normal in a study done by deCODEme in Iceland. (See the list below.)

Who	Possible Genotype	What It Means
David Duncan	CC	Baseline odds of exfoliation glaucoma
	CT	Substantially increased odds of exfoliation glaucoma
Kevin Kelly	TT	Greatly increased odds of exfoliation glaucoma

So far, Kevin has no symptoms of this disease, although in his mid-fifties he is approaching the age when it might begin to manifest. Or he may never get the disease, even with this alarmingly high-risk factor. "This is the kind of information that can be useful," he says. "If I came out baseline, I might not take it as seriously, even if it does run in my family."

I ask Kevin why he spent the money on these tests. "I am interested in finding out how to quantify who I am as an individual, the quantified self, and to see what technology can contribute to this project." He has started a blog and discussion group with *Wired* writer Gary Wolf called the "Quantified Self" (www.quantifiedself.com) to post his and others' musings and discoveries about not only genetics, but any other devices, algorithms, formulas, or knowledge that can improve one's self-knowledge.

"But what happens to this search if some of this information is faulty or not yet ready for prime time?" I ask.

"It's not useful if this information is incomplete or inaccurate," he says, "but the best way to fix this is with more and better information."

Kevin and I also check out our genetic variations on 23andMe that might be called recreational, fun, or just plain wacky. These are DNA markers for wet or dry earwax, the sprinter's gene, and a marker that increases one's chance of becoming a heroin addict. I'm not kidding. Kevin and I both are at high risk to ride the white horse, according to a 2004 study conducted in Sweden.

Here is what 23andMe says about this test:

In the brain heroin is converted to morphine, an opioid painkiller. Morphine acts by signaling through a receptor encoded by the gene OPRM1. Different versions of the OPRM1 gene are thought to affect how much morphine one needs to feel a given effect. This study of 139 heroin-addicts (primarily Swedes) and 170 non-addicts found that people with one or two copies of the G version of the SNP rs1799971 have almost 2.9 times the odds of being a heroin addict.

A 2.9 times the normal risk factor is relatively high, although 23andMe assigns the validity of this data only two stars (out of four) because the researchers tested only about three hundred people, giving these results a low statistical power to predict whether others, such as Kevin and me, will really become addicts. For this reason, 23andMe calls such findings "preliminary."

Just for the record, Kevin Kelly and I are not addicted to heroin.

Most of these somewhat vague behavioral traits and attributes appear only on 23andMe, which at the time featured results for about seventy genetic traits, compared to twenty-six for deCODEme and seventeen for Navigenics (these offerings vary slightly according to gender and sometimes ethnicity and age). This reveals a difference in assumptions and culture among the sites. The company 23andMe has close ties to Google and the Web 2.0 crowd and tends to treat nearly all genetic results as equally fascinating bits of information, listing and describing earwax right up there with age-related macular degeneration and colorectal cancer. Each of the seventy-plus traits it offers is accompanied by detailed reports about the trait, studies, risk factors, and links to more information.

"All of our information is from peer-reviewed journals," said Brian Naughton, a computational geneticist and the founding R&D architect of 23andMe. Like many of the staff at this company, Brian is young. He finished his Ph.D. recently, in 2006. "We believe that this information should not be withheld," he said, sitting in a small conference room at 23andMe, which is located in the heart of Silicon Valley just off US 101. The spartan headquarters has walls covered with whiteboards that have goals and information scribbled on them. Upstairs is a pile of workout equipment, a nod to the focus on health and lifestyle that pervades the valley's culture. "We are working hard to go through as many studies as possible. We rate the information, but we're the provider of the information, not a second-layer peer-review process."

"People don't have to buy the service," added Andro Hsu, a young Ph.D. biologist trained at UC Berkeley. "This is for people who want to build a picture of themselves. If they have a gene marker for curly hair, that helps build the picture. It's not for everyone."

At the opposite end of the spectrum of these online sites is Navigenics, which offers only information on medical conditions, not on bitter taste or heroin addiction. In between the two is deCODEme, which offers mostly disease information but tosses in some mildly recreational results, such as bitter taste. Both 23andMe and deCODEme also provide results for ancestral data. It turns out that through our Y chromosome, Kevin and I are descended from the same group of ancient humans who came out of Africa and eventually ended up in Europe. But Kevin and I differ in our lineage through our mitochondrial DNA. My mtDNA signature connects me with a group that most Europeans belong to. Kevin's mother's lineage links him up with a different group that was more recently in the Middle East and also includes Kurds, Druze, and Ashkenazi Jews.

For many other traits, Kevin and I tested the same, for better or for worse. For instance, both of us have a high probability of having blue eyes and for being lactose tolerant, things we knew even without a genetic test. And neither of us has a variation that causes some people to get flushed cheeks when they drink alcohol, although this variation occurs far more often in people with an East Asian ancestry than with Caucasians. Both of us have a proclivity to learn from our mistakes: a useful thing for writers and geneticists. Less attractive is a DNA marker linking us in another very small study to a reduction of 3 points in our IQ scores, whatever that means. We diverge, however, on a marker for risk taking: I am a medium risk taker and Kevin is a bigger risk taker. He has a gene marker that is often found in sprinters, while I have the version of the same marker that is associated with athletic endurance. (See the following table.)

Then there is one of my favorites: a marker associated with rapid caffeine metabolism (as I suck down another latte while writing this). I have a genetic variation that is linked with being able to drink coffee all day with no added risk of heart attack, although it is a study that needs to be verified. Poor Kevin has a variant that links caffeine consumption to an increased risk of heart attack—if the heroin doesn't get him, the caffeine will, although he tells me he doesn't drink anything with caffeine.

DNA Traits: Comparison of the Author and Kevin Kelly					
Trait	Gene Marker	Author	Risk	KK	Risk
Alcohol flush*	rs671	GG	Normal	GG	Normal
Avoiding errors	rs1800497	GG	Avoids errors	GG	Avoids errors
Blue eyes*	rs12913832	GG	Blue eyes	GG	Blue eyes
Bitter taste*	rs713598	CC	No bitter	CG	Bitter
Earwax	rs17822931	CC	Wet	CC	Wet
Heroin addiction	rs1799971	AG	Higher risk	AG	Higher risk
Endurance (or sprint)	rs1815739	TT	Endurance	CT	Sprinter
Intelligence	rs363050	GG	Lower IQ	GG	Lower IQ

All traits are from 23andMe.com.
* Also on deCODEme.com.

About half of these traits get 23andMe's top, four-star rating, although even this rating includes studies with only a thousand people tested to make a link between a genetic marker and a trait or a disease. Most geneticists consider a test group of only a thousand people to have a low statistical strength for common traits compared to studies with thousands or tens of thousands of subjects tested.

None of the sites says how many people have paid hundreds or thousands of dollars to use the site. I suspect as I write this that it's not a large number, just as it was rare in the mid-1980s to see someone using the first clunky, huge—and expensive—mobile phones. Even Kevin balked at paying $2,500 for Navigenics, after shelling out for 23andMe and deCODEme. "It's too much money for what you get," he says. But mobile phones got smaller, sleeker, more useful, and much cheaper.

One's genes, though, are not cell phones or computers; they are part of what makes us who we are, and they can give us clues to how we will live and die. That's why it's crucial to get this right. It's also why I suspect that Kevin Kelly and I will be exceptions, at least in the near term, in revealing and sharing our genetic data with others—even with friends. As bioethicist Arthur Caplan has told me, "Most people are nervous about revealing genetic information. They are afraid that insurers or employers or, in some cases, friends and family will find out."

I ask Kevin Kelly whether he's worried about this, and we both invoke *Gattaca*, the 1997 film starring Ethan Hawke and Uma Thurman that imagines a world where work, sports, insurance, politics, and relationships are determined by genetics. In a *Gattaca* world, if you fail to measure up genetically—say, you have a high risk of an early heart attack or a high risk of being a heroin addict—you are not allowed to hold certain jobs. You won't be elected to public office, and you are told to get lost by potential dates. In this world, forget keeping your DNA private; it's deposited everywhere you go. You kiss a date, and the residue from your lips and saliva can potentially be tested for compatibility. Tiny flecks of skin and hair (presumably with intact DNA inside recoverable cells) can be collected and analyzed to make sure you measure up or that you have the right sort of genetic future.

The movie, written and directed by Andrew Niccol, puts too much weight on the role of DNA in determining destiny. One's environment, neural circuitry, and many other factors are also great influencers—as we shall see later in this book. But he does suggest a world where our DNA is laid bare for all to see. In one scene, a flashback, the lead character, Jerome, describes his birth and the almost instant genetic panel that is taken that will dictate his future life as a genetic undesirable—which his father, Antonio, watches on a computer screen:

> Antonio turns his attention from his baby to the data appearing on the monitor. We see individual items highlighted amongst the data— "NERVE CONDITION—PROBABILITY 60%," "MANIC DEPRESSION—42%," "OBESITY—66%," "ATTENTION DEFICIT DISORDER—89%"—

> JEROME (VO)

> My destiny was mapped out before me—all my flaws, predispositions and susceptibilities—most untreatable to this day. Only minutes old, the date and cause of my death was already known.

I doubt that even in a world where DNA tests are highly accurate people will be as paranoid as those in the movie. Indeed, the plot describes how Jerome overcomes his high probability of having a fatal coronary by his passion to succeed and by sheer force of will—two factors that may actually have some genetic basis but are also among the many factors that must be considered when assessing someone's possible health future.

Another key theme in the film is the utter lack of privacy of genetic data in a society that puts up virtually no real barriers to checking and to abusing information about people's DNA. The first part of this equation, lack of privacy, is something that every geneticist and many legal experts and politicians have told me will likely be all too true in the real world of the near future. "No one can keep this information secret," says geneticist Francis Collins. But he and others believe that strong laws can be passed to prevent abuse.

Last year, Congress passed a law long championed by Collins and other geneticists called the Genetic Information Nondiscrimination Act (GINA). Prompted in part by the appearance of online, direct-to-consumer genetic testing sites, the Senate and the House passed the act by overwhelming margins, and the bill was signed by President George W. Bush. It forbids insurers and employers to use a person's genetic data against him or her—something that I'm particularly relieved to hear, given that I'm publishing my genetic results and have wondered what my insurance company might do with the information. (I got no response when I contacted my insurer to ask them.)

Even with the law, it's likely that as genetic tests become more common and less expensive, abuses will occur. Already, law enforcement authorities in Britain and in some municipalities in the United States are collecting or talking about collecting vast databanks of DNA to search for matches in crimes. This might help locate and nab criminals, but it also exposes innocent people to scrutiny by governments without their consent.

There is little to stop someone right now from testing anyone they want to in any lab that can run scans of DNA using a SNP array. Let's say I'm at a lunch for the president of the United States, and I quietly snag a glass he's been drinking from. I send it in under a pseudonym to one of the online sites and get back results ranging from a proclivity to depression to a high risk for Alzheimer's. How would this information affect the political landscape and our view of the commander in chief, even though he or she might never get these diseases? Linda Avey of 23andMe told me that her company requires so much spit that it would be a red flag to get a tiny sample or a vial that was mostly water. But the technology exists to run a genetic scan on even a smidgen of material, as long as intact cells with DNA are present.

We may see a new type of genetic paparazzi who try to snag DNA from celebrities, taken perhaps from a water glass or a fork. BRAD PITT'S GENES REVEALED! could be on a future cover of *People* magazine. And what's to stop an enterprising candidate for the senior vice president position at work from snooping on the genetics of a rival candidate for the job? "We don't want people misusing this information," says Gregory Stock of the University of California at Los Angeles Program on Medicine, Technology, and Society. "We don't want to be electing people based on genetic traits that are poorly understood and don't mean much."

Genetic information may be used to test for undesirable traits but also to single out super-athletes and geniuses. Or we might start judging CEOs on whether they have a sketchy genetic marker for intelligence or for learning from their mistakes, or one that gives them a higher-than-normal risk factor for dementia or schizophrenia—when they may never suffer from these diseases. And what about heroin addiction?

Worried about publishing my friend Kevin Kelly's genetic results in this book, I e-mailed him to ask again whether it was okay. Ever the technology enthusiast, he sent me back a one-word response: "Absolutely." As gung-ho as he is, however, none of us early genetic adopter-exhibitionists really knows whether we are being stupid, since we have no clue about what might be hidden in our genes that has yet to be discovered. Perhaps more unforgivable is that we may be releasing information that our offspring in coming generations would rather have kept as quiet as possible—if, in fact, anyone is able to keep his or her DNA secret.

Genes 'r' us

Last spring, I was in New York on a fresh spring day for the launch of Navigenics' new storefront in SoHo, possibly the first retail outlet in the world selling DNA tests for diabetes, heart attack, and celiac disease. The ultra-high ceilings, exposed brick, and chic scuffed floor (a relic of this building's past as a Lower East Side industrial shop) looked like a SoHo gallery or a designer-fashion boutique. Boxes of Navigenics kits in colorful boxes with spit containers inside were stacked neatly on a long counter, and kiosks with displays about genetics and computers with Navigenics demos were spread throughout the store.

Arriving a few hours after the store officially opened its doors at 9 a.m., I was impressed by the aesthetics of the place, which seemed like an inspired idea for a San Francisco genetics company that was trying to make a name for itself beyond the Bay Area. I wondered, though, how many people were going to walk in off the street and buy a genetic scan of their disease potential for $2,500. But it suggested what might be coming in the future. Will Navigenics storefronts one day be as ubiquitous as Starbucks, offering up menus of custom scans of this or that DNA for customers paying perhaps as little as $25 or $50? Maybe those two stores will join forces to offer customers a skinny half-decaf mocha-chai latte with extra foam while they wait for their results for 3,170 genetic traits.

At the opening party, about two hundred scientists, investors, academics, and journalists sipped wine and listened to the company's two founders and two key investors talk about genetics on a panel moderated by Greg Simon, a former adviser to Vice President Al Gore and the chair of Navigenics' Policy and Ethics Task Force. Except for Simon, who now heads a patient advocacy group called FasterCures in Washington, D.C., members of the panel (and many of the guests) come from the Bay Area—seeing them gathered for the launch of an edgy new tech company reminded me of the constant stream of high-gloss fetes during the dot-com era in San Francisco. There are some parallels between that heady era of experimentation and this new age of online genetics, as it touts another disruptive technology aimed at revolutionizing people's lives and at making money while trying to sort out exactly how to make this potent combination actually work. Like early dot-coms, the online DNA companies are testing business models, formats, and methods of delivering new information—with the added challenge that some of this information might be upsetting to people who get bad news about their health. What will work and what will crash and burn remain as mysterious and exhilarating for this new enterprise as selling socks and catnip on the newfangled "Internet" was in the mid-1990s.

A core component of the companies' dot-comish strategy is to take their products directly to consumers, circumventing physicians and a traditional health-care system that has just begun to integrate and perhaps to offer some of the tests for common diseases. This has created the sort of techno-gap Silicon Valley loves: where new technologies and discoveries offer new ways to organize, analyze, and sell data that the brick and mortar outfits—in this case, hospitals and doctors—are slower to accept. The online genetic companies have even embraced the populist ethos of the early dot-com era, with Navigenics CEO Mari Baker telling me, "We believe it is a fundamental right for people to have access to their own DNA. They will own their own DNA and their results. It's up to them whom they want to share it with; it's their call. The results will be sent to the patient, not to a doctor."

Another strong link with dot-coms and Silicon Valley that night in New York was the presence on the panel of John Doerr of Kleiner Perkins Caufield & Byers, the powerhouse venture capital firm. Doerr, a billionaire and an early funder of Google and Amazon, among many others, sits on the Navigenics board and led Kleiner's investment in the company. His partner, biotech venture capitalist Brook Byers, was also speaking alongside Navigenics' cofounders, oncologist David Agus and geneticist Dietrich Stephan. Another more recent Kleiner partner was in the audience and also said a few words: former vice president Al Gore. He told me later that he hasn't been tested yet on the Navigenics site, but he was considering it. Doerr said he had been tested, but he was mum about his results.

Navigenics' business plan differs from its competitors' by ignoring ancestry data and proclivities such as bitter taste, sprinting, and novelty seeking to focus exclusively on genetic markers associated with disease. Navigenics also charges more: $2,500 for about seventeen disease indications (this is at the time of this writing; the company expects to increase the number of diseases over time, and I suspect its price will go down), with a $250 annual renewal fee, a huge leap up from the $1,000 or so charged at the time by the other two sites. CEO Mari Baker defends this by saying, "We'll have fifty conditions offered within a year, and a hundred within three years. What value do you place on that if you find out something important about your health?" Baker, a small, feisty woman with short blond hair and warm but confident eyes, discovered on her Navigenics profile that she is at high risk for developing celiac disease—an intolerance of glutens in wheat and other grains—which runs in her family. "I have cut out glutens," she says. "That means wine instead of beer!"

She says that the site also has licensed information about diseases from the Mayo Clinic in Minnesota and is working on setting up collaborations with Mayo and other leading clinics and hospitals. Its focus is more traditional than deCODEme and 23andMe in how the site looks: it seems more inspired by WebMD, with smiling doctors and serious-looking graphics, than by, say, Google Earth. (At first, Navigenics was also the only site among the three that offered genetic counseling before and after the tests. Now deCODEme offers this service, too.) "People might be freaked out by their results," Navigenics counselor Kari Kaplan tells me. "That's why we feel that we need counselors to put results into context for people."

Navigenics also plans to include testing and information on genetic markers that indicate whether a person can respond to or metabolize certain drugs and on environmental factors such as smoking. "Environment is huge," says Navigenics' Michelle Cargill. "Africans and Asians have a low risk of heart attack, for instance; then they move to America, and their risk goes up. We're interested in adding age and ethnicity." Navigenics, like 23andMe, offers a score about the relative influence of genetics and the environment for different diseases. Heart attack, for example, comes out as 57 percent caused by genetics and 43 percent caused by environment, a determination that Cargill says is based on the best data available, which she admits is not always complete or reliable. "But it's getting better," she says.

The second major online "experiment" is 23andMe, which has been the least shy about including genetic markers for everything from avoidance of errors and heroin addiction to cluster headaches and diseases that I have never heard of, like ankylosing spondylitis, a rarely occurring inflammation of the spine and joints. Some geneticists have criticized the site for including studies of small populations that are not well validated.

"I think that many of their markers are not ready for prime time," says Michael Christman, a geneticist and the CEO of the Coriell Institute for Medical Research in Camden, New Jersey. Linda Avey, the cofounder of 23andMe, tells me that they are actually being conservative. "We could include many more," she says. "There are people who want to know this information, and they have a right to it," adds Andro Hsu, 23andMe's content manager. "This is not an iPod; it's not for everyone."

In addition, 23andMe has implemented a star-rating system, which I wrote about when I visited Kevin Kelly. This is a great idea, although the bar seems a bit low for its top-rated studies, which get four stars for testing at least a thousand people and getting independently verified in at least one additional study. "A thousand people in most cases is not enough," says Steve Murphy, a physician and geneticist and the founder of a personalized health practice called Helix Health in New York City.

The company 23andMe also has a tech-celebrity connection with its cofounder, Ann Wojcicki, who is married to Google cofounder Sergey Brin. Google is an investor in the company, and Brin has said that he wants to unleash Google's search technology to better organize genomic information, something that is desperately needed. Last spring, Google Health launched a program for patients and hospitals to organize their medical records; and one hears rumors that Google Health will one day include personal genetics as well. Some observers think, and many technophiles hope, that 23andMe may be a toe-dip in the water for Google to gauge the reception and viability of genetic information sites, but so far Google has remained silent about its genomic plans.

The third member of the genomics triumvirate, deCODEme, differs from the other two by being both a pharmaceutical development company—with two drugs for heart disease being tested in humans—and one of the leading gene-hunting research labs in the world.* Many of the gene markers used by the other two companies were discovered by deCODEme researchers; the company is also offering diagnostic tests for physicians to use for diseases such as heart attack and diabetes. Kari Stefansson is adamant that the other two companies are mere "computer companies" that know little about genetics—a not entirely accurate assessment, given that Navigenics and 23andMe both have prominent geneticists working for them and helping them assess SNPs to be included on their sites. But he does have a point, that for more than a decade deCODEme has been on the scientific cutting edge of genomics. Its business model to patent and sell the most useful of its tests as medical diagnostics ordered by physicians—as opposed to the direct-to-consumer "information" on multiple diseases offered on the Web sites—also sets deCODEme apart from other purveyors of genetic data.

---

* As of this writing, deCode was experiencing financial difficulties in the wake of the 2008 banking crisis and announced that it would be spinning off some assets.

When these sites went live, the medical establishment was largely caught off guard. "Physicians lack the training to understand this information," Francis Collins, formerly with the NIH, told me at the time, "and hospitals are not ready to know what to do with it, either." The government and regulators were also largely unprepared. In 2006, the Federal Trade Commission issued a *Facts for Consumers* report that warned consumers to be wary of companies that offer do-it-yourself genetic tests and promise to tell you that your genes can reveal a definitive risk factor for developing a particular condition. And over the years, a handful of genetic tests have been approved by the Food and Drug Administration (FDA), such as the test taken by cancer patients to see whether they are positive for a mutation of the HER2 gene: a test that indicates they can take Genentech's drug Herceptin. But the vast majority of genetic tests have not gone through a regulatory approval process.

The catch-up has begun, however. "These companies are forcing physicians and hospitals to think about what this information really means," says geneticist James Lupski of Baylor College of Medicine. His institution and several others are developing new curriculums to train doctors in basic genetics, while Navigenics is sponsoring a Continuing Medical Education course—online, of course—for doctors that by the fall of 2008 had already been completed by twenty-five hundred physicians.

Government is also stirring with the passage last summer of the Genetic Information Nondiscrimination Act (GINA) within six months of the launch of 23andMe and deCODEme, and within a few days of the launch of Navigenics. The last Congress also considered two new bills to more closely regulate genetic testing and information, one cosponsored by then Senator Barack Obama and the other by Senator Edward Kennedy, Democrat of Massachusetts. I suspect that the current Congress—and the new president—will revisit this issue.

Regulators are also reacting. Within weeks of the company's going live, an advisory committee at the Department of Health and Human Services was calling for tighter regulation of consumer genetic tests, warning that they were often marketed with little scientific evidence of their usefulness to individuals. The panel called for the FDA to require evaluation standards to prove the usefulness and validity of these tests and asked for a mandatory registry of all laboratory tests. State governments, too, are looking into setting regulations and investigating companies to make sure they comply with state laws.

Last year, New York State sent warning letters to several online genetic companies informing them that they might be in violation of state laws governing medical testing. California health officials issued cease-and-desist orders to thirteen genetic start-ups, including Navigenics and 23andMe, demanding that their tests be validated by science and approved by a physician. Both companies argued that they were selling information, not medical tests, though both were able to come to terms with the state by showing that their test results were based on peer-reviewed science and after they agreed to include physicians in their testing process.

My suspicion is that this will not be the final word for regulation at the state or federal level of these Web sites and of genetic information.

"We still have relatively little oversight for this information," says Greg Feero of the NIH. "Right now, these tests are being interpreted all over the map, which you have seen for yourself with your heart-attack results. You should not be getting contradictory results like that. There should be a *Consumer Reports*–type of rating on the clinical validity and utility, so that a doctor would feel comfortable saying, 'Let's go on to Navigenics or some other source and check on your prostate markers and diabetes markers.'"

Feero's former boss, Francis Collins, believes that guidelines are necessary, though it is unclear whether they should come from the National Institutes of Health and the federal government or from a private or nonprofit *Consumer Reports* sort of service. "We don't want to squash these new companies," he says, "but we also don't want the public turned off because they have received bad information, or if they are frightened by risks that either aren't well validated or are very low."

When things shake out, it's likely that serious medical information will be available online but will flow through physicians and the health-care system—possibly through Navigenics and similar companies. The information will be tested following guidelines that are either agreed on voluntarily by everyone or required by the FDA or some other agency. "My vision is to have all of this information available to a physician," says Baylor's Jim Lupski. "This will take some serious rethinking, for physicians to learn how to use this information for predictive and preventive medicine, but it will happen." Financially, this is where the big profits will come from, says venture capitalist Doug Fambrough of Boston-based Oxford BioSciences Partners. "To make money, you have to tap into the medical market, and you have to get insurers to pay for it," he says. Nondisease traits—eye color, ancestry, risk taking—will most likely be available sans regulation through online sites.

The closest example right now of a model linking up the Web and tradational medicine is Ryan Phelan's DNA Direct, which tested me for the BRCA genes for breast cancer. It proffers mostly pure diagnostic tests, those approved by the FDA or in wide use by physicians. The company offers seventeen online tests, with its information aimed at people who have some indication that they might contract a disease. This could be either a family member who has it or at the suggestion of a doctor, although anyone can order the tests. Costs range from $199 for simple tests to $3,456 for the most complex version of the BRCA panel. The fees also cover phone access to genetic counselors.

The online experiment is not being conducted only by for-profit companies. In Camden, New Jersey, just outside of Philadelphia, the Coriell Institute is preparing to launch a nonprofit version of an online consumer genetics site that places genetic testing in the context of tra- ditional medicine. For more than fifty years, Coriell has been the pre- mier lab for creating cell lines and preserving them and other biological samples in cryogenically frozen storage. In 2007, its new CEO, Michael Christman, left Boston University's Center for Human Genetics in part to organize the Coriell Personalized Medicine Collaborative, a research project that aims to sequence the genetic markers of a hun- dred thousand people. The Collaborative plans to post results and information on an online account such as 23andMe and the other com- mercial sites, although, like Navigenics, the Collaborative will include only well-validated markers for disease. It plans to offer this service for free, paying for it with grants that also have paid for a new state- of-the-art genotyping lab.

I visit Coriell on a drizzly afternoon in early spring, arriving in the rundown center of Camden. I pass boarded-up houses to reach Coriell's stout brick headquarters across the street from the sprawl- ing campus of the Cooper University Hospital. Cooper is part of the new genetic Collaborative, along with other medical institutions and organizations in the area. "One of the first groups we are testing is physicians at Cooper Hospital," says Michael Christman. "Most doc- tors have no idea what to do with this information, so this is a way to teach them and to provide them with an example of what the infor- mation looks like." Christman says that they will include only genetic markers that are medically actionable. "We want patients to be able to do something about their risk factor—either a treatment like a drug or through lifestyle changes such as a better diet."

"We're the nonprofit alternative," he says. "This won't be direct to consumer." This is physician-based, he adds, with an oversight committee composed of geneticists, physicians, ethicists, and community members, and chaired by Erin O'Shea, a Harvard and Howard Hughes Medical Institute chemist. The committee will decide which markers to include with an "up or down vote," says Christman.

During my visit, in a small room upstairs from Coriell's huge repository of shining silver cryogenic tanks, I give up more spit to be run on a genetic array chip—and I am still awaiting my results.

Another focus of genetic testing that has stayed under the radar are traditional diagnostics companies such as LabCorp and Quest Diagnostics, which are the largest providers in the business of testing patients for everything from cholesterol and HIV to cancerous tumors. Quest, for instance, earns $1.3 billion a year, a sixth of its revenues, from molecular diagnostic testing, which includes numerous DNA tests. I visit one of its genetic-testing facilities outside of San Juan Capistrano, California—the Nichols Institute, named for Albert Nichols, who founded the institute in 1971—which is perched in an ultramodern black-glass building that follows the contours of a ridge above a dry, rocky canyon deep inside a county wilderness area. The sign in the parking lot says to watch out for rattlesnakes.

Quest offers several dozen genetic tests, many for rare disorders such as fragile-X syndrome or Tay-Sachs disease. All tests are ordered directly from physicians and hospitals, who get the results and send them to their patients as they do for most traditional medical tests. The reports include several pages of explanatory material and an invitation for a physician who may not fully understand the test or its implications to consult with an expert. "We have a small army of counselors—genetic counselors, M.D.s, and Ph.D.s—to handle all the calls," says Joy Redman, the manager of the genetic counselors in the company's western division.

I am surprised at the number of tests offered by Quest, given all of the talk about how physicians and the medical establishment do not understand or use much genetic testing. It turns out this is not entirely true. Quest runs five hundred thousand genetic tests a year; a big seller is one that checks for cystic fibrosis. This has become a common test for parents when they get pregnant. Mom and dad are tested to see whether they are carriers; if they both are, then the parents can choose to test their unborn child, since the baby will have a 25 percent chance of having the disease. "The American College of Medical Genetics says that every couple should be tested," says Charles "Buck" Strom, the medical director of the Genetics Testing Center at Quest. He and several other senior scientists and genetics specialists are sitting with me inside the Nichols conference room to talk about their business and to give me my results for a panel of tests the company ran on my DNA.

These tests are different from my previous tests on genetic markers. Some check for SNP variations, but they also look for abnormalities in longer sequences of DNA and in insertions, deletions, or copy variations of sequences. "We can run any test here, including low-volume tests for rare diseases," says Raj Pandian, senior laboratory director of this facility. "We have the technology to run more than just SNPs." For me, they have run tests for about twenty diseases, ranging from fragile-X to tests to see whether I properly metabolize various classes of drugs. "You came out normal on every test except one," says Buck Strom.

"Which one?"

"You were heterozygous for a mutation that concerns metabolizing certain drugs." He shows me the result on a Quest report: it's for the Cytochrome P450 gene, which is important for metabolizing some drugs. "You are a medium metabolizer," says Buck Strom. "This means that you do not metabolize drugs for depression and other mental disorders as well as people who don't have your variation. These are drugs like Prozac; the class is called selective serotonin reuptake inhibitors. It's not a big problem. You probably should take a slightly higher dose of these drugs"—should I need them.

"That's one of the more useful results I've had," I say, "if I ever get depressed."

"Yes, it does show how useful some of these tests can be. If you had been homozygous, you would have a high risk of not metabolizing these drugs. They would have little or no effect."

"Do you have any plans to offer a panel like I took for healthy people?" I ask.

"No," says Buck Strom. "We offer tests that have clinical validity, that doctors order because they have read in the literature that they should, or because they know it will be useful in treating their patients. We have no plans or desires to sell these tests directly to consumers."

"We do think that they might be useful for preventing illness," says Raj Pandian.

"We are moving very aggressively to develop tests that doctors want to use to make predictions about future diseases," agrees Strom, "but we will develop these tests with physicians."

As I leave Quest, walking gingerly to avoid rattlers, I feel more strongly that a traditional medicine model like Quest's will one day merge with a more consumer-directed approach, where serious medical tests and analyses will be done with a physician's involvement, while other genetic information will be handled by finding results online.

A couple of months after Navigenics launched in SoHo, I happened to be in the neighborhood and walked over to see how it was doing. I was surprised to see that the chic storefront had vanished. A designer clothing store was in the space, with fashionable dresses and gowns hanging on racks instead of the sleek kiosks and the graphics about DNA. Later, Navigenics' cofounder Dietrich Stephan told me that the space had been rented temporarily for their opening. "Some people didn't realize this, and they think we went under or something," he says with a smile. Navigenics did not go under, but the disappearance of something so real associated with genomics, which remains highly abstract, unintentionally suggests that the business of genetics will be ephemeral for some time to come. Perhaps we aren't yet ready for DNA to be proffered like lattes, although I suspect that DNA testing will one day be as ubiquitous, perhaps even in a storefront near you.

Ready for prime time?

Several months into my genetic journey, I am sweating on a cross-trainer at the gym, pondering my results. I am now armed with thousands of genetic markers associated with hundreds of diseases and other traits, with much more information to come. I have learned my genetic proclivities for diseases that I've never heard of, discovered genetic cousins in France and New York, and connected myself genetically to a terrible lizard that died sixty-eight million years ago. But what have I really learned that is useful in informing me about my health-care future and, more philosophically, about who I am?

Like Kevin Kelly, I am mostly underwhelmed by the offerings on the commercial genetic-testing sites. Many of the results confirmed what I already know: that I'm healthy with some modest risk factors for a few diseases. The main exception is my risk for heart attack, which left me confused because of the contradictory results. I'm confident that the scientists and those who interpret genetic results will sort this out, but so far, for me, the results are jumbled enough that they leave plenty of room for me to indulge my tendency to shrug and dismiss them.

Some of these tests have proved useful and even crucial to other people. For instance, thousands of pregnant women and fetuses are tested for rare genetic disorders such as Tay-Sachs disease, which genetic counselor Kari Kaplan told me was beginning to disappear as many parents opted not to deliver fetuses that test positive for this genetic disorder. Tests such as BRCA1/2 have been life-saving for thousands of families that carry this mutation; others such as those gene tests that inform us if we metabolize pharmaceuticals properly can also prevent side effects and tell us if a drug will work for us. But for most common diseases that afflict millions of people, the science remains young and the predictive power sketchy.

"The use of this data for individuals is in many cases a little bit ahead of reality," says Francis Collins.

Harvard's David Altshuler is more outspoken. "These are not clinical tests. They shouldn't be outlawed, but there is no value in testing an individual for one million SNPs. We don't scan with an MRI for everything; those who do are scamming people. I don't order every test for every person. This is a science fiction mind-set that is not based on common sense. To think about this is clinically simple-minded. It's the idea that just knowing something is useful—well, maybe, maybe not."

Navigenics cofounder Dietrich Stephan argues that for some SNPs, the risk factors are well enough understood that physicians can start incorporating them into a diagnosis of, say, heart attack. In a recent presentation to a group of scientists (see the list on the following page), he listed the following risk factors for heart attack (I have added the bold):

LDL (bad) cholesterol greater than 160	1.74
HDL (good) cholesterol less than 35	1.46
Smokes	1.71
No exercise	1.39
9p21 (chromosome 9) DNA markers	**1.72 (high risk)**
MTHFD1L (gene)	**1.53 (high risk)**

"The genetic risk factors are similar to other known risk factors like smoking," says Stephen. "They can be used by physicians as one more element in a patient's diagnosis."

Most doctors have yet to embrace this contention. "As a family doctor, I don't think there is sufficient evidence for action for most of the genetic tests for common diseases," says Greg Feero of the NIH. "It's more important to focus my time with patients on things like smoking cessation and losing weight." Josh Adler agrees, telling me that he is not yet ready to accept that genetic markers for heart disease have an equal validation to cholesterol tests or to the impact of smoking. These tests have been proved in millions of patients for many years, he says. "The genetic tests don't have that kind of confirmation. Until they do, many doctors are going to have a hard time believing them." For me, he says, the information did not convince him to alter his diagnosis that I have a basically healthy heart, with a borderline level of cholesterol that we should monitor regularly. "These tests will be useful," says venture capitalist Doug Fambrough. "In ten to fifteen years, everyone will be sequenced, and there will be a system in place to use this information for health care."

On the cross-trainer, I wrap my hands around the grips and check my pulse: 129. Normal. Thinking about Kari Stefansson's call from Iceland warning me to go on statins for my heart, I pump harder, and harder still, and top out at 158. Also normal. I can feel my heart pumping, sending blood through my body. I feel great. I tell myself that it will take a lot more than a high-risk variation in the rs10757278 marker to convince me to take a pharmaceutical every day for my heart.

As I continue my tests for the Experimental Man project, I may yet be persuaded. But so far, the vagueness of genetic information that fails to balance all of the factors that might contribute to what my ticker may do in the future does not outweigh my lifelong belief that I am healthy.

After my workout, I'm still in gym shorts as I say good morning to my daughter in Scotland over Skype. I suddenly am reminded of when she was born, feeling for the first time her heart, then the size of an acorn, beating as I held her close. Thankfully, her experimental daughter results came out better than mine for heart disease, but there is the uncertainty of the breast cancer information and possibly other risky genes as yet undiscovered that are hidden inside her.

So much is unknown but on the verge of being discoverd. This is exciting as a science project but very personal where Danielle and my two sons are concerned—and for their children to come—although I suspect that by the time they are my age, panels of genetic risk factors will be as common as taking a pulse.

But genetics is not the full story. As we learn more about our DNA, it will be integrated with other cutting-edge technologies and data. That's what the rest of this book is about: how the programming we have been born with intersects with our environment and with that most mysterious organ, the brain—and how much science can tell me about how this is integrated into physiological systems and pathways and, ultimately, how it fits together in the entire body.

As I sign off with my daughter on Skype and watch the screen go dark, I wonder what factors I have contributed, with her mother, that protect her, or not, from the world we live in: from the chemicals, the sun, and the stress of modern life, and even from the unnatural curiosity that her father has in asking all of these questions in the first place.

2

ENVIRONMENT

Full of rebellion, I would die,
Or fight, or travel, or deny
That thou has aught to do with me.

—GEORGE HERBERT, "NATURE," 1633

Genetics loads the gun, but
environment pulls the trigger.

—JUDITH STERN, UNIVERSITY OF
CALIFORNIA—DAVIS

Light my fire

In a light Swedish accent, the scientist on the phone is asking me, "Are you sitting down?" Yes, I say, sipping a cup of coffee in a crowded food court in San Francisco International Airport, where I'm waiting for a flight.

Åke Bergman, a chemist at Stockholm University, pauses for an uncomfortable beat or two. I had asked him to help me with what would become the next phase of the Experimental Man project: how the environment outside of our bodies affects us, as opposed to our internal genetic programming. Several weeks earlier, a lab had tested levels of environmental chemicals inside my body—320 chemicals that I might have picked up from food and drink, the air I breathe, and the products that touch my skin. It was my own secret stash of compounds acquired by merely living. The list included older chemicals that I might have been exposed to decades ago, such as DDT and PCBs; industrial pollutants like lead, mercury, and dioxins; newer pesticides and plastic additives; and the near-miraculous compounds that lurk just beneath the surface of modern life, making shampoos fragrant, pans nonstick, and plastics flexible, light, and cheap.

Bergman is a world expert on polybrominated diphenyl ethers (PBDEs), a class of fire retardants that I have also been tested for. Until recently, PBDEs were added for safety to just about any product that can burn. Found in mattresses, carpets, clothing, the plastic casings of televisions and computers, electronic circuit boards, and automobiles, these chemicals have been mixed into products in order to raise the temperature at which they would otherwise ignite, which makes them harder to burn. They save hundreds of lives a year from death by fire, but they also have a tendency to break loose as gas and particles, released into the air when televisions or computers heat up and when microscopic flecks of textiles break off and products degrade.

I expected to have normal levels of PBDEs, but this pregnant Swedish pause is making me nervous. I already knew these chemicals were where I would prefer them not to be: inside my body. I also knew how much was there. What I didn't know was what this meant.

But I am about to find out.

The data I sent to Bergman comes from the relatively new science of "biomonitoring," which measures and detects even tiny levels of chemicals in people and animals. "In the past, we tended to measure chemicals in the air, food, or water, or in animals," says James Pirkle, the deputy director for science at the Centers for Disease Control's National Center for Environmental Health. "Now we are getting real human data." Since 2001, CDC's Environmental Health Laboratory periodically tests thousands of Americans in biomonitoring studies that are intended to establish for the first time national reference levels for hundreds of common chemicals, ranging from pesticides and dioxins to heavy metals, with more studies coming all the time. Based on blood and urine specimens collected by the National Health and Nutrition Examination Survey (NHANES), these reports also parse out data on chemical levels based on age, sex, and ethnicity, providing researchers, public health officials, and certain science writers with baselines to measure exposures of various individuals and populations.

Until recently, scientists lacked the technology to detect the minuscule traces of chemicals inside people. Typically, these are present in parts per billion (ppb), a standard toxicological measurement. One part per billion is like putting half a teaspoon (2 milliliters) of red dye into an Olympic-size swimming pool and then looking for traces of color. Figuring out the impact of these tiny amounts is even trickier.

In large doses, some of the toxins—such as mercury, PCBs, and dioxins—have caused horrific damage to people who were accidentally exposed to huge doses. In Bhopal, India, in 1984, a Union Carbide plant making pesticides released forty tons of the poisonous gas methyl isocyanate into the air, killing at least three thousand people within days and eventually killing a total of perhaps five thousand. Another incident happened in Minamata, Japan, where the Chisso Corporation, a chemical company, released high levels of mercury into the Minamata Bay between 1932 and 1968. This highly toxic metal accumulated in shellfish and fish eaten by locals, who suffered severe mercury poisoning. Their symptoms included numbness, muscle weakness, vision problems, and impaired hearing and speech. In extreme cases, insanity,

paralysis, coma, and death followed within weeks of the onset of symptoms. Fetuses were also exposed in the womb, resulting in severe birth
defects. More than three thousand people were officially affected, and
these and thousands more have been paid compensation by Chisso.

Many toxicologists, however, insist that the parts per billion inside
most of us are nowhere near the levels of those living in Bhopal and
Minamata Bay and are probably nothing to worry about. "In toxicology, dose is everything," says Karl Rozman, a toxicologist at the
University of Kansas Medical Center, "and these doses are too low
to be dangerous." What's more, some of the most feared substances,
such as the mercury I consumed when I conducted my fish gorge, dissipate within days or weeks—or they would, if we weren't constantly
reexposed.

Still, a few illnesses are mysteriously increasing in number of cases
as chemical levels rise in the environment. From 1987 through 2002,
autism increased tenfold. From the early 1970s through the mid-1990s,
one type of leukemia was up 62 percent, and male birth defects have
doubled. These diseases buck the general trend toward a leveling-off or
a decrease for many forms of cancer and other diseases. Some experts
suspect a link between these increases and the human-made chemicals that pervade our food, water, and air. Pediatrician and mercury
expert Leo Trasande of Mount Sinai in New York City has called the
release of thousands of industrial chemicals into the environment "an
uncontrolled experiment on six billion people." Little firm evidence
exists, however, to either implicate or show clear links between trace
chemicals and diseases.

Over the years, however, one chemical after another that was
thought to be harmless has turned out otherwise once the facts were
in. The classic example is lead. In 1971, the U.S. Surgeon General
declared that lead levels of 40 micrograms per deciliter or less of blood
were safe. It's now known that any detectable amount of lead can cause
neurological damage in children, shaving off IQ points.* From lead to
DDT and PCBs, the chemical industry has released compounds first
and discovered damaging health effects later.

On the phone, Bergman finally speaks, saying that he has reviewed
the results I sent him of a chemical analysis of my blood, which

---

*Since the 1970s, levels of lead in the blood of children have dropped by 90 percent, thanks to
bans of lead in gasoline and other products, and public health efforts.

measured levels of PBDEs deposited inside my arteries and veins. In mice and rats, high doses of PBDEs interfere with thyroid and liver function and have hampered neurological development in fetuses and newborns. In 2001, investigators in Sweden fed young mice a PBDE mixture similar to one used in furniture and found that they did poorly on tests of learning, memory, and behavior—research that was confirmed in a recent study conducted in Maine. In 2005, scientists in Berlin reported that pregnant female rats with PBDE levels no higher than mine gave birth to male pups with impaired reproductive capability.

Little is known about the impact of PBDEs on human health. Yet another expert on PBDEs and toxicology, Linda Birnbaum of the Environmental Protection Agency, told me that when a chemical such as PBDE is toxic to a wide range of animals—besides rats and mice, this one has been tested on everything from zebrafish to mussels—it is likely to have roughly similar effects on humans.

"I hope you are not nervous, but your concentration is very high," Bergman finally says. My blood level of one particularly toxic PBDE, which is found primarily in U.S.-made products, is 249 ppb—that's twelve times the mean level found in a recent Centers for Disease Control study that tested thousands of Americans. It's more than a hundred times the average found in Swedes, says Bergman. The news about another PBDE variant, also toxic to animals, is nearly as bad. My levels would be even higher if I were a worker in a factory making the stuff, Bergman says.

Primarily, PBDEs that are used in manufacturing come in three versions, called "Penta," "Octa," and "Deca." These Latin nicknames refer to the number of bromides that predominate in each chemical mixture—Penta, of course, is five, Octa is eight, and Deca is ten.[*] Pentas have been used in foam, mattresses, and furniture; Octas in electrical equipment; and Decas in hard plastics on electrical equipment and commercial textile backings. Decas tend to be expelled from the body in two or three weeks; Pentas and Octas stick around in our bodies for years. Decas have been shown to cause developmental neurotoxicity in mice and rats and may affect the immune and

---

[*] Pentas, Octas, and Decas are mixtures of different chemicals in the PBDE family. Pentas primarily contain PBDE-47, Octas primarily contain PBDE-99, and Decas primarily contain PBDE-209.

PBDE (Flame Retardant) Levels in the Author			
Chemical	Author's Levels	U.S. Mean (CDC)	U.S. 95th Percentile*
PBDE-47 (Penta)	249†	20.5	157
PBDE-99 (Octa)	40.5	5.0	42.2
PBDE-209 (Deca)	22.4	N/A	N/A

*Ninety-five percent of the people tested have a lower level.
†Levels are parts per billion (ppb).

reproductive systems. Based on animal studies, they also are a possible human carcinogen. The table above shows my results for some of the more important PBDE chemicals associated with Penta, Octa, and Deca.

Scientists have found PBDEs planet-wide, in polar bears in the Arctic, cormorants in England, and killer whales in the Pacific. In 1999, Bergman and his colleagues were the first to call attention to their accumulation in people when they reported an alarming increase in PBDEs in human breast milk, from none in milk preserved from 1972 to trace amounts in 1997. Since then, levels in people around the world have been tested, mostly in small studies, with a range of mean results, from a mere 1.5 ppb of Pentas in Japan to 11 ppb in Australia. In Nicaragua, children living in a dump registered alarming levels up to 639 ppb. In Oakland, across the bay from me, two small children tested by the *Oakland Tribune* for a story had blood levels significantly higher than mine: 490 ppb in a five-year-old girl, and 838 ppb in an eighteen-month-old boy.

In 2004, Penta and Octa manufacturers voluntarily halted production in the United States after being slated for banning in states such as California and Washington, although Decas have continued to be made at the rate of about fifty million pounds a year, mostly for television and computer casings. In 2004, the European Union banned Pentas and Octas; in 2008, they also banned Decas. Manufacturers are developing alternatives that seem less dangerous, such as chlorine-based inorganics and safer bromides. Birnbaum says it is also possible to use nonchemical fire barriers, such as metal, in computer casings.

This is all very interesting, I'm thinking, but how did these chemicals get in me?

Bergman says that the main source of exposure comes from breathing dust. PBDEs attach to dust in one's house, office, and car. This may explain the high levels in the Oakland children. Presumably, they inhale and ingest PBDE-laden dust while crawling on the floor. Once inside the body, the Pentas and the Octas accumulate in fat and, in women, in breast milk. Another source of exposure is by eating animal fats, although Birnbaum thinks the main source is dust.

But I am not an infant crawling on the floor, I point out to Bergman. He launches into a series of questions to find out why I apparently have enough flame retardants in my body to slow a small fire, should I ignite.

Have I recently bought new furniture or rugs? No. Do I spend a lot of time around computer monitors? Yes, but my current computer, I'm told, is PBDE-free. Do I live near a factory that makes flame retardants? Nope, the closest one is more than a thousand miles away. One curious possibility could be my proximity to the World Trade Center disaster just after September 11, 2001. I was not in New York City on the day of the attack but was there a few days afterward, staying within a mile of ground zero when the air was still acrid with chemical fumes from the destroyed buildings. PBDEs from materials in the buildings were measured at high levels in this noxious cloud, which contained numerous compounds that have caused long-term damage to workers and rescuers and to residents of lower Manhattan. But it's unclear whether my few days of exposure could have caused my high levels.

Then I come up with another idea.

"What about airplanes?" I ask.

"Do you fly a lot?" Bergman says.

"I fly about two hundred thousand miles a year," I say.

"Interesting," Bergman says, telling me that he has long been curious about PBDE exposure inside airplanes, where plastic and fabric interiors are drenched in flame retardants to meet safety standards set by the Federal Aviation Administration and its counterparts overseas. Since 2004, Boeing and other plane manufacturers say, they have been phasing out PBDEs in new airplanes, but these chemicals remain on some older planes.

At the time of the call, Bergman was hoping to run tests of PBDE concentrations in airplane air and inside frequent fliers' bodies—a project he since has implemented, publishing the results in a paper in the fall of 2008. To find out the exposure levels of people on airplanes, his high-flying experiment studied nine test subjects who took long flights of nine to eleven hours. These passengers took air samples on their flights and had the levels of PBDEs tested in their blood before and after the flights. The scientists found that the air on board was thick with PBDEs at high levels. The "after" levels in the passengers' blood also showed significant increases, although they were far less than mine. "The findings from this pilot study call for investigations of occupational exposures to PBDEs in cabin and cockpit crews," wrote Bergman and his team in their study.

This leaves me with an intriguing hypothesis, although it will take further study to clear up the mystery of where I picked up this substance that I had not even heard of until I began working on this project.

So I'm left with the question: How worried should I be about PBDEs?

"I wouldn't be panicked about it," says Linda Birnbaum. "Animal data from rats, mice, fish and other organisms suggest that the main concern is reproductive. I'd be much more concerned if you were a woman." Bergman agrees that my white maleness reduces the risk, adding a guess that "Any level above a hundred parts per billion is probably a risk to newborns."

This makes me feel a little better, but what about my daughter and the other three billion women on the planet?

Birnbaum can't say for sure, though any dangers to humans, male or female, from tiny levels would not show up for years if at all. "This compound is not highly toxic in an obvious way, like mercury," she tells me. "The effects are more subtle." In animals, detrimental effects such as thyroid and endocrine disruptions and developmental issues manifest later, perhaps in a lower sperm count or in reduced ovarian function. "We don't know exactly how this would work," she says. "It's hard to tell doctors, 'Here is what to look for in your patients.'"

Can I do anything?

Birnbaum says to eat foods that are low in animal fats because this is one source of PBDEs. "People should keep their homes and offices as clean and as dust-free as possible," she says. We can also buy goods from companies that claim to no longer use PBDEs. These include Dell, Canon, Hewlett-Packard, Ericsson, Mitsubishi, and Sony.

Still, even with the bans and the phaseouts, an enormous mass of products using PBDEs, from automobiles to bedspreads, remains on

Earth. They will be used for years to come. This may be why levels of bromides have been increasing exponentially in people and animals, with the levels doubling every three to five years. Bergman's breast-milk study in Sweden suggests that PBDE levels have been doubling in people approximately every five years. These upward ticks may slow with the bans but not anytime soon.

As I hang up with Åke Bergman, my flight is announced. I find myself strolling onto the plane eyeing the familiar sleek plastic bins, seat covers, and carpet as if they are something outlandish and possibly dangerous. I look at dust on people's shoes and feel the low vibration of the floor as the engines idle, waiting to take me across the world on a journey where I'll breathe recirculating air that on my last flight I hadn't given a thought to.

Three-thousand-mile trail of blood

The experiment that leads to Åke Bergman's phone call starts with yet more blood—my blood—and a journey it takes from New York City to Vancouver Island, Canada. But before becoming a long-distance traveler, my plasma has to be extracted from my arm. This occurs at the Mount Sinai Medical Center in upper Manhattan in New York City, where I find myself facing down another needle, this one held by phlebotomist Ron Leon, a bald nurse with thick biceps who looks like he can throw me to the ground in an instant should I misbehave but will feel horrible if he causes the slightest pain when drawing blood. Also standing by my side is Leo Trasande, the Mount Sinai physician and expert on mercury and other toxins who tested my mercury levels after I caught and ate the halibut off the coast of California. Leo has volunteered to be the expert doctor overseeing this portion of the Experimental Man project.

I am providing much more blood than I did for the genetic tests—thirteen vials—to run tests on levels of PBDEs and hundreds of other chemical pollutants. These results will produce an internal snapshot of my chemical body burden at this very moment, 9 a.m., on an unusually hot April day in New York City. One major difference between this test and those that assessed my DNA is that genes are mostly static.[*] They are my underlying software, which was programmed when I was conceived, whereas these chemicals are dynamic and change constantly as exposures go up and down, are dispelled or stored by my body, and interact with other chemicals, natural and artificial. They also interact with genes by turning them on or off or by damaging them—as I shall describe later.

"This won't hurt," insists Leon.

With my usual aversion to needles in my own veins and to seeing my own blood, I don't watch as Leon pricks me so expertly that I barely feel it. He then reassuringly begins to count the vials: "That's one, that's two . . ."

Trying to be the tough journalist, I grit my teeth and say nothing, though I swear I can feel the blood flowing out.

By the time the count reaches eight or so—I try not to listen—I feel woozy. By ten, the room is fading into a deepening fog. Sweat on my forehead is turning cold. Calmly, I inform Leo Trasande that I am about to pass out.

"Just three more," says Leon in a concerned but firm voice.

"Are you okay?" asks the ultra-thin Trasande, with large glasses and short-cropped red hair. "We can stop if you're having problems."

"No . . . ," I mutter, trying to push through the fog as another tube of blood fills. But the fog is winning.

A sharp jolt and pain in my nose: smelling salts from Trasande.

"Whoa!" I say, my head abruptly clear.

"All done," says a triumphant Leon. He shows me the vials piled in a plastic dish as the smelling salts wear off, and I swoon into the waiting hands of Leo and Ron.

After a speedy recovery, I learn that the blood will be taken to a back room and processed before being sent on ice (with my urine samples) three thousand miles away to a lab called AXYS Analytical Services, one of the few labs in the world that is capable of running the sensitive tests that are required to extract parts per billion of chemical toxins.

A few weeks later, I fly to Vancouver Island, Canada, and follow my trail of blood to a room-size freezer at AXYS. I stand among frozen samples of everything from rockfish caught off the coast of

---

[*] Recent discoveries suggest that genes can change over time as a result of environmental and other factors.

Alaska to polar bear tissue collected on Arctic ice floes. Upon arrival from New York, my blood was stored in this deep freeze, which is part of a small complex of low-slung buildings surrounded by pine trees. Nearby is Victoria, a small city on the southern shore of this huge island. Victoria is close to the city of Vancouver and to Seattle but seems remote.

I'm getting a tour of AXYS with the then business manager, Laurie Phillips. She is showing me the journey of my blood through the detection process, which the company has just finished. After my tour, the lab will provide me with results.

Founded in 1974 to run environmental tests for the government of Canada, which was opening up new oil and gas fields in the Arctic, the company expanded over the years to now test fish, reptiles, mammals, soil, water, and people for over six hundred different chemicals. "Our business is finding very small levels of chemicals," explains Phillips. "Sensitivity is our specialty." AXYS has tested thousands of people's blood serum, most of which passed through this freezer.

Few people, however, have had as many chemicals tested for and detected as I have. Out of the 320 chemicals AXYS looked for, it found detectable levels for 185 of them—pesticides, PCBs, dioxins, plastic additives, and, of course, flame retardants. But even this hefty count is hardly all of the man-made chemicals known to exist in the environment. The EPA has listed tens of thousands of chemicals since the agency was created in 1970; toxicologists say there are about 80,000 commercial chemicals in existence. I would like to have been scanned for all of them, but the tests are too expensive. Those that I did have tested were priced at about $25,000, which AXYS discounted to $15,000 for this project.

Nor was I tested for chemical cocktails: the blending of chemicals that becomes more potent and behaves differently from when the chemicals are on their own. Obviously, chemicals such as pesticides, PCBs, and phthalates do not occur inside us in isolation. "Combinations of chemicals may have additive effects, or they might be antagonistic, or they may do nothing," says James Pirkle of the CDC. "We don't know much about this yet, although we're beginning to study it."

In the AXYS freezer, set to -22 degrees centigrade, Laurie Phillips hands me a thick overcoat with a faux fur–lined hood as she points to piles of plastic and metal containers. The metal containers are stainless steel for samples that have to be as free as possible of phthalates, the pervasive additive to plastic that is used in making most coolers. Rogue particles of phthalates might contaminate a specimen and throw off the result. "In here we have bird livers, fish blood, and fish tissue," says Phillips, a short woman with long, dark hair and a no-nonsense manner. "We have a study looking at chemical levels in edible fish parts and another one asking how great blue herons get dioxins in their bodies. We are testing mice for exposure to metals, using hairs on the mice to check for trace metals. We have water samples and pieces of tundra." The lab's clients include, besides Canadian regulatory agencies, local authorities and companies testing to see whether they are in compliance with environmental protection laws and safety rules in plants and in facilities that expose workers to chemicals. Academics, environmental groups, nonprofits, and the occasional journalist also send samples. So do law enforcement agencies when they go after polluters for breaking the law.

The blood trail next takes us from the freezer to one of two large lab complexes. First stop is the biohazard room, where all specimens with even an infinitesimal chance of containing harmful toxins if released into the environment are treated. "That would be human samples, sewage samples, and certain soils," says Phillips. My blood is probably not a major threat to AXYS workers or to the people, fauna, and flora of Vancouver Island, but, taking no chances, AXYS technicians dressed in hazmat suits worked on my specimen in this sealed room with reverse airflow and special filters to prevent the release of any unexpected dangers.

My hazard-neutralized blood was then taken to a clean room that is as free of contaminants as possible to start the complicated process of separating out each toxin from the stew of other chemicals in my body. The air runs through special filters, and everything is glass or steel. (Glassware is sterilized at 350 degrees Fahrenheit for eight hours and rinsed in a cleaning solvent prior to use.) The use of plastic products is again carefully controlled because of phthalates. Computers are a no-no because they might throw off microflecks of PBDE flame retardants. "We also worry about people's deodorants and hand creams that could show up in a sample," says Phillips. "People joke about naked analysts around here so that nothing's coming off their clothes. We have fresh air that is flowing up through holes in the floor past people's faces and drawing stuff away." But they can't eliminate every trace of phthalates and other chemicals even in here, so the lab creates "blank" samples that do not contain a specimen but are used to detect whatever minuscule amounts of contaminants in the room might slip past the precautions. Levels that show up in the blanks are subtracted from the specimen levels to get the true amount of a chemical tucked in my blood or in other specimens.

In another part of this lab, flasks with liquids colored brown, purple, and green are bubbling on burners, looking like a mad scientist's concoctions in an old horror film. My blood was put in flasks and dissolved in more solvents, then cleaned up to remove unwanted chemicals, a process called "fractionating." "You end up with an extract, and the extract has in it the dioxins and pesticides, plus other stuff we still don't want," explains Dale Hoover, an AXYS executive who has joined us. "We narrow and narrow it down to the compounds that we want and get rid of everything else."

Now the extracted chemicals are moved to the final stage in the process, where the specimens are injected into a $500,000 mass spectrometer. These devices look like elongated washing machines turned on their sides, with long tubes attached that scan the chemicals at almost the atomic level. Some mass specs turn a specimen into gas; others smash it apart. The machine then measures the mass of each particle, which has a unique signature for each compound. "We know what the mass of dioxin is," for example, says Hoover. The scanner records the amount of dioxin or DDT in blood.

In a conference room a safe distance away from the biohazard room and the mass specs—we're back in the world of plastic chairs and flame-retarding computer monitors—Laurie Phillips and Dale Hoover show me my results, printed out on Excel grids. One set of three pages lists my raw PBDE results, giving data for forty-six variations of this chemical. This is weeks before I will speak with Åke Bergman or have even heard of him, and I have only a dim idea of what I am looking at, so I ask, "Do these levels seem normal?"

"Your brominates seem high compared to other serum samples that I've seen," says Hoover.

"What does this mean?" I ask.

"Our job is not to analyze these results," he says, offering to refer me to some studies and suggesting that I talk to this Swedish fellow named Åke Bergman.

We look at other results, from bisphenol-A to DDT. Mostly, they say, I have normal levels, with perhaps some anomalies besides PBDEs to check out. Thumbing through the pages of complicated numbers and designations, I groan at the daunting task of making sense of these results, which seem as indecipherable as my raw genetic data. Like genomics, environmental biomonitoring produces heaps of information in search of interpretation. I just hope that the tools to analyze this data are more user friendly than those that are available for most of genetics. I already know there is no 23andMe or deCODEme with consumer-oriented graphics and explanations, although I suspect there will be soon.

For the moment, I am faced with ferreting out information myself about my accumulation of these chemicals and whether they are dangerous. It seems like the best way to do this is to go on a journey into my past and my present to find out where and how I might have been exposed—and to talk to experts and locals along the way.

Idyllic childhood in Kansas, except for the toxic waste dump

My own exposure, like yours, started in the womb at the moment of conception, with whatever toxins happened to be present in the sperm and the egg that created me. As the cells divided and my organs formed, my mother was already downloading environmental chemicals through the placenta and the umbilical cord into my body. These undoubtedly included mercury, which even today annually puts 60,000 newborns at risk for adverse neuro-developmental effects from in utero exposure, according to a National Research Council report. A 2005 study coauthored by Leo Trasande puts the number of children whose IQs are lowered by mercury exposure at 316,000 to 637,000 a year. I can only imagine the impact of mercury and other chemicals swirling through the fetal me. I seem to have turned out okay, though, with most faculties and organs intact, including those that are responsible for reproduction.

This is a miracle, considering when I was born. The year was 1958, a dozen years before Congress passed the Clean Air Act, and fourteen years before it passed the Clean Water Act. This period was possibly the most polluted in U.S. history, if not world history, given the numbers of people affected, the number of novel chemicals present in the environment, and the global reach of many pollutants. Lately, however, China, India, and other rapidly industrializing countries are trying hard to do worse. The late 1950s and 1960s were the height of the Age of Lead in gasoline and paint and the dawn of the Age of Plastic, not to mention tens of thousands of other chemicals dumped raw into the air and the water.

I was exposed to even more chemicals as I grew up in a small, rural community just outside of Kansas City, Kansas. The area around my house was idyllic: woods, limestone river bluffs, streams, and a big lake. But my house was situated within a few miles of a major industrial corridor along the Kansas River—which for us was out of sight and out of mind—where factories made cars, soap, and fertilizers and other agricultural chemicals. When we drove toward downtown Kansas City, we passed the plants, plunging into a noxious cloud that engulfed the car with hazy effluents and an awful chemical stench. Flames rose from fertilizer plant stacks, burning off mustard-yellow plumes of sodium that smelled like rotten eggs, and animal waste was dumped directly into the river. Older friends from that era remember a near-permanent haze from coal-burning power plants and blood from slaughterhouses in this cattle town that sometimes turned the rivers red.

"A lot of the plastic-making facilities would have caused odors in those days," says Denise Jordan-Izaguirre, the senior regional representative in Kansas City for the federal Agency for Toxic Substances and Disease Registry (ATSDR), "and you had the Amoco refinery, which really stunk up the place."

I remember one day when the older brother of my then best friend fell into the Kansas River while we played on its sandy banks near our town, a few miles downriver from some of the factories. We had been warned to stay out of the river because of its strong currents, but that wasn't the problem. This boy came right out but soon developed rashes and second-degree burns from something in the water. He had to be taken to an emergency room.

In the nearby farmland, trucks and crop dusters sprayed DDT and other pesticides in great puffy clouds. Every year in the spring, a truck drove slowly through our town, spraying DDT to eradicate mosquitoes. We kids were warned to stay indoors. Instead, we rode our bikes through the mist blowing out of the back of the truck, holding our breath and feeling very brave.

This was during the early days of the environmental movement, when organizations such as the Sierra Club were beginning to point out that the unfettered dumping of chemicals, litter, and other pollutants into the environment was out of hand. This message didn't sit well with many Americans, who were used to disposing of their waste directly into waterways, the air, and the land. Until then, nature had mostly absorbed this influx of garbage, or at least there wasn't yet enough of it to do serious, large-scale harm to the environment or to humans. But as industry expanded and chemicals became pervasive in the early and mid twentieth century, nature was being overwhelmed. This was most evident in places such as Los Angeles, which had smog so dense when I was a boy that it was like a heavy, noxious fog. In New York City, where my mother's parents lived in the mid-1960s, I remember their worrying about breathing the air in Manhattan, which they said was the equivalent of smoking two packs of cigarettes a day.

In Kansas City, a visit to the clippings archives of the *Kansas City Star* gives me a brief flavor of the attitudes of that period and the battles that started in the 1960s to clean up my hometown. I remember this personally because my mother, a prominent nature photographer and artist, got involved in the environmental movement early in a region of the world where this was not popular. I grew up reading Rachel Carson and books such as *Limits to Growth* and watching my mother deliver lectures and write books and articles as a fiery advocate for preserving the Tallgrass Prairie. This extraordinary ecosystem (buffalo, wolves, antelope, and diverse grasses and plants), which began just west of our house, had been assaulted for over a century by the encroachment of ranchers and farmers and more recently by pesticides and fertilizers. In her view, this was destroying a bionetwork that once covered an enormous swath of land from Canada to Mexico and had nearly as much floral diversity as a rain forest. One of her early presentations was a slide show and a talk documenting litter, garbage, and obvious visuals of factory effluents and fouled water that she delivered to local civic groups in our conservative county in Kansas. Audiences were stunned—less by what she was showing than by her audacity at suggesting that their beloved region was anything but pristine, and that their businesses and industry might be less than model citizens.

In 1967, a reader wrote to the *Kansas City Star*, worried that a prominent creek—Brush Creek—running through her affluent neighborhood was a "polluted stream" because the city's sewage system was allowed to overflow into it during rainstorms. In an article responding to the letter, the *Star* reported that county health officials told its reporters that the state of Kansas had approved this arrangement and that the term *"polluted stream"* was "a matter of degree of impurity or contamination." Dr. Bruce Hodges, then the county health director, assured everyone that the creek had recently passed an inspection by the chief sanitary officer. "Although the water contained elements that made the water unsafe for drinking or swimming, by Health Department standards it could not be termed polluted," Hodges told the paper.

Three years later, in 1970, as pressure built in Washington for Congress to do something about America's befouled water and air, the unusually enlightened Democratic governor of Kansas, Robert Docking, called for the creation of a state Department of Environmental Resources. According to the *Star*, Docking wanted the new agency to "work to identify and overcome all that degrades our earth, our skies, our water and every living thing so that by the end of the environmental decade of the 1970s [we] may see our environment immeasurably better than at the beginning." In the 1970s, industries lining the river near my home were repeatedly fined for violations of pollution standards. In 1971, the *Star* reported that seventy-five industries had received a hundred citations and were "under notice to bring their plants into compliance with state clean air standards for particulate pollution." Sulfur dioxide levels created by burning coal and petroleum were two to three times over the state limit. Even one of my favorite barbecue places in Kansas City, Gates Bar-B-Q, was cited in 1975 for smoke and particulate violations.

This general toxic soup was bad enough. Worse was my discovery that just a mile from my childhood home is a major EPA Superfund site, on the National Priorities List for hazardous places—a spot where a dump once sprawled across a limestone bluff above the Kansas River. We boys used to play there on hot, muggy summer days, entering from a stand of woods that divided our community from the river. The dump was a motherlode of old bottles, broken machines, steering wheels, and other items only boys can fully appreciate.

For years, companies and individuals in this corner of Johnson County had dumped thousands of pounds of material contaminated with toxic chemicals here. "It was started as a landfill before there were any rules and regulations on how landfills were done," says

Denise Jordan-Izaguirre of the ATSDR. "There were metal tailings and heavy metals dumped in there. It was unfenced, unrestricted, so kids had access to it."

Kids like me.

Even if I hadn't played in the dump, now called the Doepke-Holliday Superfund Site, it was situated just half a mile upriver from a county water intake that supplied drinking water for my family and forty-five thousand other households. "From what we can gather, there were contaminants going into the river," says Shelley Brodie, the EPA remedial project manager for Doepke. In the 1960s, the county treated water drawn from the river but not for all contaminants. Drinking water also came from twenty-one wells that tapped the aquifer near Doepke. I read the voluminous EPA reports about the site and saw listed numerous chemicals that have shown up inside me, including pesticides and heavy metals. I ask Jordan-Izaguirre whether the dump might have been a source of my exposure. "Sure, it's possible, some of the metals or some of the pesticides. It would be hard to section out what you were exposed to then, as opposed to other places you might have lived."

Doepke-Holliday was assessed by the EPA and cleaned up between 1983 and 1995. Instead of removing the refuse, however, they buried everything in a lined landfill and sealed the site under an impermeable cap designed to keep the material away from the air and water. "It's an engineered cap," says Shelley Brodie, "It's high-density polyethylene. What you do is grade the site so there will be less runoff. You end up putting in compacted silt and a drainage net, and then you put the impermeable liner over that, and then you put six to twelve inches of compacted soil and grow vegetation. As it continues to rain or snow, the moisture will go down the beginning layers and affect the vegetation, which is what we want. Keep it covered with healthy vegetation."

In 2005, the second five-year review of the site declared it safe and well managed. "The sad thing is if you did dig it up," says Brodie, "you can almost still read the newspapers in there. It's kind of entombed. But that's the whole point, that no water goes in that will mix with the waste and generate leaching that will expose the environment to the waste."

Incredibly, there was a second Superfund site just a quarter mile in another direction from my house. At night while I was growing up, we often heard the *pop-pop-pop* of gunfire in the distance, from a shooting range tucked in the woods over the hill across the street where we lived. It was a place where farmers and sportsmen hung out, drank, and shot skeet. Over the years, long after the range was abandoned, lead from the shotgun shells and the white paint on the shattered clay pigeons scattered on the ground had built up dangerous levels of this heavy metal. It was a quick operation to clean it up, says Jordan-Izaguirre. Crews came out and carted away the shot and the smashed clay. "We did it because kids were getting on the site as the area became more developed. It's attractive for the kids to play with the shells."

Yes, it was.

Some thirty years after leaving Kansas for college, as I research this project, I find myself on a cold, bright day in January standing above the river I knew so well, near the site of the dump. The river in the winter is muddy brown and sluggish with a low level of water. The cottonwoods are still there and the cattails and the tangles of roots and fallen trees jutting out from sand banks. The Santa Fe Railroad's tracks run along the bank where we kids placed pennies on the rails and waited for trains to come and cut them to shreds. I wanted to climb up one of the bluffs north of the dump to see what remained of the woods that we used to explore, but the area is more developed now, and there is a fence surrounding the property. An interstate was built just after I left home that cut off access to the dump from my boyhood house.

I arrive at the Doepke site near sunset and check in with the company that still runs an active landfill on part of the property. Every few minutes, a garbage truck appears and heads up the bluff, its engines roaring in low gears on the steep incline. As the sun begins to dip down, I make my way to the top, finding a quiet, grassy field where heaps of junk, gravel, barrels, and old tires once stretched off in all directions. I am standing on the EPA's cap, where the grass is cut low and is brown in the winter, and ravines filled with gravel run here and there for drainage. I get out of my car and stand beside what looks like a park and allow myself for a moment to marvel that this place has been cleaned up. I wonder how my mother would feel if she saw this. In fact, I was told that the community in the 1980s and 1990s, after my family moved away, had rallied to support the EPA in cleaning up this Superfund site.

"I have to admit, we were a little taken aback at all of this pollution right in our backyard," says a neighbor across the street from my old house. He has grown old and white-haired since I last saw him. When I was a kid, this man had been one of the people who was angry at my mother's slide shows. "You know," he says, "your mother was right all along."

Back in my car, I do a quick tour of the rest of the area, driving past the cleaned-up shooting range and heading down the river on the interstate toward the row of factories that is closer to downtown. With a brilliant Midwestern sunset now blazing orange and yellow in what seems like a much bigger sky than we have in California, I pass a pressing-steel plant; the Kaw Power Plant, with its three tall silver smokestacks; a Sinclair gasoline terminal; and the huge Colgate-Palmolive plant with its fields of pipes filled with chemicals lining the river bank.

In the waning twilight, I get out of the car near the power plant on the edge of the river and take a deep whiff of the air—and smell perhaps a faint scent of mud and decaying leaves, but none of the vile aromas I remember. I climb down a low embankment to the river, step out on a spit of gravel, and stand above the brown, sluggish water. Do I dare? Denise Jordan-Izaguirre told me that fish were back in the Kansas, and that it was clean from most chemicals. I reach out my hand, remembering how small it was when I was last this close to this river. Back then I wouldn't have taken the risk. Now I put my hand into the cold water and it feels like—water. No burns, no smell, no funny films of frothy chemical bubbles. I realize that however cynical one gets about our government and efforts to fight pollution, and more recently global warming caused by human activity, the water running through my fingers and the air I am breathing suggest a modicum of hope for the future.

I have traveled to eastern Kansas to do more than revisit scenes from my youth. I am here to investigate what chemicals I might have ingested. In part, my AXYS test results read like a chemical diary from forty years ago. For starters, I wonder whether all of those bicycle rides behind the DDT truck were responsible for the traces in my blood of this now-banned pesticide. I also have detectable levels of other childhood pesticides, such as the termite-killers chlordane and heptachlor.

Have some of these lingered since my childhood?

To find out, I visit two toxicologists at the University of Kansas Medical Center in Kansas City, Kansas: Karl Rozman and John Doull. Rozman worked for years as an adviser to the U.S. Occupational Safety and Health Administration (OSHA) and is an expert on DDT and other pesticides. Doull was the coeditor of a major textbook on poisons called *Casarett and Doull's Toxicology: The Basic Science of Poisons*. The two of them meet me in Rozman's office, which is filled with thick test books and scattered papers and journals.

"The levels you have are about what you would expect decades after exposure," Rozman tells me as he thumbs through my AXYS results. I ask about DDT, and he looks up the number.

"Your level of DDT is slightly high," he says, "but that's not the most important one to look for. DDT breaks down into metabolites that persist a long time, and they are what is important."

The main metabolite of DDT is called DDE, he explains, and my level is 256 ppb, slightly less than the CDC mean of 295 ppb. "Having that level suggests you were exposed as a child," he says.

I also have slightly elevated levels of the pesticide family of chemicals that replaced DDT after it was banned in the early 1970s, when I was a teenager. Rozman explains that these replacements, called organophosphates, are in some ways worse than DDT. "The issue with DDT was the persistence in the environment," he says. "That's what made DDT cheap and attractive as a pesticide: it had a low level of toxicity that lasted a long time, so you didn't need multiple sprayings. The organophosphates are highly toxic when applied, and quickly dissipate, in about thirty days, and need repeated applications. That's the trade-off."

"About fifty children a year die from organophosphate overdoses," says Doull. A few crop-duster pilots have died spraying these chemicals when they turned back into a cloud of chemicals they had just sprayed, which knocked them unconscious, causing them to crash. "So the advantage of DDT was that acutely almost nobody ever died," says Rozman, which is probably why we were able to ride around in the plume behind the DDT truck. "But with these types of compounds, the chronic effects are of concern because they accumulate and cause damage to animals and possibly to humans over time."

The long-term effects of pesticides from my youth are suspected in a higher than normal rate of leukemia and other cancers among farmers in Kansas. This is according to a study out of Kansas State University that perused cancer death records between 1980 and 1990, the decade after I left home. Another more recent study out of the National Institutes of Health (NIH) has linked people using DDT and similar pesticides for more than a hundred days in each person's lifetime to a 20 to 200 percent increased risk of developing diabetes, depending on the pesticide.

This risk factor from a chemical toxin has interesting implications if scientists are able to factor it in with a person's genetic risk factors for diabetes—a notion we will address later in the book.

The highest levels of pesticides from my youth, however, came from a fungicide called hexachlorobenzene, or HCB—at 10.57 ppb. It was used in seeds for crops to fend off fungi until it was banned in 1966 as a carcinogen in animals, causing increases in liver, kidney, and thyroid cancers, and as a "probable carcinogen" in people. HCB has a half-life in humans of about twelve years, says Rozman, which allows him to calculate what my levels were as a child. "The half-life is a very important concept," he explains. "The half-life is the time it takes to get rid of fifty percent of the compound from your body. Now, with the half-life of twelve years, that means that over this time, you would come down from, say, an exposure of one hundred to fifty. In another twelve years, it would be twenty-five, and in another twelve years, it would be twelve and a half. You can do the calculations the other way around, so with hexachlorobenzene, about forty years ago, when you were ten years old, your levels would have been about eighty."

"Was that a dangerous level?" I ask.

"That is far below the threshold. Fortunately for you," says Rozman.

"That's assuming I'm not getting any more exposure now."

"Yes, which is probably a wrong assumption because there will be trace amounts in the environment even though the chemical was banned."

"But if I had been a farmer handling seeds soaked in HCB forty years ago, I might have had an even higher level."

"Yes, that is actually what happened. With hexachlorobenzene, we do have experience in many thousands of people—farmers—getting exposed in an accident that happened in Turkey in the 1950s. The United States provided grain for Turkey, which was treated with hexachlorobenzene to keep out fungus when it was being stored. There was starvation in that particular region, and they baked bread directly out of these seeds. They were supposed to use it as seed to plant crops, not to use directly as food. I believe three thousand people were exposed and many, many died. Because there was a huge dose, hexachlorobenzene started accumulating in their systems, and this particular compound after, I don't know, several months or half a year, particularly in younger children, caused liver disease."

"That's a significant and tragic exposure," says Doull. "I want to emphasize that these levels in you are rather trivial, even if they're higher than normal. You need to be careful how you talk about this. People who read your book and say, 'Well, there's a chemical there and it's a bad chemical, it can cause birth defects or cancer or whatever, and, gee, I don't want any of that stuff there.' But we don't know whether these chemicals are dangerous at these very low amounts."

Rozman and Doull also talk about mixtures of chemicals, telling me that similar chemicals can have an additive effect. "So you have to add them up. With dioxins, some of these all react by the same mechanisms, so they all have to be added up."

I hastily pull out my dioxin levels, and Rozman assures me that even added together, they are unlikely to do me any harm.

I'm happy to hear this, although Doull tells me that it's not always a matter of simply adding up additives. Some chemicals are more potent than others, which has to be factored in. "Chlordane, for example, compared to DDT. You could say chlordane has five times greater action than DDT, so you would have to adjust the amount to get the same effect." By action, he means the toxic impact on a tissue, a cell, or an organism in causing damage or disease. Some compounds synergize others—that is, their joining initiates a new reaction that can cause an effect that is many times more toxic than the two chemicals when separated. "Those things you need to worry about," says Doull.

The work of cleaning up the area where I grew up goes on, says Denise Jordan-Izaguirre. "In Kansas City, we've cleaned up most of the sites we know about," she told me in a recent e-mail. "We're still finding old plating facilities so you get metal shops and we've done removals. I was just at a meeting, there are six thousand sites in Missouri they've identified where there was mining, milling, refining, or smelting of lead. It's a huge problem for both groundwater and surface soil."

My level for lead is 3 ppm, I tell her.

"This suggests you may have been exposed to some lead. You don't do anything until it's twenty-five. At twenty-five, they'd start monitoring you, over fifty they'd remove you from your job, over seventy-five they would chemically remove it. A level of three, though, isn't all that bad."

For that, I'm grateful, although I remain amazed that my generation so far seems to have endured the deluge of chemicals thrown our way as children. As we reach our forties, fifties, and beyond, most of us continue to be reasonably healthy. This is probably because of improvements in health care and treatments that have caused life span to nearly double in the United States since 1900.

People born in the late 1950s and the early 1960s have produced the usual complement of great artists, entrepreneurs, inventors, scientists, lawyers, investors, and engineers. Though, of course, one has to wonder whether an Albert Einstein, a William Shakespeare, or a Franklin Roosevelt never happened because of a few downward clicks in IQ due to lead exposure or hexachlorobenzene or to the synergistic impact of dioxins mixing with sulfuric dioxide being illegally pumped into the air when I was child living near the river we called the Kaw.

Hotspot on the Hudson

Kansas was not the only place where I was exposed to toxic chemicals. At age nineteen, I left Kansas for college at a place and a time that put me at the height of another human-produced substance that was detected inside me: polychlorinated biphenyls. PCBs once were used as electrical insulators and heat-exchange fluids in transformers, fluorescent lights, and other products and can lurk in the soil anywhere there's a dump or an old factory. But some of the largest releases took place along the banks of New York's Hudson River in the northeastern part of the state from the 1940s to the 1970s, when General Electric used PCBs as coolants in transformers it was manufacturing in the towns of Hudson Falls and Fort Edward.

About 130 miles downstream from the GE plants is Poughkeepsie, where I attended Vassar College in the late 1970s and the early 1980s. PCBs were at their highest concentrations in the river just as these chemicals were being banned in 1979. Until then, GE had legally dumped excess PCBs into the Hudson, which swept them all the way downriver to Poughkeepsie, one of seven municipalities that draw their drinking water from the Hudson.

PCBs, oily liquids or solids, can persist in the environment for decades. In animals, they damage the liver and cause cancer. Some PCBs chemically resemble dioxins, the highly toxic by-product that's most famous for being in the defoliate Agent Orange. In lab animals, they cause other injuries that include reproductive and nervous system damage, as well as developmental problems. In separate accidents in Japan and Taiwan during the 1960s and 1970s, several thousand people were exposed to rice oil contaminated with PCBs. They suffered from low birth rates, spontaneous abortions, birth defects, and low intelligence.

In 1984, a two-hundred-mile stretch of the Hudson, from Hudson Falls to New York City, was declared an EPA Superfund site, the largest in the country, and plans to rid the river of PCBs were set in motion. GE has spent almost $400 million to date on the cleanup as it prepares to dredge up and dispose of PCBs in the river sediment under the supervision of the EPA. It is also working to stop the seepage of PCBs into the river from the now largely abandoned factory in Hudson Falls. The Fort Edward plant is also being cleaned up as it continues to manufacture electrical equipment. The EPA puts the total price tag of the cleanup, when complete, at about $750 million.

It's possible that the drinking water in my dorm and on the Vassar campus contained traces of the PCBs that poured into the river far upstream, although the main source of human exposure is eating fish caught in polluted waters. PCBs collect in a fish's fatty tissues, prompting New York State to ban fishing in the upper Hudson decades ago. As far as I know, I didn't eat any Hudson River fish during my college days, and experts tell me that PCBs are oily and nonsoluble and too heavy to be carried easily in water. I have, however, registered detectable levels for dozens of the 209 cogeners of PCBs, including the highly toxic, dioxinlike PCBs that are created when PCBs are burned and interact with the environment. My levels are mostly very low, although it's hard to compare them to the CDC's report on national PCB levels since apparently the detection at AXYS is more sensitive than that at the CDC's lab. For many of the cogeners inside me, the NHANES study reports a "<LOD" = less than the limit of detection.

But PCBs, like many other persistent chemicals, are ubiquitous on Earth, meaning that I could have gotten my body burden almost anywhere. "PCBs are everywhere," says Leo Rosales, a former EPA official then based in Hudson Falls, "so who knows where you got it?"

On a cold, sunny day in January, I visit Hudson Falls and Fort Edward, after driving up from Albany on US Route 4 along the wooded banks of the upper Hudson. The area is primarily a mix of farmland, stands of leafless forest in the winter, and industrial towns that have seen better days. Both towns are still dominated by the two large General Electric plants that sprawl along the river banks, one in the heart of Fort Edward, right on the main drag along Route 4, and the other a few miles to the north on the edge of Hudson Falls, slightly northwest of Main Street (Route 4). I pass by modest wood-frame houses and run-down storefronts, some of them closed. Pickups and older SUVs carry yellow ribbons in support of the U.S. troops in Iraq.

I stop off at a small building on Route 4, the EPA field office for the Hudson River project, and meet up with the EPA's Leo Rosales. Young, with short-cropped hair and an earnestness that reminds me of Peace Corps workers I met when I was a reporter overseas, Rosales shows me displays and posters in the office's front area explaining the problem and the GE-EPA cleanup project.

"What is the danger of these PCBs to locals?" I ask.

"The impact on people is not definitive," he says. One study in Hudson River communities found a 20 percent increase in the rate of hospitalization for respiratory diseases, while another found no increase in cancer deaths in the contaminated region. The effect on wildlife has been devastating, however. "In one case a few years ago, the [New York State] Department of Environmental Conservation people did a bird study, on the spotted sandpiper," says Rosales. "They tested the eggs to see if any PCBs were there and were surprised there were high levels of PCBs in the eggs. Levels of fifty-two parts per million. Keep in mind that the legal definition of hazardous waste for PCBs is fifty ppm. So that bird egg is a hazardous waste. After testing that bird egg, you will have to put it in a hazardous waste landfill. We tell that message to people, it's not only the fish you might eat, it's also the wildlife, it's the environment as a whole that's being impacted by this. People may not be in imminent danger, but this chemical is not being diluted in the river, it's not burying itself, it continues to be a threat."

Some meetings with community leaders have been held here, he says, and not all have gone smoothly. The proposed solution of dredging several tons of river bottom and transporting it to a site out of town to clean it up is not popular with everyone. "There is a silent majority of local people who want the project," he says, "a vocal minority does not." Locals are concerned about the pollution in their midst—some, as I would soon discover, are also frightened, others bitter—but many remember GE fondly and consider the dredging to be either dangerous because of all the PCBs being moved around or a waste of money and effort for a threat they believe is minimal. Some local politicians, farmers, and business leaders blame the EPA's project for calling attention to PCBs and driving away business from the area. Home owners are upset because they can't sell or refinance their homes around the plants; several of them sued the city of Fort Edward when it reassessed the value of the homes and upped their taxes.

I climb into Rosales's car, and he takes me on a tour of the area. The Fort Edward plant covers thirty-two acres and still has a large "General Electric" sign and logo on the side facing Route 4, locally called Broadway. One of the houses we drive by has a sign in its window: WE OPPOSE DREDGING.

The original plant was built in 1942 to produce bomber turrets during World War II. In 1946, GE moved in and began to manufacture capacitors using PCBs. The plant still employs about three hundred people, according to GE—fewer than before.

We drive slowly past the plant's main parking lot, which is mostly empty of cars. Its asphalt is cracked, and the painted parking space lines have faded. Rosales tells me that leakage from the plant has caused tens of thousands of pounds of PCBs to gather in a pool below this lot, collecting in a small geological depression. Most of it is still there. One resident later told me that the factory floor used to get thick with PCBs, "like maple syrup," and workers cleaned it up with degreasers and solvents and washed it down drains in the floor. It has collected underneath the plant in a layer of porous sand over clay.

Across the street from the parking lot and the factory, Rosales and I pass by streets of simple one- and two-story homes. Outside are kids' bicycles, family cars, swing sets, and a doghouse or two. I see one mother carrying a baby and getting into her car; down the street, two boys are walking toward the river.

In 1982, PCBs and other contaminants were found in the water that came up from the wells that were used by most of the people living in these homes. In 1983, authorities replaced the wells with municipal water hookups, although for nearly forty years the water for drinking and everything else had come from this aquifer hard by the PCB pool, below the lot, and was untreated. This legacy has left some of the residents furious—including Dennis Prevost, a retired army officer and public health advocate. He blames PCBs for the brain cancers that killed his brother at age forty-six and a neighbor in her twenties. "I grew up a block from the Fort Edward plant, across the street from the parking lot," he later told me. "My family has suffered health effects from living near the plant."

Ed Fitzgerald, an expert on PCBs at the State University of New York at Albany, and others cite the cancer investigation that showed no abnormally high rates in the area, but Prevost claims that this study did not take into account people who moved away from the area and then died elsewhere from cancer. Fitzgerald responds that he has explained to Prevost and other residents that the risk from the wells was probably small because PCBs tend to settle to the bottom of an aquifer, "so it's unlikely that the drinking water would have high levels." Eating contaminated fish is a more likely exposure route, he explains. "In the seventies, fish contained hundreds of parts per million of PCBs. These levels have gone down, but they are still high. Some lighter PCBs also become airborne and may have caused some exposure, but they degrade faster. The state conducted a big study; asked batteries of questions; took extensive tests, blood tests, and indoor air samples; and determined that the risks were not abnormally high."

Heading away from the river toward a narrow canal built in the nineteenth century, the Champlain Canal, Rosales pulls off the paved road onto a frozen dirt road to show me the then empty 110-acre field where the dewatering plant for the cleanup would be located. (The plant has since been built.) This is where PCBs would be removed from the sludge dredged from the river bottom. "This site is unprecedented," says Rosales, as we get out and trudge around in a few inches of snow. The plan was to build huge, airplane hangar–size buildings where the soil will be washed and a water treatment plant that will clean the contaminated water, handling approximately two million gallons of water a day. Once treated, water will be discharged to the nearby Champlain Canal, which will be monitored for pollutants. GE planned to build a road to access the dewatering site, along with seven miles of new railroad track to connect the site with the main rail line.

"Two full trains a day of PCBs will head out for disposal," says Rosales. The waste-disposal facility is located west of Andrews, Texas, near the Texas-New Mexico border. This 1,340-acre facility opened in 1997 as an endpoint for hazardous material collected by the EPA, the Department of Energy, and the Army Corps of Engineers. I was told by officials at the EPA and locals that no one in New York wanted this stuff anywhere in the Empire State. In Texas, New York's problem will be stored in lined landfills that will ultimately be capped and sealed—and then watched for decades, if not centuries, to make sure they are not leaking, or until other technologies come along that can neutralize these chemicals. At least, that's the idea.

"Really, what we are doing here is accelerating the recovery of the river," says Rosales. "So, that way, the river would cure itself a lot quicker than if left undone, with no cleanup."

As we leave the site, we pass scattered farmhouses, and Rosales tells me these people are worried about living so close to the dewatering site. "They say, 'EPA, you've been telling us for so long that PCBs are hazardous, now you're gonna stage this in front of my house?' They're very concerned about this. But we keep telling them the risk is from eating fish, not from being close to a dewatering site. It's going to be secure and monitored at all times, where the river really isn't."

Back in Fort Edward, as I leave town, I take a drive across a bridge to Rodgers Island, overlooking the back side of the GE plant. On the north end of the island is a cluster of houses, a picnic ground, a beach, and a recreational center with a pool, a baseball diamond, and swing sets. A few years ago, GE tested PCB levels in the soil around the houses and found levels high enough to take emergency action. "They scraped up all the dirt and brought in clean fill," Rosales had told me. "The swing set is on high enough ground to be PCB free," he added.

The air is cold and damp as the sun drops over the river. A thin layer of wet snow covers the ground down to where the water gently laps the blackened sand of the beach. I walk down to the edge to reach the water as muddy gunk sticks to my shoes. A short distance away, the GE plant stands quiet and still. At my feet I see little pellets the size of very small buckshot washing up on the sand. Are these PCBs? Rosales had said they are here—the heavier ones in the muck, lighter ones still floating in the air. "You will be breathing them down on the beach," he said, adding that "touching it, putting your hands in it, swimming in it, even breathing it for a short period of time may not necessarily be hazardous."

This did not seem reassuring as I squatted down, determined to touch the little pellets to see what they felt like. I reached out my hand, just as I had done at the bank of the Kansas River near where I grew up—but this time I pulled my hand away.

Whose body burden?

Where I now live in San Francisco, I'm encountering a newer generation of industrial chemicals—compounds that are not banned and are increasing year by year in the environment and in my body. Sipping bottled water after a bike ride, I could be exposing myself to bisphenol-A, an ingredient in rigid plastics that are used in products that range from water bottles to safety goggles. Bisphenol-A causes reproductive system abnormalities in animals, which recently has led to calls to ban it at the federal level and in some states, although, as of this writing, no action has yet been taken. My levels were so low, they were undetectable—a rare moment of relief in my toxic odyssey.

And that faint lavender scent as I shampoo my hair? Credit it to those ubiquitous phthalates, molecules that not only add flexibility to plastic, but also help dissolve fragrances into shampoos and to thicken lotions we lather on our hands and faces. The dashboards of most cars are loaded with phthalates and so is some plastic food wrap. Heat and wear can release phthalate molecules, and humans swallow them or absorb them through the skin. Because they dissipate after a few minutes to a few hours in the body, most people's levels fluctuate during the day.

Like bisphenol-A, phthalates disrupt reproductive development in mice. An expert panel that was convened by the U.S. government's National Toxicology Program concluded that although the evidence so far doesn't prove that phthalates pose any risk to people, it does raise "concern," especially about potential effects on infants. "We don't have the data in humans to know if the current levels are safe," says Antonia Calafat, a CDC phthalates expert. I scored higher than the mean in five out of seven phthalates tested. One of them, monomethyl phthalate, came in at 34.8 ppb, in the top 5 percent for Americans. Leo Trasande speculates that some of my phthalate levels are high because

I gave my urine sample at Mount Sinai in the morning, just after I had showered and washed my hair with phthalate-rich shampoo. A few hours later, my levels would have been less.

My inventory of household chemicals also includes perfluorinated acids (PFOAs)—those tough, chemically resistant compounds that go into making nonstick and stain-resistant coatings. 3M used them in its Scotchgard protector products until it found that the specific PFOA compounds in Scotchgard were escaping into the environment and phased them out. In animals these chemicals damage the liver, affect thyroid hormones, and cause birth defects and perhaps cancer, but not much is known about their toxicity in humans.

I could go on, but I'll stop here and ask a basic question: Why do we tolerate these chemicals?

In some cases, we don't. DDT and PCBs were banned a generation ago. More recently, Penta and Octa PBDEs have been banned or phased out in the United States and Europe, and Decas have been banned in Europe and in Washington State.

But it is not always clear that banning a suspect chemical is the best option. Flaming beds and airplane seats are not an inviting prospect, either. The University of Surrey in England assessed the risks and benefits of flame retardants in consumer products. The report concluded, "The benefits of many flame retardants in reducing the risk from fire outweigh the risks to human health."

After all, every industrial chemical that is not an unintended by-product of manufacturing was created for a purpose.

These products allow us to live as we do. They are the basis of our civilization and our lifestyle, which is built on a superstructure of chemicals. From 1935 to 1982, DuPont's advertising slogan was "Better Things for Better Living . . . Through Chemistry," because people loved what chemicals do. For much of the twentieth century, chemists and chemical companies were heroes. Swiss chemist Paul Hermann Müller was awarded the Nobel Prize in 1948 "for his discovery of the high efficiency of DDT as a contact poison against several arthropods." DDT was considered a miracle chemical for ridding locales of the scourge of malarial mosquitoes, from the American South to East Africa—until Rachel Carson's 1962 book *Silent Spring* launched the antichemical and wider environmental movement by presenting evidence about how DDT was harming the environment.

No one celebrates chemicals anymore, a point acknowledged when DuPont dropped "through chemistry" from the company slogan in the 1980s and in 1999 adopted the more nebulous "The Miracles of Science." Yet as a society, we continue to love the products and the lifestyle the chemicals afford. When I give talks about my toxic testing, I sometimes ask audiences to raise their hands if they would be willing to give up such chemically laden items as deodorant, cell phones, and automobiles. Most people answer yes, they could see forgoing deodorant (I quickly ask them to put their arms down); fewer reply yes to cell phones and cars. Then I ask about books, and there is a nervous twitter. I explain that books are loaded with chemicals in the paper-making process, the ink, and the cover. No hands go up, although presumably, these are bookish crowds.

My point is that all of us have been complicit in this chemical soup that has settled on our planet, in plants and animals, and inside each of us, although this realization might make us squirm.

Chemical companies make sizable profits from giving us what we want. In 2007, U.S., European, and Canadian companies shipped almost $1.5 trillion worth of chemicals, a 9 percent increase over 2006 for U.S. companies, which earned almost 7 percent profits. In 2008, the dramatic ups and down in the cost of oil for making petrochemicals caused costs to spike sharply upward mid-year, affecting sales and profits, although it's still safe to say that few human activities match the scale and effort that are put into making chemicals. Naturally, the industry wants to keep prices down by using the cheapest materials possible and avoiding costly systems to reduce pollution from its manufacturing and from its products.

When I was a boy in Kansas, industries that made or used chemicals were only lightly regulated and were subject to few environmental curbs. As I saw when I returned to my boyhood home, a dramatic and surprising shift has taken place in the most overt forms of pollution along the river near my home, although in retrospect it's obvious why it happened. By the 1970s, enough Americans could no longer tolerate the chemical chowder in our air and water that they forced political action from Congress.

The regulations that emerged were not, however, based solely on the risks and dangers of chemical pollutants. Congress also mandated that most environmental rules use calculations to weigh the risks and benefits of chemicals and manufacturing practices for everything from pesticides and mercury emitted by coal-burning power plants to plastics and plastic additives. Risks and benefits can include not only the potential toxicity and the health impact of pollutants, but also the impact on the costs and the financial viability of companies, public power utilities, and government agencies.

Back at the University of Kansas, I talk to Karl Rozman and John Doull about this. Both have served as advisers on expert panels that assess and establish regulations and risk-benefit processes. I ask them how it works.

"We have methods for evaluating risk," answers Doull. "We can run tests on toxicity in animals and collect data on the health impacts of particulates, for example. This is an evolving science, but we have developed methods. What we really don't have are good ways to evaluate the benefits of chemicals. We tend to do that economically. We say, Well, you know, it reduces your costs to feed people to use pesticides; it enables us to produce food for the whole world. But how do you measure that against the risks of the pesticides? We don't have good yardsticks for this."

"There is an example along these lines in Africa right now," I say. "Some public health people want to reintroduce DDT to go after mosquitoes because malaria cases are rapidly rising, and the parasites are becoming resistant to most treatments. They would like to spray small amounts in doorways of homes, very targeted, though not from airplanes in huge clouds like they used to."

"It's all about your benefits and risks," says Doull. "You have to weigh putting some people or animals at risk against the benefits and ask the question: do those benefits really justify imposing that kind of risk or not?"

"This is what happens when new drugs are tested," adds Rozman. "Is a new drug that might treat cancer, but has some toxic side effects, too dangerous, or does the benefit make the risk acceptable?"

What about assessing an *individual's* risks versus benefits for having trace levels of chemicals in his or her body? I ask. That is, the benefits the person gets from using a product that is either convenient or vital to his or her life, such as a car if the person has to drive to work, versus the possible harm from effluents coming from that car?

Rozman and Doull tell me that the EPA and other government agencies tend to work on risks as they pertain to populations of people—say, for all Americans, or on local populations that may have been exposed to unusually high amounts of chemicals, such as Kansas farmers and former GE workers in Hudson Falls, New York. Public health experts also collect data on vulnerable segments of the population: old people, who more easily succumb to smog; women who are pregnant or are in their childbearing years; and fetuses and infants. Leo Trasande tells me, though, that no specific federal laws exist to protect the most vulnerable from most chemicals. "The food quality protection act [of 1996] requires the EPA to create laws protecting infants and children from the dangers of pesticides," says Leo. "There are no laws like this for other chemicals."

Beyond this, he says, neither the government nor scientists have focused on the impact of average exposures on individuals—in large part because the technology has not existed. Public health officials have traditionally focused their efforts on protecting populations, not individuals, confirms Jim Pirkle of the CDC, although he suspects this will gradually change as scientists learn more about how minute levels of chemicals cause disease, and as more people are tested for their personal body-burden levels of chemicals. Information on individual genetic susceptibility will also play a larger role, as scientists learn more about envirogenetics.

The paucity of information makes it difficult for an individual to assess properly his or her risk-benefits for using everyday chemicals. But even if we did have a baseline of information, it wouldn't matter since our exposure to most chemicals is involuntary—which is frustrating. "I think that people inherently view risks that are involuntary differently than voluntary," says Karl Rozman. "Driving a car is voluntary, and, if you decide it's too risky, you don't drive, whereas these chemicals in your body are involuntary."

For a few chemicals, such as mercury, we can, to some extent, control our exposure. For instance, after my experience on the *Osprey* catching and eating the halibut and also the store-bought swordfish, I have limited the amount of big predator fish I eat. This is not a huge sacrifice for me, since I don't much like big fish, anyway. If I get a hankering for fish flesh, I can eat smaller fish that contain less mercury. Since getting my phthalates' results back, I am also shying away from the lavender and minty scents in shampoo and deodorant, and I try to keep my bisphenyl-A exposure from those crinkly plastic bottles to a minimum.

I could try to remove chemicals by using controversial techniques such as chelation therapy, where chemicals are added to the body to flush out heavy metals, such as mercury and arsenic. Most physicians counsel against chelation, however, except in cases of very high exposures. The process, they say, either doesn't work or can also flush out desirable metals and chemicals that the body needs and can cause other complications.

Ultimately, we could just ban everything, though there is this little wrinkle—the fact that we humans like using some of these chemicals, whether or not we admit it. Take plastics. Without them, many items that we take for granted would either be impossible to make, such as lightweight cell phones and computers, or would cost a bundle.

I have a confession: as a consumer, I like plastic. Within three feet of where I'm sitting right now, plastic has made possible my laptop, iPhone, water glass, pens, backpack, watch, speakers, sunglasses, and running shoes. As an environmentally minded person, I'm appalled by all of these chemicals on the loose, but does this mean giving them up?

During one of my phone calls (on my plastic cell phone) with the CDC's Jim Pirkle, I ask him whether he thinks the levels of chemicals that his agency is measuring inside Americans' bodies are dangerous.

"I don't want to answer that, because we don't have the data," he says, echoing others. "If you had asked me that question in the fifties and I said that lead levels at twenty parts per million were okay, I would have been wrong." Today, any detectable level of lead is considered unsafe for children.

When rivers saturated with chemicals were causing rashes and burns on children's skin, and the air was unbreathable, the personal risk-benefit assessment for millions of people was clear. They demanded action, and action was taken. But it's harder to know about the chemicals of today, many of which are invisible to the eye and the nose.

This lack of knowledge allows for a range of opinions from experts on how harmful these chemicals are in tiny amounts. "Chemicals are not all bad," says Scott Phillips, a medical toxicologist in Denver who was referred to me by the American Chemistry Council (ACC)—the lobbying group in Washington that represents the chemical industry. "While we have seen some cancer rates rise, we also have seen a doubling of the human life span in the past century." Representatives of the ACC insist that they support new scientific efforts to tease out the toxicity of chemicals produced by their industry, although they have opposed legislative efforts at the federal and state levels to require more thorough testing of chemicals for toxicity. No new, comprehensive laws regulating the industry have been passed by Congress in more than thirty years. Yet ACC toxicologist Richard Becker contends that "As new science comes forward, we'll welcome it to make regulation more efficient. The problem is right now we don't have all the information."

Leo Trasande disagrees, insisting that the unleashing of so many chemicals into the environment without knowing their impact is an uncontrolled experiment on all of humanity—not to mention on the ecology of the entire planet.

The key is to know more about these substances, so that we are not blindsided by unexpected hazards, says former California state senator Deborah Ortiz, who chaired her state's Senate Health Committee and has authored and supported bills to ban PBDEs and to monitor chemical exposure in humans. "We benefit from these chemicals, but there are consequences, and we need to understand these consequences much better than we do now," she said.

In 2006, the European Union launched an effort to do just that when its Parliament passed the REACH act, standing for Registration, Evaluation, Authorization, and Restriction of Chemicals. REACH requires companies that want to market new chemicals in Europe after June 2007 to prove first that the substances they market or use are safe or that the benefits outweigh any risks. Provisions in the law also require some testing of existing chemicals. The law, which the chemical industry and the administration of President George W. Bush opposed, also encourages companies to find safer alternatives to suspect flame retardants, pesticides, solvents, and other chemicals.

The basis for REACH is a concept called "the precautionary principle," the idea that chemicals should be tested and better understood for possible toxicity before they are approved. In Europe, this protocol is replacing the system that is still in place in the United States. "Essentially, this is a standard of innocent until proven guilty," says David Sherr, a professor of environmental health at Boston University. Currently, only a small percentage of the eighty thousand chemicals used by industry in the United States have been tested for toxicity. New chemicals are generally approved without a test, unless there is a reason to suspect they might be toxic. "This is a difficult standard," says Trasande, "since many of these chemicals are new, and their toxicity is unknown."

For now, the U.S. government has no plans to enact a stateside version of REACH—although this might change in the new Obama administration—although American consumers are likely to benefit from the EU's action, as chemical companies develop new standards and chemicals to comply with the new law in Europe. Their legislation also is likely to launch new and innovative efforts to produce safer, cleaner chemicals in general, a process that is already beginning with the nascent "green chemical" movement. China and other fast-developing nations are also considering REACH-style laws, although it's unclear how well they will be enforced.

As I type these words on my plastic MacBook, sipping coffee from a treated–paper cup, I can't help feeling just a smidge of anxiety about my own relationship with the chemicals in my life. Realistically, I'm not going to give up my Teflon pans, chemically treated and dyed blue jeans, and most other products that make life convenient. Yet I wonder whether this is more of a devil's bargain than I would like to admit. On a global scale, the indiscriminate use of pesticides, plastics, and flame retardants may be a gathering disaster on the scale that Jared Diamond wrote about in *Collapse: How Societies Choose to Fail or Succeed,* which chronicles how societies from Easter Island in the Pacific to the ancient Maya came crashing down because of overpopulation, overuse of resources, and pollution. Diamond calls this ecological suicide "ecocide" and offers at least an implicit warning that this sort of collapse could happen again unless we are clever and vigilant.

The question is, can humans have our chemically laden cake and eat it, too? Politics and specters of global collapse aside—this book is about individual health—at least part of the answer lies in each person's individual biology and his or her ability to metabolize and fend off the harmful effects of the environment. This, then, is the next stop in the Experimental Man journey: to link up the first two sections of the book, the environment and genes, to find out more about protections or sensitivities I may have been born with to cope with what nature and humans are lobbing at my body as I make my way through life.

Do my genes protect me?

Near Acadia National Park in northeastern Maine, I feel the frigid wind blow off the ocean at the Mount Desert Island Biological Laboratory (MDIBL), a cluster of buildings surrounded by a quiet winter forest by the sea. My San Francisco leather jacket isn't nearly warm enough when I follow Carolyn Mattingly, a tall, slender scientist wearing a much more appropriate thick overcoat and knit cap, as we scurry from Mount Desert's offices to her lab in a separate building. Mattingly is part of a team at the MDIBL that runs an NIH-funded database connecting data and studies that investigate how environmental chemicals interact with genes and proteins, and whether there is a connection to any specific diseases. Called the Comparative Toxicogenomics Database (CTD), her team's effort is the most comprehensive Web site of its kind and by far the most comprehensible for a nonscientist.

Most genetic databases online are massive, unruly sites that are designed (or, more accurately, have proliferated like algae in the ocean) as repositories of raw data and studies downloaded by scientists whose federal grants require them to make public their findings. Good and bad studies and data are deposited indiscriminately, leaving nonscientists and scientists alike facing a Babel of information that is difficult to understand and to evaluate. To use the CTD, you punch in the name of a gene, a chemical, or a disease—"mercury," for instance. Up pops a menu offering links to whatever genes and diseases interact with this chemical, plus a list of peer-reviewed studies and media articles. For mercury, the site during the summer of 2008 tells me there is information on 292 genes that have been tested for a response to mercury in fourteen organisms, including mice, rats, dolphins, cows, ducks, and humans, with links to dozens of diseases, from cancer to neurodisorders.

Mattingly talks quietly. She is shy and, like many scientists, responds to visitors as if she's been interrupted from intensely absorbing work or just woken up from a sound sleep in the middle of a dream. She tells me that the scientists on the CTD team have curated more than

122,000 chemical-gene and chemical-protein interactions involving more than 4,000 unique chemicals—including foods, vitamins, and naturally occurring compounds such as cholesterol—and more than 13,500 genes in some 200 species. The team also integrates data from other sources on more than 60,000 chemicals.

The database is not set up for a search like mine. I want to link together my personal genetic and chemical information to create an individual profile of how they interact. Nor is the fledgling field of envirogenetics ready to make much sense of such an investigation. "We are very early in trying to sort out these connections and interactions," says Mattingly.

But I've come here anyway to get at least a rough idea of how my DNA might protect me, or not, from mercury, PBDE flame retardants, pesticides, and PCBs. These are just a few of the millions of interactions occurring inside me right now between environmental factors—everything from the sun's UV rays to what I had for breakfast—and my genes. Because I could write several volumes if I tried to cover all of these topics, I'll concentrate only on chemical pollutants, although keep in mind that much more is going on for each of us as we simply live each day.

The idea that chemicals and other environmental factors affect people differently is hardly new, as anyone with an allergy or another obvious sensitivity can attest. "We know that people metabolize environmental chemicals in their bodies differently," says the CDC's James Pirkle. But the detailed study of how genes and specific environmental factors interact on a molecular level *is* relatively new, in part because researchers first needed to understand the basics of genetics. With this task well under way, after decades of emphasis on genes, scientists are now beginning to link together specific environmental factors and their impact on genes and on disease. This effort is partly driven by the unexplained escalations in some diseases that may be caused by a combination of toxins and high-risk genetic variations. "Recent increases in chronic diseases like diabetes, childhood asthma, obesity, or autism cannot be due to major shifts in the human gene pool," says geneticist Francis Collins. "They must be due to changes in the environment, including chemicals, diet, and physical activity, which may produce disease in genetically predisposed persons."

As Kansas University's John Doull tells me, there are three major factors to consider in assessing these interactions: "You have the agent, the chemical, that is capable of producing adverse effects. And you have the individual, his genetic profile, and the biological system in him that's potentially going to be damaged. The third thing, of course, is the exposure: the dose and timing. When you look at those three things: the agent, the subject, and the exposure, the most important factor is the susceptibility of the subject to have this kind of damage produced. For some exposures with certain doses and timing, we have evolved excellent defenses. For others, some of us, or all of us, are vulnerable."

Most people and animals react similarly to chemical exposures, says Karl Rozman. "But you always get a few outliers who react very badly." He tells about the tragic case of a student at Kansas University who was prescribed codeine for pain—codeine being a chemical, too, that happens to have been approved for human use. "She apparently took the right doses because that was all that was missing from the bottle," he says. "But she had lethal levels of codeine. It turned out that she did not have the enzyme that metabolizes this chemical. You and I probably would metabolize codeine with a half-life of 2.6 hours. I calculated the half-life in her was about nine hours, so instead of getting rid of it, as she took more, she built it up."

This case is doubly horrific because a genetic test now exists that could have predicted this young woman's susceptibility to codeine and many other drugs from a variety of classes, including antidepressants, antipsychotics, beta-blockers, pain medications, and some anticancer drugs. The culprit DNA here are variations in what is called the Cytochrome P450 complex, or CYP450: a grouping of proteins that among many other things metabolize codeine and other drugs and chemicals. Some people have genetic variations that keep them from producing certain CYP450 enzymes—genotypes that can be detected using a patented genetic test called AmpliChip CYP450 from Roche Diagnostics.

Thankfully, I came out normal when I took the Roche test, making me a pharmaceutical company's ideal candidate for a slew of drugs. My CYP450 profile also suggests that I have some general protection from unwanted chemicals that wind up inside me, in that I metabolize a long list of them at a normal rate.

Following Carolyn Mattingly through the snow, we enter a building tucked into the woods. Before we look at my data, Mattingly is showing me her "wet lab" research—what she does when she is not working on the CTD database. In a small workspace in this freezing-cold building, she is testing tiny striped zebrafish for reactions to arsenic, an industrial by-product that is prevalent here in Maine and New England. The fish are smaller than my little finger, zipping about in mini-aquariums that are about to be saturated with arsenic, a chemical that is also inside my body. Her fish work is relevant to me and to other humans because zebrafish—a favorite animal for testing toxins—contain many of the same genes that we humans have. "Zebrafish give us some idea of which genes are affected by arsenic," she says.

"Should I worry, as a human, if I have a susceptibility gene that has been found in zebrafish?" I ask, half kidding.

"There are similarities between genes in fish and in humans," she says, "but animal models only take us so far in understanding what happens in humans, although they provide some meaningful clues."

Later, in a conference room overlooking thick, leafless trees, Mattingly opens her laptop and delivers my results for about forty toxins that registered inside me at higher-than-average levels: my now familiar complement of DDT, PCBs, flame retardants, and metals. In all, the database registers 343 genes that are affected by or associated with these chemicals.

She warns again that this information as it applies to me, an individual, is incomplete. "We don't really understand how many of these chemicals work and interact at a very basic level in cells and in the body, so being able to know how they respond to different people with different genetic variations is problematic at best."

I tell Mattingly that I understand the caveats and accept them. My goal is to forge ahead to see what, if anything, can be applied to individuals at the embryonic phase of this science.

First on the list is Mattingly's favorite toxin, arsenic. The database lists 1,400 genes that are affected by arsenic, based on tests run on several types of animals, including humans. When I tested my arsenic level, it came out as 12 parts per billion, safely under the danger threshold of 23 ppb, which is good news since I have mutations in 22 out of the 142 genes in humans that might make me susceptible to arsenic exposure. These include a variation in the ABCB1 gene that may inhibit my ability to expel arsenic. ABCB1 is in a family of genes that provides resistance to a wide range of metals and other chemicals, including, for some people, beneficial chemicals in drugs used for chemotherapy and other treatments.

Next are the PBDE flame retardants that appear in such elevated levels in my body. The CTD lists only seventeen genes that have been tested for interactions with PBDE-47, the ingredient in Penta flame retardant that I scored the highest on. I am hoping that I have DNA that protects me from this chemical or that at least keeps me from being susceptible. But Mattingly has found no SNPs in my personal data that match up with the genes, and, anyway, none have been tested in humans. Only two animals have been listed as being tested for genetic reactions to PBDEs: *Mytilus edulis*, the blue mussel, and *Oncorhynchus mykiss*, the rainbow trout. For mussels and trout, the news is not great. Apparently, PBDE-47 affects the activity of all seventeen genes, which could play havoc with their functions, though the details are poorly understood.

The database also has no directly relevant data on genes negatively impacted by the PFOAs that are found on my Teflon pans. This is probably because concern about this family of chemicals is relatively recent. Out of more than three hundred genes found on the database in relation to PFOAs, only three were in humans, and these were tested on human cells in petri dishes. Using human cells to test for toxicity to chemicals is a new development in environmental toxicology, but it is imperfect, as I will explain later, since a cell isolated from an organism can tell a scientist only so much about processes in a body that work as complex and interactive systems of cells and other molecular structures.

For PCBs, the vast majority of studies also concentrated on animals, although there were a few interesting results on human cells. A number of studies have shown some link between PCB exposure and genes that are complicit in breast cancer. These include a Danish study in 2002 indicating that certain PCBs—including several that showed up in me—reduce the expression of the BRCA1 gene. As we learned earlier in the book, BRCA1 acts as a tumor suppressor for breast cancer, playing a role in how cells develop and in repairing DNA. The study concluded that PCBs therefore "have the potential to affect breast cancer risk." My daughter has not been tested for PCB levels, but she and I both have some potentially deleterious genetic variations in our BRCA1 genes that are associated with a higher risk of breast cancer. (We also have contradictory markers that confer a normal or low risk.) As of now, information does not exist that would link up our BRCA results and other markers associated with breast cancer with PCB exposure, although I'm sure this linkage will one day be available. This does make me wonder, however, if PCBs, which were prevalent and on the rise when my grandmother contracted breast cancer in the sixties, played a role in her disease.

In other tests, David Sherr of Boston University and others have shown that PCBs, dioxins, and other pollutants seem to activate the aryl hydrocarbon receptor (AhR), which, among other things, can cause cells to become cancerous. One cause of cancer is when healthy cells grow out of control and lose their identity: they "forget" that they are programmed to be liver or heart cells and go rogue, becoming "immortal" by not dying in the usual manner and by continuing to replicate into more rogue cells. Sherr's lab has found that many cancers are accompanied by elevated levels of AhR. "In a normal cell, AhR causes the cell to grow if needed," says Sherr. "It's a basic part of life for many organisms. Certain chemicals make the AhR think it's being activated. The right kind of PCBs turns on the AhR and it becomes a persistent signal and can cause cells to go cancerous." Sherr says that certain gene variations may be at work inside some people, putting them at higher risk for AhR activation by chemicals. "It would help tremendously to know more about these genes and who is at risk," he says.

Mattingly checks on DDT and finds close to three hundred references to studies in her database, including a few on humans. One 2004 study in France investigated the impact of various doses of DDT and the pesticides Aldrin, Chlordan, and Dieldrin on two Cytochrome P450 genes and found that these chemicals cause an increase in the activity of CYP3A4 and CYP2B6, which produce enzymes that metabolize a broad range of drugs and chemicals. This and other studies suggest that these pesticides alter not only specific genes, but also pathways of genes that interact as a system inside our bodies that can also be altered by interacting with chemicals.

Much of this data is also too preliminary to link up with my specific genetic variants to see whether I am personally more or less susceptible to DDT and other pesticides. The number of studies trying to make these links is ramping up quickly, however, attested by the rapid increase in papers listed in the CTD database. Within a few months of my first accessing the site, the number of studies investigating the gene-chemical-disease connections with DDT rose from about thirty to almost three hundred. (Some of these studies are not new; they simply had not yet been included in the database.) Not that volume alone will produce meaningful results. Indeed, research focused on the affect of chemicals on individuals remains scant, as scientists try to understand the basic mechanisms of how these chemicals work and interact.

Mattingly also goes over my results for mercury, picking up on the earlier research I conducted after catching my halibut with Josh Churchman on the *Osprey*. The CTD reported on almost three hundred studies investigating more than 250 genes in species such as zebrafish, cows, dogs, dolphins, rats, mice, chickens, pigs, and humans. The human data suggested preliminary findings linking mercury exposure to cancer, cell death and proliferation, and diabetes. Mattingly hasn't found any links between the genes in the CTD and variations in my genetic data that might increase my risk of acquiring one of the diseases or the conditions caused by mercury—which could be good news, although the database, of course, isn't set up to catch individual genetic variants.

Nor does the CTD contain every study on mercury's impact on humans. ("We are trying to keep up!" says Mattingly.) One series of studies on mercury-gene interaction in humans that is not yet in the CTD database is the one I discovered just after my fishing expedition on the *Osprey*—those conducted ley Karin Broberg and her team at Lund University in Sweden. In her latest paper, she has identified four SNPs that may make a human more or less susceptible to the harmful effects of mercury exposure. These SNPs seem to affect the production of glutathione, the chemical in cells that metabolizes and facilitates the body's removal of toxins such as drugs, poisons, and environmental heavy metals.

Scientists have long known that the elimination of methyl mercury varies greatly among individuals, ranging from 30 to 70 days, with extreme cases up to almost 190 days. These differences are most likely genetic and probably have to do with a person's ability to churn out enough glutathione to more quickly clear out the toxins, especially in the liver and the brain. "These genetic variations can change an individual's susceptibility to carcinogens and toxins, as well as affect the toxicity and efficacy of some drugs," wrote Broberg's team in their paper, which was published in 2008.

For the study, the Lund University researchers went to Västerbotten, a region in northern Sweden, to test 292 subjects who reported on a survey that they ate fish (large and small, lean and fat) at least two or three times per week. The team tested mercury levels in the subjects and genotyped them for the four SNPs, coming up with significant differences in the levels of mercury for people who had a similar intake of comparable fish. The differences were more pronounced when subjects had a combination of certain higher-risk SNP variations.

Using my genetic results, listed in the table on page 161, I was able to test only two of the four SNPs in Broberg's study. For those, I came out normal, meaning that my cells, in the absence of other data from additional SNPs, probably expel methyl mercury more in the usual 30- to 40-day range than in the more dangerous 75- to 190-day end of the spectrum. However, when I checked my family's results for these two SNPs, my mother and my brother came up with a higher-risk variation but not the highest. This upset my mother. She has always watched such things carefully, although I suggested that at age seventy-six, she hasn't suffered any noticeable harm from possible mercury retention, nor is she a big fish eater. My brother's response showed a flash of the humor I love in him. He said: "Let's go eat fish!"

Carolyn Mattingly runs one more test for me. Because genes and SNPs—those single-letter changes in our DNA that can cause one person to be at higher risk than another for a disease—do not operate in isolation in our bodies, scientists are beginning to develop software and computing power to track a change caused by a mutation or a chemical in one gene with what happens to other genes that interact in pathways that regulate, say, inflammation or metabolism. For instance, if mercury or PCBs cause a gene to ramp up or down its activity, this can affect its ability to activate or otherwise influence the next gene, which can cause an entire pathway of genes or proteins to be altered or to malfunction. Imagine a line of cars on a long bridge moving at a steady, if frustrating, twenty miles per hour, when a driver with allergies suddenly has to sneeze. He tries not to, but it cannot be helped: Achoo! The sneeze causes him to slow down ever so slightly, which affects all of the cars behind him. The delay ripples down the line, causing a cascade of cars to reach the other side of the bridge later than they would have without the sneeze. Something like this happens in genetic pathways when one gene is affected by, say, the introduction of mercury from eating a halibut with butter and basil, as I did after my fishing trip aboard the *Osprey*.

			Envirogenetic Interaction of Mercury: The Author's Results	
Chemical	Gene	SNP	Author's Results	Impact if High Risk
Methyl mercury (MeHg)	GSTP1-105	rs1695	AA (low risk)	⬆ Risk of asthma, adverse drug response, some cancers
MeHg	GSTP1-114	rs1138272	CC (low risk)	⬆ Risk of some cancers
MeHg	GCLM-588	rs41303970	N/A	⬇ Glutathione production
MeHg	GCLC-129	rs17883901	N/A	⬇ Glutathione production

The understanding of connections between environmental toxins and pathways is in most cases very early, but Mattingly was nice enough to send me two diagrams of gene pathways that look like the electrical schematics on the back of the fuse box in my garage. These make little sense to a nonscientist like me, but give some impression of what these systems of genes look like.*

The first diagram shows a field of about thirty-five genes connected in a grid that are influenced by my forty top chemicals. Mattingly created the diagram using a commercial software package that demonstrates how genes interact with one another and how affecting one of them can affect others. For instance, a gene called CFB, which is involved in immunological responses, reacts to mercury exposure by decreasing its activity. According to the diagram, this alteration in CFB expression might cause other genes "downstream" to be altered, too, causing complications in critical pathways and possibly triggering a disease response.

Mattingly's second diagram details the interactions of genes involved in my respiratory system. This schematic includes more than

---

* To view the diagrams, go to www.experimentalman.com.

thirty genes that have something to do with diseases that range from asthma and migraines to heart disease. In this system, I checked out two genes that are affected by environmental toxins inside me. For one gene, I have a high-risk variant; for the other, the normal variation.

For the first gene, ADRB2, my variant in very early and sketchy studies suggests a slightly elevated risk of migraines, asthma, and perhaps heart disease. (These studies do not take into account environmental factors.) At the same time, according to other studies, exposure to certain heavy metals such as cobalt and copper also causes this gene's activity to increase. The combination of these two factors zapping this gene is not known, but it's possible that this one-two punch could cause changes up and down the pathways that include ADRB2.

Mercury and other metals have been shown to turn down the activity of the second gene, called F5, which controls part of the process of blood clotting. Having a certain SNP mutation in the F5 gene can also cause thrombosis and increases the risk of stroke. For this gene, I have the normal, or lower-risk, variant, although if I had the higher-risk version, having metals in my body might have amplified the deleterious effect both in the F5 gene and in the gene pathways that it's a part of.

As the afternoon wanes, I say good-bye to Carolyn Mattingly and make a dash through the snow to my car. My head is spinning with what she has been explaining to me. Without question, the diagrams linking interactions with genes, chemicals, and disease pathways will be a crucial part of the future of personalized medicine, although it is far too early for them to be of much use to anyone, let alone individuals. They are also extremely difficult to understand for the nonscientist. Yet even in this crude form, the tables and diagrams contain potentially powerful information about the affect of chemicals on the delicate and dynamic balance of forces inside our bodies.

Like the man in the car on the bridge who couldn't help sneezing, I feel as I head away from Acadia that I have just learned what could happen if I slow down or speed up the actions of my genes even slightly. This suggests that Mattingly is on to something much more important—and complicated—than genetics alone and that this will be playing out to affect us all during the next few years.

With more snow threatening and the sky turning dark and pregnant with moisture, I feel something that is new to me in the Experimental Man project: fragile.

Immortal cells bathed in mercury

A day after visiting Carolyn Mattingly, I happily find myself in a balmier clime near Washington, D.C. I'm heading to another embryonic envirogenetics project, in Rockville, Maryland, at the National Institutes of Health's Chemical Genomics Center, the NCGC. In Maine, I had learned about possible interactions between chemicals and genes; now I was about to find out how the soup of trace chemicals inside each of us might impact the world within our cells: the genes, the proteins, the mitochondria, and other structures that are vital to the smooth operation of life.

The NCGC uses modern "high-through-put" technologies developed by the pharmaceutical industry and researchers looking for compounds for potential drugs. These drug candidates are first tested for toxicity and to see whether they cause a desired reaction in a cell or a tissue. Thousands of tests are run at a time, in the hope that one or two will work against a target disease. The same technology can be used to test environmental toxins against human cells, says NCGC director Chris Austin. He has joined forces with scientists at the EPA and the National Toxicology Program (NTP) to run for the first time thousands of common toxins through his center's new $40 million high-through-put lab.

A tall, long-limbed neurologist turned researcher with light brown hair, a sturdy jaw, and marquee good looks, Austin walks me through labs where robotic arms load and unload tiny well-plates of samples, with each well the size of a pencil point and filled with cells, genes, or other targets being exposed rapid-fire to different chemicals at varying doses. The facility is small, by drug-industry standards, but the machines are state-of-the-art in a lab built amid the rolling hills north of the Potomac River, a rising hub of biotech companies and labs.

"We have started with three thousand chemicals," Austin says as we walk, explaining that the project is testing cells in vitro—out of the body—from the liver, the brain, the blood, and other key organs in humans and in rodents. Austin's team wants to see how the cells respond to chemicals that include many on my list: pesticides, phthalates, and the rest. "We're studying how chemicals interact with different components in cells," he says. "We're also looking at how different SNPs cause an increase or a decrease in a person's risk for developing diseases that are caused by chemicals interacting with genes."

Someday, scientists hope to eliminate or greatly reduce the use of animal testing for toxicity and replace it with cell models, says Austin. "For these early tests, we are still trying to set up methods and to see how cells react to different doses and chemicals to see whether we get any useful information. We want to understand the mechanism that determines why or why not these chemicals are toxic. It's very preliminary." Even if it works, the in vitro data will not be as useful as in vivo—in body—data right away, he explains. "One reason is that the in vitro cells act differently in a petri dish than they do in a human. But we're trying to get as close as we can to what really happens in a person.

"We want to find out whether we can predict reactions in cells and animals using this data," he says. Eventually, they hope to use the method to develop computer models that can predict outcomes of different chemicals in human and animal cells before a new chemical is approved for use in a product. Further out in the realm of possibilities are tests on individuals to measure levels of toxins in cells and to determine personal variation in how cells respond, although this is highly speculative. "I have no basis for knowing whether that would work," says Austin.

"We're seeing some interesting reactions in the cells," he continues, "compounds turning on and off genes and pathways, and we have lots of ideas. But we're trying to figure out what to do with this data beyond the science. There are questions of politics, and even the law if people decide, say, to sue chemical companies based on our results. If we publish this data prematurely, will people draw unfounded conclusions about these chemicals? We're struggling with this. Everyone out there has an agenda. Let's say that these chemicals turn out to be nontoxic in trace quantities—that would be what the science says, although some people won't like this. Same thing if we find that these chemicals turn out to be toxic. We are the dispassionate seekers of the truth, though we are not unaware of the politics and emotions that go along with our work. I fear that an emotional reaction for some may prevent the proper intake of the data."

He continues to point out machines and processes, but of course all of this talk about testing an individual's cells has given me an idea.

"Hey, what about testing my cells using your process—is that possible?"

For a beat or two, Austin stops talking and flashes a look that has become familiar during the Experimental Man project: of half surprise and half incredulity. Then he smiles broadly, looking like Gary Cooper as he understands what I'm asking: to use my cells as the target for thousands of chemical toxins, and to use this experiment to tell the story unfolding in his lab.

"Sure, why not?" he says. "We order anonymous cells from a supplier right now, but I can't see any reason why we couldn't use yours."

He warns me that the tests won't tell me much, however. "As soon as your cells leave your body and we put them into a petri dish, they begin to behave differently than they do inside you. In your body, they are dynamic, working with other cells. This is the problem with this testing. We need to figure out how to stop these changes, and we will, but right now it's a bit artificial. But you have to start somewhere. The Human Genome Project took years and started small."

Over the next few weeks, we devise the experiment and bring on board two researchers at the NCGC who are working directly on the cell assay project, Ruili Huang and Menghang Xia.

The first step was for me is to give more blood, from which technicians will extract a cell type called human vascular endothelial cells, or HVEC. This is the easiest cell on their test menu for a live human to access. To extract my HVECs, I went to the Coriell Institute in Camden, New Jersey. This was the same place I went to have a panel of SNPs tested for Coriell's nonprofit Web site to determine genetic risk factors for certain diseases. Coriell specializes in preparing cell lines for labs to use all over the world. After relieving me of a few HVECs, it took them three months to prepare and grow a batch of cells that would become stable and "immortal"—that is, they would grow indefinitely if properly fed and maintained.

This is probably the only chance for a small (very small) part of me to live forever.

While we waited for the cells to complete the immortalization process, there was protocol to deal with. Whenever researchers want to test humans, they are required to file for an IRB—Institutional Review Board—at their institution or sponsoring agency to approve the experiment. Chris Austin wasn't sure one was required, since these cells were outside of my body during testing, but, to be on the safe side, he asked for an opinion about my experiment from the NIH's ethics consultation service. Every medical and research institution has an ethics service available twenty-four hours a day to weigh in when ethical issues arise, such as when to withhold care for the terminally ill, or whether it is appropriate to use an experimental treatment on a patient. This NIH committee, like most, was composed of physicians, ethicists, religious representatives, and nurses.

A few weeks later we got our response from the committee.

They said no.

Chris Austin and I were surprised, although I told him that this was not the first time that objections had been raised to the Experimental Man project. Much of what I proposed when I approached researchers and institutions was unprecedented: testing a healthy individual using diagnostic and analytical tools and technologies that were usually applied to the sick or were so new that many experts considered them too early to be of use for individuals.

I had given my argument to counter these objections for my project many times: that my experiment is marking a moment when new methods and technologies are on the cusp of being ready for healthy individuals, some more so than others. My effort is to describe these technologies and to assess their current usefulness—or their lack thereof—and their potential for the future. As I have discovered in my testing so far, much is not yet ready, although I believe that the exercise of explaining the technologies and the potential by testing a real human might help people understand the science and its implications. It might also help encourage a greater emphasis by the research community on applying these technologies to individuals.

The committee's objections were typical of what I had heard before. Its official "Analysis and Recommendations" said:

> While performing tests on the journalist's cell-line could be a part of an interesting journalistic story, the tests would not be of value to the kind of research performed by Dr. Austin's research unit. Given this fact, the consult team had several concerns which suggest that it would not be advisable to comply with the journalist's request.

The objections raised were the possible misuse of taxpayers' money; the danger that the public might be misled about preliminary research; and the idea that if I was allowed to have such a test run, it would be unfair to deny others. The committee also suggested the following if the tests were allowed:

Understanding of results: in addition to concerns about the public being misled by the reported results of the test, Dr. Austin should make sure that the journalist understands the risks and benefits of the tests, and of the reported results.

IRB review: According to our interpretation of the title 45 CFR (Code of Federal Regulations) part 46, the tests that were to be performed on the journalist do not constitute research, and so might not have required IRB approval. However, we strongly advise that if Dr. Austin were to perform the tests on the journalist, he gets prior approval from OHRP (Office for Human Research Protections).

Lack of implied endorsement of journalist's publications of the results: If Dr. Austin were to comply with the request, then he would need to consider how he can guarantee that the public would not be misled by the journalistic report, and that his own scientific credibility be maintained, without implying that he endorses the report.

Two additional points made by the committee were to be sure that the tests, if done, were conducted according to federal criteria governing lab work on humans and a concern that Chris Austin could be a party to litigation if the tests and results were misrepresented and members of the public or others were confused.

I appreciated this input from the committee and the glimpse into the inner workings of ethical considerations at a medical research facility such as the NIH. It is crucial that such a mechanism exist to protect patients and the public and to properly discuss and vet new technologies, methods, and experiments. I was also grateful that the committee members left the door open for Austin, his team, and me to further discuss the request with the committee—which we did, in the ethics department's conference room in Bethesda.

Five members of the committee joined Ruili Huang and me in person, and Chris Austin on speakerphone from California, where he was traveling. I started the meeting by giving a brief presentation of the Experimental Man project, suggesting that it was primarily an educational effort, while submitting that one crucial role of the media is to explain and assess the activities of the government, including the NIH. With the public and even members of Congress and other leaders sometimes confused by aspects of this new science, I argued, it is vital to find creative methods for describing it.

The committee parsed the issues down to three: the ethics, the legality, and the scientific merit of my proposed experiment. No one thought there was a serious ethical breach involving some outlandish use of human experimentation, since my cells were outside my body. After some discussion, they agreed with my contention that the experiment could be described as an educational activity, rather than research, and that this was perhaps a beneficial use of public funds if it contributed to a greater understanding of the NCGC's work.

The issue of legality, they said, was not their purview.

As for the scientific merit, they turned to Ruili Huang and asked whether my cells had any value for her experiments. She answered that it was too early in the research to know—they were still in a proof of concept phase, exploring what happens when cells were subjected to chemicals. They had not yet set up a rigorous parameter for which cells to use. "I think they might be useful, but we will have to wait and see," she said.

This impressed the committee members, leaving two final concerns: how could they be sure that my report wouldn't mislead the public or damage the reputation of Chris Austin and his team?

Normally, journalists do not allow subjects to read or have a say over their stories—this is the freedom of the press not to be impeded or dictated to by the government or others—but in this case, I said, I would work with Chris Austin—which I did—to ensure that the material was as accurate as possible and would not damage his reputation.

In the end, the committee members gave their nod to the project, with the proviso that we seek an opinion from the NIH legal office and that we check in with the IRB and the OHRP.

Unfortunately, this process was going to take too long for the results to get into the book, although Chris Austin pledged to continue the process and to conduct the tests once we received final approval. Any results will be posted on the online version of the Experimental Man project.

But I can report on what the results of my cell assays are likely to be. Huang and Xia had already run screens of 1,408 chemicals chosen by the NTP against anonymous human (and rodent) HVEC cells. My HVEC cells are likely to react in a very similar way.

The team has tested some fifty cellular functions, such as mechanisms that kill the cell; that damage genes, proteins, and pathways;

and that cause the membrane of the cell to degrade. The fifty "assays" included tests on markers for the BRCA genes involved in breast cancer, the Tau protein associated with Alzheimer's disease, and targets that deal with oxidative stress. Out of the 1,408 chemicals tested, 428 showed some toxicity in the cells. Some chemicals were toxic across all cell types and species, and others were selective, says Huang. "Some of the toxins appear to be more heritable than others," meaning that genetic components played a role. "We want to match toxic response to genotypes," she says. All cells were exposed to fourteen different doses of chemicals administered over the course of forty hours. Of course, the chemicals varied widely as to potency over a given period of time, with some being highly toxic in small doses over a short stretch, and others that are far less noxious in the short term but that will persist in the cells over time and perhaps gradually cause damage.

Huang emphasizes an important caveat with these tests: that the chemicals were directly applied to cells, rather than working their way in as part of a dynamic process in the body. This direct cell immersion in the lab was likely to cause reactions that might not happen in real life, where mechanisms outside the cell could prevent all or some of the chemicals from entering the cell or might otherwise influence a single cell's exposure. "The body does different things to compounds that don't happen when we work in vitro," says Huang. "In some cases, the compound we're testing is not toxic, but the body converts it into something that is."

Huang shows me the results for two of my favorite chemicals that are among the 1,408 tested: mercury and PBDEs.

Mercury was so toxic that it killed nearly all of the cell types within forty-eight hours, including the anonymous HVEC blood cells that were tested by the team. "It kills the cells so quickly, it's hard to measure what happens," says Huang. Mercury also caused significant changes in a majority of the assays, including the production of glutathione, but it didn't affect the activity of the breast cancer genes or Alzheimer's. Huang says the researchers are conducting additional tests on glutathione pathways for mercury, including an investigation into the susceptibility of brain cells. "That's one of our pathways," she says.

The PBDEs exhibited a more subtle impact. For the fourteen doses applied over forty-eight hours, the HVEC cells showed no impact at all for the Penta or the Octa PBDEs, and there was negligible impact on any of the fifty assays. But this is probably because these chemicals are not acutely toxic and cause whatever damage they inflict slowly over time by persisting inside an organism for many years. "The PBDEs we tested were selectively active across the same cell lines," says Huang. "They were more toxic in human cell lines than in the animal's."

"This shows the difference between a chemical like mercury with a short-term toxicity that is high," says Ruili Huang, "and one that has a lower toxicity in the short term." She tells me that much work is needed to determine doses and timing for each chemical and what damage occurs, if any.

"I think we will one day know what these chemicals do to cells in much detail," she says.

I ask Chris Austin whether these tests can tell us much about the tiny levels of mercury and PBDEs and other chemicals that most of us carry around.

"Not yet," he says, "but that's the idea, to know this."

I ask him whether this sort of science is what legislators in the EU had in mind when they passed the new REACH laws mandating that chemicals be tested for toxicity before they are approved.

"Probably," he says, "although the Europeans are at about the same state of science that we are, so it will be a while. But we're headed in the right direction."

Meanwhile, my cells at Coriell remain immortal and waiting.

The rise of envirogenetics

I'm sitting in an Indian restaurant in Silicon Valley with a biotech entrepreneur, a physician researcher, and a recent Ph.D. from India. This team is cooking up a stealth company that will take what deCODEme and the other genetic testing companies are doing one step further by combining the exposure of an environmental toxin with a person's genetic risk factors. Their nascent commercial effort also plans to conduct original research on patients to establish, they hope, a strong link between several SNPs that make one susceptible, or not, to the detrimental effects of a chemical that I have become all too familiar with: mercury.

Admittedly, I had something to do with this commercial juggernaut unfolding over tandoori and samosas, which makes me feel a little uncomfortable describing it in this book, since the ringleader is my girlfriend, Lisa Conte. But I can't not report it, since what she is doing is as much an example of what's coming in the future for envirogenetics as 23andMe and deCODEme are for genetics.

Lisa is the founder and CEO of Napo Pharmaceuticals, a small biotech that develops drugs from medicinal plants collected in the rain forest and other regions of the world. I have been telling Lisa about my reporting on mercury and genetics, and, like a good entrepreneur, she immediately saw a potential opportunity of the sort a writer like me doesn't much think about: that offering consumers a series of tests, including genetics, to determine their sensitivity to mercury and to other chemicals might have a market.

Lisa's efforts at Napo put her in touch with patient groups and with environmental and global health advocates who might be interested in such a test. An activist in global health—her company's lead drug, now in FDA human trials, treats diarrhea, a scourge that kills upward of 2.7 million children a year—Lisa believes that an envirogenetics test linking mercury and genetics would be eagerly embraced by mothers like her, who are worried about mercury during pregnancy. So she assembled a team, which includes those sitting here in the Indian restaurant, to create a Web site that not only offers tests for mercury levels and possible genetic proclivities, but will also provide information on what a variety of experts say about its impact, pro and con. This would include a section on the controversial claims that mercury used in preservatives in certain vaccines has triggered autism in some infants. Some parents of autistic children blame this preservative, called thimerosal—which has since been removed from new vaccines—whereas a number of public health experts insist there is no proven link.

"Pregnant women are confused about the risks of mercury," says Lisa. "They are told to be careful, to eat only a certain amount of tuna and large fish, that there might be danger in eating more. Women are also confused about the possible link with autism. What if we can offer a test that tells them they and their children are genetically more, or less, susceptible to mercury?"

I shared what I had learned about genetic testing: that many of the SNPs associated with diseases or traits have not been medically validated, and many of them need to be tested on larger populations, and so forth. This is why Lisa is planning to run tests similar to those conducted in Sweden by Karin Broberg on the glutathione SNPs, but with larger numbers of people and testing more SNPs. Meanwhile, she and her team that day at lunch—epidemiologist George Gellert and geneticist Pradeep Babu—are developing a Web site called MyMercuryRisk that they hope to launch in the spring of 2009.

I expect that existing direct-to-consumer genetic sites will also one day offer gene-environment tests, although as of this writing they have not. As the science improves, and statistics are worked out to determine risk factors for interactions, it's possible that envirogenetics will eclipse "pure play" offerings from genetic companies. Much needs to be worked out and refined, however, before this information is useful in a broad sense for many chemicals and environmental factors, even if a few sites like MyMercuryRisk could be offering clinically validated tests within a year or two.

As a distinct field, envirogenetics got its first significant boost—and funding—when Francis Collins teamed up in 2006 with David Schwartz, the former director of the National Institute of Environmental Health Sciences, and convinced Congress to spend $40 million for a new Genes, Environment, and Health Initiative. Since then, the GEI has funded a range of projects that investigate priority toxins such as mercury, ozone, diesel exhaust, and pesticides, and the effects of other environmental influences like diet and stress on disease. The project is also sponsoring the development of new biomonitoring technologies, including better ways to track everything from psycho-stress events and blood-cortisol levels to chemicals that dissipate quickly in the body, such as phthalates. "We're trying now to get relevant data on gene-environment interaction, to match exposure data with genetic data," says Brenda Weis, a project manager for GEI. "The initiative is still new. No one knows exactly what we will find or how the data will come out."

One GEI initiative at the Harvard School of Public Health is looking into how genes play a role in people's individual responses to mercury and selenium exposure. Selenium is another chemical that appears in fish and may mitigate some of the harmful effects of methyl mercury, although this is controversial. Harvard researchers plan to investigate how genetic variation affects individual responses to chronic zinc, chromium, and scandium exposure. This study will tap into medical information already collected as part of the Nurses'

Health Study, which has followed the health of 120,000 nurses in the Boston area and beyond since 1976. Dietary information will be taken from questionnaires that are completed by participants, and levels of mercury and selenium will be measured from toenail clippings. Genetic information will be acquired from blood and cheek swabs.

"Given the biologic relevance of mercury and selenium for human health and prior candidate gene studies demonstrating heritability, we anticipate discovery of major novel genetic regions that will greatly advance our understanding of the intersection between genes, dietary habits, and metabolism," reports the Web site for the project.

Chris Austin believes that a much larger effort is needed, something akin to the Human Genome Project: perhaps the Human Envirogenomics Project? He and others believe that the only way to create meaningful envirogenetics data is through a large prospective cohort study, collecting DNA samples and information about exposure to a variety of environmental factors from five hundred thousand to a million participants who are followed for a number of years. This study would require a huge investment of time, effort, and money and could cost as much as $3 billion—close to the cost of the Human Genome Project—possibly more, according to a report issued by the Secretary's Advisory Committee on Genetics, Health, and Society at the Department of Health and Human Services. The committee concluded that the health benefits would be enormous and worth the cost.

This project would provide valuable data for government regulators and public health officials—and, perhaps, for individuals who are trying to make the personal cost-benefit decisions I mentioned earlier regarding our use of chemical-laden products. The results would also help fill the current vacuum of information about who is most susceptible.

"A comprehensive study of this sort might tell us everything is okay," says David Sherr of Boston University, "though I suspect that it will tell us that some of these chemicals are not safe even at trace amounts, which would push the government and the chemical industry to make changes."

When I had all of my environmental tests done, I went back to my internist, Josh Adler, with the results of the top forty environmental toxins in my body, and he e-mailed me back this response:

On the chemical report, I am somewhat dismayed that there are so many chemicals in measurable levels in humans (you included). Given that we are talking about parts per billion, I suppose it really should not be surprising. Anyway, other than the somewhat less toxic [of the] PCBs, your levels were within the standard range (whatever that means) for humans. But you were surprisingly high on those other PCBs.

Any idea where they may come from in your case? As far as assessing your health risk based on these data, I don't think it helps much. But, with a particular chemical that is in higher concentration than the rest, it prompts me to ask what the source might be. I certainly would not have asked about PCBs in your environment without these data.

On the SNPs, like most of the other SNP data, there is not enough influence [in the risk factors] to change what we do. We can't see mercury in the brain or liver on any type of scan, so this would require a periodic biopsy for monitoring. The risk/benefit ratio does not work out here, given the potential dangers of the biopsy. But does this suggest you should limit your consumption of San Francisco area fish? Probably a good idea.

Josh didn't catch the significance of the flame-retardant levels, not being trained to look for this. When I mentioned it to him, he said he didn't know what to make of it since the effects of PBDEs even at my levels are unknown.

In the end, Josh pointed out that with all of my testing, he has found no indication that my chemical stew is causing grievous harm. Yet who knows whether the toxins may still have a subtle affect, slowly breaking down my body's defenses.

Josh also wanted to know whether the testing had caused me psychic harm, frightening me or causing anxiety.

I ponder this one night sitting on my flat-topped roof in San Francisco, where I have a great view of the bay and the extraordinary hive of human activity in the city laid out below, much of it dependent on chemicals that have shown up inside me. Interstate 280 and the Bay Bridge swim with tides and eddies of automobiles; a partially shutdown power plant releases a steady white plume of effluents from a tall stack. Far to the north, a massive oil refinery and a storage depot spread across a hillside. Ships move up and down the bay, and jets roar overhead. For an urbanite in the twenty-first century, I find this scene both breathtaking and reassuring. I don't feel frightened or anxious, but I do feel an edge of unease, mostly because I lack the information to know what all of this is doing to me, if anything. Efforts like the GEI and MyMercuryRisk may soon tell me more, but right now, I'm left with perhaps more ambivalence than I was with my pure-play genetic testing.

Deep down, however, I suspect that there are trade-offs for our spectacular civilization that we have barely begun to understand, even as our technology is beginning to provide clues.

One thing that sets this information apart is that we have done it to ourselves. We were not born with these chemicals; we made them.

A lasting impression for me as I climb down from my roof is wondering how I would feel if my envirogenetic profile were different—if I had DNA that made me either susceptible to, or more resistant to, toxins. So far, I have been lucky in the few tests I've taken, although, for all I know, there are genetic variations lurking within that are waiting to be triggered by the next breath of air or the next bite of broccoli. Or maybe it's my children who will have their genes affected by some new chemical that is being introduced even now, which will present dangers previously unknown to our bodies as they have evolved.

Then again, up on my roof, I'm sitting near a huge earthquake fault line that could obliterate this beautiful city in a matter of minutes—a less than cheery thought that snaps me back into my usual fatalism: that when my time comes, it will come.

If I say that enough times, clicking my heels, I may also end up back in Kansas.

3

BRAIN

If the human brain were so simple
that we could understand it, we
would be so simple that we couldn't.

—EMERSON M. PUGH

The incredible shrinking brain

You want the good news or the bad news first?" asks a grinning James Brewer, suggesting that the bad news isn't all that bad.

"I'll take the bad," I say.

"Your brain is shrinking."

We're sitting in a cafeteria at the San Diego convention center, attending the largest neuroscience meeting in the world, a gathering of more than thirty thousand brain researchers in this rapidly expanding field. A neuroscientist from UC San Diego, Brewer looks like a large and curious toddler with a balding head, bright eyes, and babylike cheeks. He is poking at a plate of soggy Chinese noodles and delivering some of my first brain test results: an assessment of the size and the health of neuro-anatomical features with multisyllabic Latin names, such as the cerebellum, located at the back of my head just above my neck, a center for controlling my sensory perceptions and motor abilities; and the hippocampus, situated in the bottom section of my brain, which is key in short-term memory and spatial navigation. He has prepared a map of my brain, he says, homing in on certain features of my neural landscape to tell me whether I'm normal.

But I hadn't expected this: not me, a man who considers himself healthy and ageless, at least in his own, er, mind.

"People's brains begin to shrink when they get into their forties," he explains. "Your shrinkage is about average for someone your age."

A few weeks earlier, my brain had been scanned some five hundred miles to the north, at the University of California at San Francisco (UCSF). This was the first time I'd had my head in a magnetic resonance imaging (MRI) machine, one of those devices that looks like a giant steel doughnut with a long gurney that slides a person lying on his back into the hole. Once inside, the machine bombards whatever part of the body one is investigating with magnetic energy that allows a three-dimensional picture to be taken. I once had an MRI taken of my lower back when I had disk problems; now it's my brain.

"Just lie still for this," says UCSF neurologist Adam Gazzaley. "We're taking an anatomical study of your brain and downloading it into a computer."

That sounds a little creepy, I'm thinking, as they slip pads around my head to keep it still. But no one is downloading the actual contents of my brain: my thoughts, emotions, knowledge, and experiences. The technicians are creating a map of what my brain looks like: those regions and structures that James Brewer will analyze, and also other scientists who have agreed to test my brain for how it reacts to everything from fear and anxiety to belief in God.

Brewer's test is called a "structural" MRI scan, which will create the schematic of my anatomy. Later, I'll be scanned for brain function: what goes on in my brain when I think. This functional MRI— fMRI—scan will check for blood flow in my brain that increases in certain areas when I react to or record an experience, initiate an action, or ponder something. Blood is needed to provide fuel to the neurons at work, and an fMRI scan picks up signatures of metal ions in blood and tissue. These areas of blood flow are displayed by a computer as colorful splotches on an fMRI image that pinpoint the location of everything from emotions to decision making and show how these foci in the brain interact in pathways and processes.

Applying fMRI technology to a healthy individual is far more nascent than applying and trying to make sense of genomics or even envirogenomics. "These tests are so early, they are almost totally irrelevant to individuals," warns Judy Illes, a neuroscientist and bioethicist at the University of British Columbia, "though I think there is so much that seems tantalizingly close to being understood." NIH neuroscientist Eric Wassermann agrees. "Data for brain scanning are noisy for an individual," he says. "What we have is mostly group data relevant to an age group or people with or without a neuro disorder." Yet Illes and Wassermann are enthusiastic about my proposed experiments, seeing them as an opportunity for a nonscientist to describe what, if anything, MRI scanners and other brain-reading technologies can tell us about the three pounds of mushy tissue between our ears, home to that mysterious realm called the "mind."

Given the complexities of the brain and the hundreds of scans and other neuro-tests one could take, in this narrative I can only dip into a sampling of what goes on in my gray matter, a prospect that I find myself approaching with far more fascination and anxiety than I felt when having my genes and environmental influences investigated and analyzed. Within the hundred billion neurons and trillion or so glia cells (*glia* is Greek for "glue"; the cells that hold neurons in place and feed and protect them) lie my hopes and fears; love, hate, anger, passion, and memories and even these words now streaming across the page. Without my specific blend of cells and tissue, genes and experiences, there would be no Experimental Man—no individual person to take these tests and to try to understand what they mean.

Each of our brains contains secrets: memories we would rather dispose of, proclivities that we are not proud of or that we struggle to control. I have talked about my tendency to get anxious when I was younger and my efforts to suppress this inclination, which still lurks deep in my mind to occasionally make me distressed and nervous about saying the wrong thing or doing something embarrassing. These fears are common and usually banal: I still get flummoxed at times in the grocery store about the question of paper versus plastic. (Plastic is easier to carry and doesn't rip, but paper is more natural; then again, paper destroys trees . . . so if I pick plastic, what will the cashier think of me, if he or she even cares?) But what if my anxieties are not normal? Will my brain "out" me by revealing that despite all of my work over the years to convince myself I'm a steely eyed customer fazed by nothing, I'm really a nervous wreck waiting to happen?

Another cause for concern is the astonishing fragility of the brain. Stories abound about head injuries that caused people to lose critical attributes that made them who they are. For instance, in 1933, the composer Maurice Ravel was swimming off the coast of France when he suffered a stroke. As science writer Steven Johnson relates in *Mind Wide Open*, Ravel survived, but for the next four years before his death the composer suffered the horror of only being able to imagine new symphonies and concertos in his head. His impairment had destroyed his ability to translate fresh compositions into musical notes. I have two close friends roughly my age who suffered minor strokes. Both survived, but for one, who is in radio and has a beautiful baritone voice, it took several months for his voice to return. We feared it was lost.

Another friend, the writer Cathryn Ramin, wrote a lovely book about how her once-sharp memory began to fade in her forties.

She tracked down this premature loss to a knock on the head when she was nine years old, delivered by her younger brother as he was whirling around in circles for the fun of it with a broomstick handle. "I was in the wrong place," she wrote in *Carved in Sand: When Attention Fails and Memory Fades in Midlife*. "The impact knocked me flat. For the next three weeks, my eye sockets and forehead turned every color in the rainbow." Cathryn had forgotten about this incident until a neuropsychologist suggested that a long-ago head trauma can cause forgetfulness. "It was a sobering thought, the vulnerability of the head, how often we injure it—and how little attention we pay," she wrote.

And yet, unlike my heart or liver, I have some familiarity with my brain. At age fifty, I am intimately familiar with many of its reactions, processes, and ebbs and flows. But this knowledge is limited. I can know only those parts that I am conscious of. Imagine a man in a dark house with certain rooms and passageways lit up with lights of varying intensity, from bright to dim, that cannot be moved and provide illumination only under certain circumstances. Those areas that are frequently lit are well known, but what about the rest of the house? And how would such a man even know he is in a house if being inside of it is all he has ever known?

I'm not going to spend much time noodling about such philosophical conundrums: about whether my mind is something separate from my body, as the seventeenth-century philosopher René Descartes posed, or the more modern notion that everything that happens in our heads can be reduced to a physical explanation (philosophers call this "materialism"). But I do want to make a point that I know when my brain is tired and when it's full of passion; when it loves, is angry, or is thoughtful. At times, it's firing on all cylinders; at other times, I can't make it do what I want it to. I also have moments when things do not feel concrete, when I agree with the poet Wallace Stevens that "Reality is a product of the most august imagination."

Centuries from now, we may be able to scientifically unravel what all of this means, pinpointing every neuron and glia, proton and electron, and quark and photon, and how they work and interact in the brain with all of its enormous complexity. Yet many neuroscientists suspect that even if we do decipher the mechanics of our gray matter, we still may not understand the feature of the brain that makes humans unique: consciousness.

This reminds me of the supercomputer in Douglas Adams's *The Hitchhiker's Guide to the Galaxy*, which spends millions of years trying to discover the meaning of life and one day gives its response: "42." This answer is meaningless, however, without knowing the question.

For now, I'm having just a simple structural MRI test to map my brain, which is about to start as I slide into the coffin-size tunnel at the heart of this huge magnet. I can hear the great machine now, despite the plugs in my ears: the loud whirring noise like a great windmill spinning. This is the magnet, weighing several tons, powering up inside the metal casing that is wrapped around my head. Then comes an odd series of clicks and what sounds like steel grating against steel.

The brain is roughly divided into three sections that might be described in evolutionary terms as old (in or near the brainstem in the back), newer (generally in the middle of the brain), and newest (mostly in front and on the top). The late neuroscientist and psychiatrist Paul MacLean, formerly of the National Institute of Mental Health, has described these three "sub brains" as reptilian (basic functions and emotions), limbic (the mammalian brain that nurtures offspring and establishes ties with mates and groups), and neocortical (the rational brain). These three brains are sometimes at odds with one another, as might be expected from an organ cobbled together by evolution for a variety of functions and environmental conditions over the eons. But the three brains are also connected and often work together. For instance, a woman steps on a nail, feels pain, and immediately pulls back her foot: that's a reptilian reflex. Just behind the woman, her son is about to step on the same nail, and she pivots and pushes him away to safety: that's limbic. The woman picks up the board with the exposed nail, smacks it on a rock to bend it back, and drops it in a Dumpster: that's a mix of the limbic wanting to protect others and the neocortical figuring out the most logical way to eliminate the risk.

Within these three regions are structures associated with specific functions, such as the hypothalamus, which controls body temperature, hunger, thirst, anger, and fatigue; and the cerebral cortex, which plays a key role in memory, attention, thought, language, and consciousness. Some parts are named after the scientists who characterized them, as stars are named after their discoverers. Two of these areas are important to me as a writer working with words: Broca's area, the speech-producing part of the brain in the left frontal lobe named after Pierre Paul Broca, who

Age-Related Atrophy Report

PATIENT INFORMATION

Patient Name: Duncan^David	Date of Birth: 1958-03-08	Sex: M
Patient ID:	Accession Number:	Ref. Physician:

MORPHOMETRY RESULTS

Structure Name	Volume (cm^3)	% of ICV (Normative Range*)	Percentile vs. Normative Range*
Hippocampus	7.07	0.42 (0.37-0.50)	27
Lateral Ventricles	32.80	1.94 (-0.59-3.04)	88
Temporal Horn of LV	2.12	0.13 (0.00-0.18)	92

AGE-MATCHED REFERENCE CHARTS*

Author's results: MRI dementia test. Image of the author's brain to determine the size of neural structures and any evidence of dementia. The author's results were normal, with no sign of disease.

discovered this region in 1861; and Wernicke's area, the seat of language comprehension that was discovered by Carl Wernicke in 1874. These men discerned their namesake areas by studying patients who had deficits in speech or comprehension due to location-specific injuries revealed on autopsy. (For diagrams that identify different parts of the human brain, go to www.experimentalman.com.)

When I finish my scans, Adam Gazzaley slides me out of the tube and leads me to the control room on the other side of a thick pane of glass. "Let me show you your brain," he says, pointing to a monitor filled with the colorful image of a head cut in half from top to bottom. I can see a crinkly looking cerebral cortex on top highlighted in brown: the cerebellum is in back and the thalamus looks like a small eggplant (see the above image).

Brewer calls up an image of a diseased brain with Alzheimer's, and it's chilling to see the differences—now sections of the brain are visibly misshapen and atrophied.

Brewer tells me that abnormalities in other parts of the brain can offer evidence of other diseases that are either present or beginning to develop that may manifest in the future. For instance, patients with schizophrenia have distinctive aberrations in the left superior temporal gyrus and the left medial temporal lobe, while those with attention-deficit/hyperactivity disorder (ADHD) have smaller brain volumes in several regions. And people who drink moderate to heavy amounts of alcohol have brains that shrink even faster than the contraction by merely aging.

Brewer assembles data on several other anatomical features in my brain, providing me with the relative percentages for each structure. Nothing stands out in terms of disease, he explains, an outcome that more or less matches my genetic profile for diseases of the brain. I've listed my gene variants in the table on page 186—a few of these were also cited in the Gene section of the book—although there is no scientific link that I'm aware of between these variations and telltale anatomical variations from MRI scans.

As you can see from the table, genetically, I have a higher-than-average risk for three diseases: multiple sclerosis, amyotrophic lateral sclerosis (ALS, or Lou Gehrig's disease), and Parkinson's disease. This looks frightening, although at my age these ailments would probably be apparent by now, if I had them. Nor is there any evidence in my scans that I suffer from them. My genetic results for other disorders such as ADHD and schizophrenia also agree with my brain anatomy: nothing to worry about, inasmuch as any of these tests can tell me much. Nearly all of this information—anatomical volumes from MRI scans and genetic variation data—is tentative and gives only a fuzzy picture of what may be happening in my brain. The best diagnostic tool remains the fact that I feel good and have no obvious symptoms for any of these diseases.

One finding in Brewer's brain volume table stands out: that my left hemisphere is bigger than my right. I ask whether that means anything, and he says that some people think that writers are more left-brain oriented, since the left brain contains the main centers of language, although most scientists consider the left brain-right brain dichotomy more pop science than real science. Inasmuch as it can be believed, left-brainers are said to be more anal and obsessed with tidiness, details, facts, logic, and reality. I think of myself as having a healthy respect for all of these attributes, but I also associate myself with many of the characteristics linked to the right brain: imagination, "big picture" thinking, feelings, use of symbols and images, philosophy, impetuousness, and risk taking.

The Author's Profile of Selected Neuro-Genetic Variations

Trait	Gene	Gene Marker	Author	Risk Factors
ADHD	CLOCK	rs1801260	AA	Normal
ALS	DPP6	rs10239794	CT	1.3
Alzheimer's disease	APOE	rs4420638	AA	0.51
Bipolar disorder	DJKH	rs1012053	AA	Typical
Cluster headaches	HCRTR2	rs2653349	GG	Average
Multiple sclerosis	IL7R	rs6897932	CC	1.8
	IL2R	rs12722489	CT	1.06
	HLA DRB1	rs9270986	AC	2.92
	HLA-DRA	rs3135388	CC	0.66
	IL2RA	rs12722561	AG	0.85
Parkinson's disease	COLL11A1	rs1676486	GG	1.0
	ADH1C	rs283413	CC	High
Schizophrenia	GNB1L	rs2269726	CT	Slightly lower

The left-right sensibility comes from an observation that I now can easily see in the images of my own brain: that it is split into two sides that closely resemble each other in a kind of mirror image. Yet there are differences, including, as Brewer noted, the location of most language functions on the left side. Left- versus right-handedness, feet, ears, and eyes also tend to emanate from different hemispheres, with roughly 95 percent of people who are right-handed having their language functions in the left hemisphere, which means 5 percent do not. Damage to either side of the brain or operations that remove one side of the brain can profoundly impact a person's ability to speak and understand language and perform other localized functions, although there are cases where injured people—especially children, whose brains are still forming and plastic—have learned to relocate partial functions to other regions of the brain.

I engage in two other tests to further elucidate which side of my brain may be in charge. The first is a visual test that shows the silhouette of a woman spinning on her left foot. According to the test, right-brained people see the figure spinning clockwise, and left-brained people see her spinning counterclockwise. Viewers are then encouraged to see whether they can make the figure turn in the opposite direction.

I stare at the figure and can only see her spinning clockwise, which makes me Mr. Right Brain. I look hard again and cannot make the lady spin the other way. Wondering whether this is a trick, I ask several people in a café near my house, appropriately named the Thinker, to look at the image. The yoga instructor, the photographer, and the software engineer I speak with see what I see: she is spinning clockwise. Except for the engineer, this does seem like a particularly right-brain group. But then something odd happens. Days later, when I have the figure up in a corner of my computer screen while working on something else, I catch her spinning out of the corner of my eye, and she is moving counterclockwise. Moving my eyes to look at her full on, she does a switcheroo back to going clockwise.

I take one more left-right test offered online, put on the Web by Intelegen, Inc. It asks eighteen questions (see a portion of the test on page 188).

I score a perfect 18 out of 18 for being left-brained. This confirms Jim Brewer's scans of my gray matter, but contradicts the dancing lady—and my own heartfelt sensibilities that I am not Felix Unger, the super-stuck-up character played by Tony Randall on *The Odd Couple* on television and by Jack Lemmon in the 1968 movie version.

Arguably, my belief that I use both sides of my noggin equally is supported by a 2002 study at Duke called HAROLD—hemispheric asymmetry reduction in older adults—that suggests older brains have less of the left versus right dichotomy going on. I'd like to think that this is because older brains learn to use both sides. In some cultures— not ours—this might be one reason why elders are considered wise. Not because they actually *are* wise, but because they have learned to use their brains more efficiently, even as their brains atrophy. I suggest this interpretation to Adam Gazzaley, and he says that most people do not interpret HAROLD in this way; he also says that my brain is still young compared to those in this study.

"Don't worry too much about this," counsels Judy Illes. "Most of what you hear about the differences between left-brain and right-brain people isn't very good science. The reality is that most of us are both left and right: we use both sides, even though one is almost always dominant over the other."

I ask Brewer and Illes whether brain volumes can provide any clues about people's personalities or their proclivities: attributes or characteristics that might identify, say, a person's inclination to be a writer. Brewer had quipped that my bigger left brain made sense because I'm a writer, a remark that he later said didn't mean much, other than a writer uses language perhaps more than other people, and language is centered more in the left than in the right side of the brain. "There isn't anything that I know of out there to tell you from an MRI that you are a writer or have a tendency to be a writer," he says, although he mentions that there is a well-known study in Britain that found a distinctive anatomical signature for London cabbies.

In 2000, researchers at University College London reported on a study of sixteen cabbies whose posterior hippocampi—associated with memory—were significantly larger than those of control subjects. The cabbies had actually grown more brain cells in their swollen hippocampi, apparently because they were required to learn and hold in

---

1. When you walk into a theatre, classroom, or auditorium (and assuming that there are no other influential factors), which side do you prefer?

   ☐ Right   ☐ Left

2. When taking a test, which style of questions do you prefer?

   ☐ Objective (true/false, multiple choice, matching)

   ☐ Subjective (discussion)

3. Do you often have hunches?

   ☐ Yes   ☐ No

4. Do you have a place for everything and keep everything in its place?

   ☐ Yes   ☐ No

A sample of four questions from the Hemispheric Dominance Inventory Test.

their heads thousands of streets, alleys, and courts. Anyone who has been in a London cab knows that drivers do have a remarkable ability to take one to "that restaurant on Oxford Street, you know, the Indian place in the alley by that big book store near Oxford Circus?" The researchers found that hippocampus volume correlated to the amount of time each person had spent as a taxi driver. "These data are in accordance with the idea that the posterior hippocampus stores a spatial representation of the environment and can expand regionally to accommodate elaboration of this representation in people with a high dependence on navigational skills," says the study. "It seems that there is a capacity for local plastic change in the structure of the healthy adult human brain in response to environmental demands."

In articles about the cabbies' remarkable hippocampi, the lead researcher, Eleanor Maguire of University College London, worried that the advent of GPS in taxis might change this unique cabbie brain. "We very much hope they don't start using it," she told a newspaper. "We believe this area of the brain increased in grey matter volume because of the huge amount of data they have to memorize. If they all start using GPS, that knowledge base will be less and possibly affect the brain changes we are seeing."

I am unable to make a direct comparison of my hippocampus with those of the London cabbies—the researchers used a cross-section measurement, and my data was given to me as a total volume—but Brewer tells me that my hippocampus falls within the "normal" range and is unlikely to be as expansive as a cabbie's. So now I know: if anyone were to look at my scan to see if I'm a London taxi driver, the answer would be no.

Other studies have found anatomical differences in jugglers, psychopaths, and criminals, although much of this work is too early for definitive identifications of people's tendencies, much less their jobs, although Judy Illes believes it won't be long until MRI scanners and other devices will provide some useful information to profile people's penchants. "We will have information on electrical signals and blood flow and genes that will give a signature of proclivities," she says. "You might have a signature that indicates you're creative and organized in a way that suggests you could be a writer. Someone else might have very different readings—a banker or a visual artist."

I asked Illes whether this technology could be abused by employers scanning employees to make sure they have the best brain profile to drive a truck or to engineer a bridge, or by overzealous law enforcement officers delving into people's heads looking for criminal intent.

Illes agrees there could be abuses, but she believes it won't be a huge problem. "I suspect that there will be laws to protect people when the technology is ready," she says.

I think about the recent explosion of interest in genetic testing—how the technology for running genetic scans has become cheap and easy enough to allow for companies such as 23andMe and Navigenics to offer genetic scans even before the information is entirely ready. The sudden appearance of these companies in late 2007 and 2008 did prompt a speedy passage of the Genetic Information Nondiscrimination Act in Congress, where it had previously languished for more than a decade, although whether this will truly protect people from prying employers and insurance companies as genetics becomes more precise is yet to be seen.

I know of no move to introduce a Neuro Information Non-discrimination Act in Congress, although I suspect that there will be something like this proposed in the next few years. I am more troubled than Judy Illes is about the possibilities of scans being used to identify proclivities that might go beyond being helpful for predicting disease and career counseling. I doubt that there will be a *Gattaca II: The Brain*, where society is enthralled with brain scans and tests to tell people how to live, work, and love. Yet I fully expect to see tiny, mobile brain scanners planted everywhere—in homes, shopping malls, office lunch rooms, airports—that may reveal far more than even the most accurate genetic tests.

Imagine a world where people check on scans of their brain to gauge their attitude before a big meeting, or check on their kids before a test. Scanners in a clothing store might identify for a clerk exactly the pair of shoes I crave and what might entice me to part with my money.

If this is our future, then we must be very careful as a society how this information is used, especially over the next few years as the technology is perfected and the science remains inexact and the results sketchy.

Remember the moon over the mountain, forget the blonde

My first ever memories are of a small house that my parents bought in Kansas City in 1962, when I was four years old. I remember a backyard with a swing set painted blue and red and the small upstairs room with yellow walls that I shared with my brother, where I was sure there were fanged, life-sucking creatures under the bed. Most vividly, I remember one day helping my father use a push lawnmower to cut the grass. Closing my eyes, I can see myself standing in front of him pushing hard, feeling dad's strength (he was thirty-one years old), and hoping that this moment would never end. Some forty-six years later, at the other end of the spectrum of memories, my most recent recollections as I write this are mundane: before sitting down at my desk, while my family is sleeping, I brewed a pot of French Roast in my morning coffee ritual and glanced at the newspaper on a Sunday morning when the 2008 Olympics ended and Barack Obama announced that Joe Biden would be his vice presidential running mate for the upcoming presidential election. In between these memories at the ages of four and fifty are a lifetime of triumphs, regrets, smells, loves, travels, sadness, boredom, stimulation, fun, anger, and hope; the births of my children and the deaths of people I loved.

How is it possible that I can remember any of this? I'm also wondering, with a tiny little tug somewhere in my mind, about the effects of aging: do I remember things as well as I used to? I now know that my brain is shrinking but seems clean of dementia and other obvious neuro-maladies. Beyond this, I think I'm as sharp in recollecting some things and as dull for others as I always have been. Or is this part of my semidelusion that I am so healthy, I don't have to worry about such things?

On a perfect day in San Francisco, with the temperature in the upper sixties and a hazy-warm sun shining, I'm in another lab at UCSF attached to a machine that is going to test my ability to remember. I'm wearing a tight plastic cap covered with electrodes, with dozens of wires connected to an electroencephalogram (EEG) that measures electrical activity in the brain. A few minutes earlier, students of neuroscientist Adam Gazzaley, who helped with the anatomical study, lathered my head in a gooey gel and strapped on the electrodes. It looks like a cross between a swimmer's cap and a device a mad scientist might use to zap my brain.

Gazzaley's experiment is setting out to prove that the brain doesn't lose its ability to remember as it ages but instead loses its ability to filter out unwanted memories. He is demonstrating that the frontal lobe—the neocortical (new) part of the brain—has a major influence on what is remembered by the hippocampus in the older (reptilian) part of the brain. Neuroscientists call this "top-down modulation," a process that uses the frontal lobe to guide both what people remember and what they don't want to or need to burn into their memory cells.

The brain does the same thing with hearing. Young ears (and brains) can easily discern the voice of a friend sitting across a table in a noisy bar; as ears and brains age, they are less able to pick out the friend's voice from the barrage of other sounds. "We think it's a filtering problem," says Gazzaley.

According to the abstract in Gazzaley's study:

Electroencephalography (EEG) was used to examine the relationship between two leading hypotheses of cognitive aging, the inhibitory deficit and the processing speed hypothesis. We show that older adults exhibit a selective deficit in suppressing task-irrelevant information during visual working memory encoding, but only in the early stages of visual processing. Thus, the employment of suppressive mechanisms are not abolished with aging but rather delayed in time, revealing a decline in processing speed that is selective for the inhibition of irrelevant information.

EEG spectral analysis of signals from frontal regions suggests that this results from excessive attention to distracting information early in the time course of viewing irrelevant stimuli. Subdividing the older population based on working memory performance revealed that impaired suppression of distracting information early in the visual processing stream is associated with poorer memory of task-relevant information. Thus, these data reconcile two cognitive aging hypotheses by revealing that an interaction of deficits in inhibition and processing speed contributes to age-related cognitive impairment.

As the students strap me in, Gazzaley explains that I'm an anomaly for this experiment. At age fifty, I fall in between his two test groups: the young'uns, ages nineteen to thirty, and the seniors, ages sixty to seventy-seven. (I actually fall closer to the older group, but who's counting?) This makes me wonder as they ask me to lean forward to fit my chin into a viselike contraption that will hold my head still: will I test closer to the young subjects or the old?

The experiment has subjects watch photographs of faces and landscapes shown for one second each on a monitor. The screen then goes blank for eight seconds, and participants like me are shown either a face or a landscape and asked: was this one of the images you just saw? I click on a box in my hand: yes with my left thumb, and no with my right. To test the idea of filtering, Gazzaley asks subjects in some of the tests to ignore the faces; in others, to ignore the landscapes. He also runs a series of photographs that are only passively viewed, to ascertain a baseline of what lights up in the brain when a person isn't trying to actively suppress or enhance faces or landscapes.*

I am feeling a touch of performance anxiety. James Brewer had already knocked down my sense of agelessness a notch or two by informing me that my brain is shrinking. Now I am subjecting myself to a test that might reveal me to be prematurely entering into codgerhood. My DNA testing already exposed a few genetic clues about memory, intelligence, and longevity, although most of these gene tests are preliminary and need to be validated. This made my mostly good DNA results suspect, although there is the genetic marker I share with *Wired* editor Kevin Kelly that suggests we're both lacking in intelligence.

For this test, researchers in the Netherlands genotyped 300 children and 276 adults, with a median age of thirty-seven, while asking them to complete "standard intelligence measures." They homed in on a SNP in a gene called SNAP-25, which "plays an integral role in synaptic transmission," according to the abstract of the study. SNAP-25 is expressed in the mammalian brain in the neocortex, the hippocampus, the anterior thalamic nuclei, and other locales associated with intelligence. "Recent studies have suggested a possible involvement of SNAP-25 in learning and memory, both of which are key components of human intelligence," says the abstract. In my test for the relevant SNP, I came out with high odds of having a lower IQ, by a whole 3 points. If I had been AA, rather than GG, I would have had high odds of being 3 points higher on my IQ. (See the table on page 194.)

The other results indicate—all of these come from 23andMe—that I learn from my mistakes; I have a slightly increased ability to remember things; and I have higher odds of living to age one hundred. This last test was run on 213 Ashkenazi Jews who had lived to at least age one hundred, and 258 controls. The centenarians tended to have the CC version of the rs2542052 SNP, while the controls who were less than a century old did not.

---

* Go to www.experimentalman.com for a slide show, photographs, and brain scans detailing Adam Gazzaley's experiments on memory.

The Author's Preliminary Genetic Results: Intelligence, Memory, Learning, and Longevity				
Trait	Gene	Gene Marker	Author	Risk Factors
Avoiding errors	DRD2	rs1800497	GG	Learns to avoid errors
Intelligence	SNAP-25	rs363050	GG	Lower IQ (3 pts.)
Longevity	APOC3	rs2542052	CC	Higher odds, living to 100
Memory	KIBRA	rs17070145	CT	Increased memory

As for the 3 points shaved off my IQ score, I'm not sure that it matters.* Over the years, I have scored well enough on IQ tests. For this book, I took several online tests and got various scores that seemed to improve the more I took them, suggesting that IQ may have as much to do with familiarity with the tests as with intelligence. My average is down slightly from scores I had as a child, which I can blame on either the deteriorating ability of my right brain to answer the math sections of these tests or to my shrinking brain. I should add, however, that a 3-point difference may mean little to me as an individual, but it could be significant for public health experts who are looking at the intellectual health of populations.

"You do know that there are no actual genes for intelligence, memory, or avoiding errors," warns University of British Columbia neuroethicist Judy Illes when I mention these studies. "This is basically reckless of these companies to be suggesting there are."

In the true spirit of the Experimental Man project, I also tried to see whether I could assemble any data implicating environmental factors into how my brain remembers and learns. One answer is that I have an overwhelming slate of factors that I might incorporate, though nothing definitive: everything from knocks on the head as a child to my

*See "You show me yours, I'll show you mine" in section 1 for more on this genetic test and for a comparison with the DNA results of writer and technology guru Kevin Kelly.

two fish gorges that resulted in spikes of my onboard mercury, some of which may have ended up in my brain. For the latter, I appear to have some genetic protection from mercury damage to my gray matter, but who knows?

I'm now ready for the test, and the students move into a nearby room, where they will monitor my session and tend to the EEG machine. They turn out the lights in my room, leaving the computer screen glowing a few feet in front of my face, and we begin.

The first round is remembering faces and ignoring landscapes. I quickly realize that the best way to remember faces is to focus on one or two distinctive details: thick eyebrows, glasses, big lips. For landscapes, it's the spray of water from a stream, a V-shaped notch in a rocky ridge, a full moon. Being the competitive sort, I am highly motivated to get the right answers, and I get mildly ticked off when I miss one. As the test wears on, however, I begin to get sleepy. This has long been a hazard for me: getting drowsy when I'm sitting still and not receiving the steady stream of mental stimulation I normally get as a journalist and as a person who must always be doing something, even if it's reading or pacing around in a room. I am also running an ongoing life experiment to see how little sleep I can get and still function, which allows me to nap on demand. By the middle of the ninety-minute experiment, I'm struggling to stay awake. But somehow I hang on, drifting off and missing my cue to press the button "yes" or "no" only a couple of times.

Even in my groggy brain, memories are being made as I stare at the photos. Exactly how memories are made is not entirely understood, but scientists think that while the mechanisms of memory occur in many places in the brain, some types of memories are focused on specific locales. For example, the hippocampus seems to be a focal point for spatial and declarative learning, and the temporal neocortex (located under the temples) for long-term memories. The brain records different types of memories: sensory memories such as an itch on my face in the middle of this experiment, which comes and goes and is forgotten; short-term memories that I retain longer, such as the faces I'm looking at, which I learn long enough to take the test and then will most likely jettison; and long-term memories, those faces and landscapes that make enough of an impression that if you show them to me again hours or days or perhaps months later, I am likely to recall them.

Scientists believe that specific memories and learning are recorded via changes in neuronal synapses. This is the process of neurons activating and creating an electrical spike called an action potential that sends signals through pathways in the brain that cause something to happen: a muscle to move, a thought to register, or, in this case, an image of a face or a landscape to imprint. Somewhere in my brain, neurons and systems of neurons activate and allow me to recall the face of a girl with short blond hair and a landscape that reminds me of an Ansel Adams photograph, with a cliff face and a moon overhead. I have forgotten the other images I saw during the test, but I still remember these two over a year later.

I don't know why I remember these pictures and not the rest—possibly it's because the girl was cute, and I have always loved Ansel Adams. Nor can I tell you why I remember even the most obscure and useless dates and historical facts, but I am terrible at remembering people's names, even those I have known for a long time but haven't seen recently. I also recall objects that are colored wine-red and bright green but have less interest in yellows and blues. I don't like the smell of most perfumes, but if I catch a whiff of Chanel No. 5, I instantly think of my mother, who wore this scent when I was a boy. Each of us has a unique menu of things we remember and things we forget.

After the test, I wash the goo out of my hair and join Adam Gazzaley and his students in the control room. They replay some of my data. It appears as jerky lines running across a monitor, like multiple lines from a polygraph test, showing the activity of certain neurons as I remember or suppress the faces and the landscapes. These readings are like musical scores of the electrical signals in my brain, a highly complex and coordinated orchestra and choir of millions of neurons firing off and receding, performing a neural symphony in each of our heads that is greater than any ever written for an orchestra.

In the future, there may be that distant time we discussed earlier when all of the voices and the instruments of the brain will be understood and their notes recorded on a magnificent score processed by computers with a billion or a trillion times the memory and speed of today's greatest supercomputers. (Or maybe the supercomputers will be our brains, augmented by hardware.) But will we ever understand the meaning of how all of these voices work together?

That same day, after lunch, Adam Gazzaley takes me to UCSF's brain-imaging center and places my head into the doughnut hole of the

MRI that will also run in this session the structural-anatomical study analyzed by James Brewer. Once Brewer's images are done, Gazzaley will direct the machine to take a battery of face and landscape memory tests. These will be similar to the tests run while my brain was hooked up the EEG, but this time the magnetic waves emanating from the machine will record blood flow in my brain, rather than measure electrical signals. The experiment is to see how I compare with Gazzaley's younger and older groups as my neocortical brain in the frontal lobe tells the memory centers in the back of the brain to remember faces and disregard landscapes, or vice versa (see the below image).

A few weeks later, I meet with Adam Gazzaley in his hyper-modern office in the new Genentech Hall at Mission Bay. He tells me that he sees nothing alarming or unusual in my results. "You aren't abnormal, and you have no evidence of disease that affects

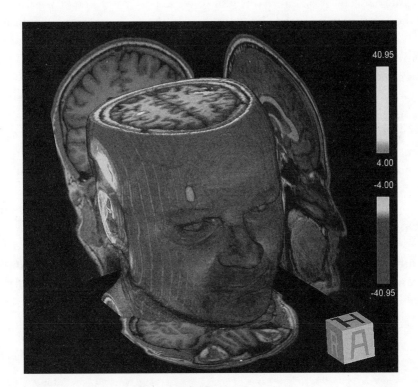

The author's brain caught in the act of remembering.

your memory," he says, pulling up some slides of my results on a large computer screen at his desk. FMRI images of my head from the front, the top, and the side pop up, with red splotches that Gazzaley says show the activation of regions in the front of my brain interacting with regions in the memory centers in the back, which also light up.[*]

"This is what's supposed to happen," he says.

Gazzaley also displays the EEG data and identifies one spike as showing how I did in remembering faces and another in ignoring faces.

He shows me my scores compared to the older and younger groups. "You did great on reaction times and accuracy," he says, "better than even the younger group." This is surprising, given that I was close to falling asleep for part of the experiment.

I am obviously pleased with these results, although Gazzaley has not yet delivered the key results from the experiment: how well I remembered and suppressed faces and scenes. On this, I didn't do quite as well. "You came out where we expect you to be at your age, about halfway between the younger and the older groups." Like the older group, I remembered things fine, but my brain betrayed me by revealing that I wasn't always filtering out what I was supposed to ignore as well as the younger group.

"It means your brain is aging," says Gazzaley. "Sorry about that."

I have similar results for the fMRI scans, with mostly great marks for reaction time and accuracy but a score that is slightly worse than the younger group on the suppression of unwanted memories.

My top-gun scores for accuracy and reaction times almost make up for Gazzaley's rather blunt declaration that my brain not only is shrinking, but it also can't filter out unwanted input as well as it once did. Gazzaley tries to reassure me that my brain still isn't all that old, and that I appear to be doing fine. "You really have nothing to worry about," he says.

"For now," I say.

---

[*] For more images of the author's brain, EEG readings, and tables of results from Adam Gazzaley's experiments, go to www.experimentalman.com.

A brain half my age

I might be aging, as Adam Gazzaley contends, but I want to get a second opinion about what exactly the inexorable march of time is doing to my brain. So a few weeks after hearing my results from Gazzaley's tests, I meet up with Pat Turk in my San Francisco office. He is a director of business development for Cognitive Drug Research (CDR) Ltd., a British firm that uses automated tests to gauge how well we remember and how well we think. I had run into another CDR executive at a meeting in Boston who offered to conduct a battery of tests to check out my attention, memory, and executive function but not by having my brain zapped by a magnet or by listening in on the electrical chatter in my head. CDR will offer a nonscanning series of tests taken on a computer that will compare my results to those of people my age and of individuals who are younger and older, using a company database of more than six thousand people tested.

Most subjects in the database were assessed for cognition and memory during clinical trials conducted by pharmaceutical companies or by academic researchers testing new neuro-drugs for everything from nicotine addiction to Alzheimer's disease. Results from the computer tests are recognized "endpoints" by the Food and Drug Administration to determine whether new brain medications work, which means that the drug industry has billions of dollars in potential revenues riding on them.

I'll also get a "brain-age" score that correlates my results with the ages of other test-takers, offering yet another opportunity for information I'd probably rather not know about how I'm aging.

Pat Turk and CDR's CEO and founder, Keith Wesnes—I will later talk to him by phone—explain that cognitive and memory tests have been administered using pen and paper for at least eighty years; researchers started using computers in the 1970s and 1980s. Wesnes is a psychologist and a neuroscientist in Britain who launched the company in 1986 as an outgrowth of computerized testing programs he developed for his academic experiments on cognition.

Besides drug trials, CDR has tested patients who have had open-heart surgery to see whether this operation affects their cognitive performance. Other tests include checking patients' scores when they take cholesterol-lowering drugs, after they are treated with chemo-therapies for cancer, and when they return from the war in Iraq after suffering head traumas. Tests have been run on children eating sugary breakfast cereals versus more healthy breakfasts—not surprisingly, the sugar caused a decline in performance—and on the cognitive impact of being obese. A competitor of CDR, CogState of Australia, offers a product called CogState Sport that monitors an athlete's recovery from concussions. Customers include the Rugby Football League in Britain and the Australian Football League.

Wesnes tells me that his latest project is to create a 23andMe for brain tests. He believes that the cognitive-testing market is on the verge of expanding into an online business that will offer tests directly to consumers, businesses, or anyone else. For healthy people, the sites could be used to run self-experiments, such as comparing one's cogni-tive abilities while listening to, say, Carlos Santana versus listening to a Bach concerto, or before and after a bike ride. The price won't be outrageous—perhaps $40 or $50, says Wesnes, although he has not yet done a business analysis to set prices.

We ran a mini-experiment on this idea as part of my column on the *Condé Nast Portfolio* Web site. In a story about CDR and my tests, we offered, with CDR's help, a short version of its standard tests for anyone who wanted to take it. Almost fourteen thousand people took the test, and dozens wrote comments. Some people felt satisfaction that they came out with younger brains than their real ages, and a number of people were either upset or unsettled when their brains' ages were older than their real ages. Pat Turk responded to these comments by writing in an e-mail: "The brain age score reflects a person's performance on our 3 tasks as compa-rable to that typically seen for a healthy subject of that chronological age. In other words, if a 60-year-old person has a brain age of 45, that person (while chronologically older) would have the cognitive performance of a typical 45-year-old. So in that regard, a lower brain age is reflective of better cognitive function. It is not solely related to reflexes, as the responses are based upon the person's level of attention and speed of processing information."

Keith Wesnes added in an e-mail, "Our database has shown that as people get older, their speed of response on these attention tasks declines. This happens in a linear fashion from 18 to 80 years and is primarily due to failings in the ability to process and react to information. As the declines occur over even quite small age differences, e.g., less than 5 years, these do reflect a general pattern in the population, but of course there is every likelihood that different people decline at very different rates."

Out of 13,836 tests taken on the Web site, men outnumbered women by almost two to one, reflecting either the Web site's reading demographic or a greater paranoia among men about how smart they are. But men were faster in their response times, albeit only by 9 milliseconds on average. (Reaction time when choices needed to be made: men—443 milliseconds; women—452 milliseconds.) The median age was 35.7, with a range that included someone who was 103 years old (biological age). The test scores brain age only up to 80 years old. Those claiming to be 18 to 25 years old averaged 64 milliseconds faster than those between the ages of 60 and 70 years old —a very significant difference, according to Pat Turk. Possibly, this reflects more time spent on computers playing video games as well as evidence of cognitive decline. Below is a table that tells testees how they fit in with the thousands of other people who took the test.* (The numbers in each column show the percentage of people of that particular actual age whose "brain ages" were in the age bands of 20 to 80. For example, 8.6 percent of people 50 years of age had a "brain age" of thirty.)

"Brain Age" Test Results: 13,836 People Tested						
	Actual Age					
Brain Age	20	30	40	50	60	70
20	60.8%	52.7	41.2	25.3	27.5	11.1
30	7.3	9.9	11.8	8.6	3.4	3.7
40	7.3	6.5	6.7	14.2	9.4	7.4
50	6.4	8.2	9.7	12.6	9.4	7.4
60	5.2	6.8	4.6	6.5	7.7	7.4
70	3.5	4.1	6.7	11.1	11.2	7.4
80	9.3	11.6	19.0	21.3	31.0	55.5

---

* Take the test yourself at www.experimentalman.com.

This online test raises some interesting ethical questions, including the possibility that employers and others will use these exams to discriminate against people, or that people will get upset about their results. "For those not in clinical trials, these tests should be taken for fun," says UC Berkeley neuroscientist Silvia Bunge, who studies memory and cognition. "They are not to be taken seriously." Yet I expect similar issues to come up for a direct-to-consumer neuro-testing site that arose for online genomic testing sites, such as: Would a trained professional be delivering results? How valid are the findings? Also, perhaps it's not too premature for Congress to be considering that Neuro Information Nondiscrimination Act I mentioned earlier.

Pat Turk plans to return a few weeks later to run a second part to our experiment: what will happen to my results if I add an environmental factor that is not uncommon in many people's lives. This is not some pollutant or frightening toxic chemical but one of my favorite beverages: red wine.

That test will come later. For now, I'm sitting down at my own desk in front of Turk's computer with a cord running to a small box outfitted with "yes" and "no" buttons for me to punch with my thumb.

Creating as much of a cliniclike setting as possible, Turk writes down the placement of the computer on my desk and how I'm holding the button box so that we can replicate these same conditions for future tests. He then reads me the instructions for the first test, explaining that I will be shown fifteen words on the screen at the rate of one every two seconds. Turk will then give me a pencil and paper and ask me to write down as many as I can remember just afterward to test my memory.

"Are you ready?" he asks.

I hesitate for an instant, wondering, again, why I am subjecting myself to this. Do I really need to know that (a) I am a moron, or at least not as quick or smart as I hope I am? or (b) that my brain age is older than my real age? I have to admit that these are not trivial worries; a great deal of my self-image and sense of who I am hinges on my being reasonably smart. Whether I am or not may be open to question, but that is a core self-belief. To have it disturbed might cause a small smudge on my self-esteem or worse.

"Let's do it," I say, taking a deep breath (I read in the CDR material that increased oxygen improves scores, so I suck in as much as I can to hopefully flood my brain).

On the screen, the words pop up in a rapid, random progression: POET, BEACH, ATTENDANT, JURY, CAVE—there are fifteen in all. I'm watching them tick by, one by one, feeling slightly panicked that I will forget them. A mere thirty seconds later, Turk hands me a pencil and asks me to write down as many of the words as I can quickly recall. I have sixty seconds.

I quickly write down three words. Then one more. Unbelievably, that's all I can remember as the seconds tick by. I close my eyes trying to will the appropriate neurons to imprint the words I just saw. At about fifty seconds I suddenly remember one more word, for a total of five.

"That's terrible," I say. "I can't believe I'm not remembering more. They are right on the edge of my mind."

"Actually, that's pretty good," says Turk.

"Really?" I ask, sure that he's humoring me.

"Let's move on," he says, running me through about twenty minutes of additional tests. Here is a sampling:

Picture Presentation: A series of twenty pictures is presented on the screen at the rate of one every three seconds for the patient to remember.

Simple Reaction Time: The patient is instructed to press the "yes" response button as quickly as possible every time the word "yes" is presented on the screen. Fifty stimuli are presented, with a varying inter-stimulus interval.

Spatial Working Memory: A picture of a house is presented on the screen with four of its nine windows lit. The patient has to memorize the positions of the lit windows. For each of the thirty-six subsequent presentations of the house, the patient is required to decide whether the one window that is lit was also lit in the original presentation. The patient responds by pressing the "yes" or "no" buttons as appropriate, as quickly as possible.

Delayed Word Recall: The patient is again given one minute to recall as many of the words as possible from the original word presentation.

Word Recognition: The original words, plus fifteen distracter words, are presented one at a time in a randomized order. For each word, the patient is required to indicate whether he or she recognizes it as being from the original list of words by pressing the "yes" or "no" button as appropriate, as quickly as possible.

Picture Recognition: The original pictures plus twenty distracter pictures are presented one at a time in a randomized order. For each picture the patient has to indicate whether the patient recognizes it as being from the original series by pressing the yes or no button as appropriate, as quickly as possible.

I feel better about my performance on the rest of the testing, but I don't know for sure until I receive my results from CDR by e-mail, along with a call from Keith Wesnes. I get both with some trepidation. What if I have an old person's brain? It turns out, my brain either fooled the test, or I needn't have worried. Wesnes's comments:

This 50 year old male performed comfortably in the higher end of the normal range for the attention and working memory tasks. Episodic memory was generally in the normal range, except for word recognition, which was not reflective of other scores and should be repeated.

Wesnes also delivers my brain age: a surprising twenty-five years old! "You were quick in responses and usually pretty accurate," says Wesnes. "But that doesn't mean that you weren't sharper in your twenties. We are all on our own scale, and we all decline with age."

If I must decline, this is the sort of decline I'm okay with.

Although I don't want to take this test again, in case my score comes out worse, sure enough, Pat Turk returns to town a few weeks later and I find myself once more face-to-face with the CDR exam. We planned to conduct the wine test, but Turk insists that I first take the test stone sober. I do—and become extremely annoyed at myself when I can remember only three words out of fifteen in the first part of the study. I'm not sure why, although I remember four when Turk asks me to recall the words after thirty minutes, at the end of the test. This suggests that on this day, my immediate recall is like jelly, but my delayed recall is better. But I still can't match the five words I got right during my previous test—possibly because it's early evening rather than early afternoon when I took round one of this test. "The time of day can influence performance on cognitive tests," says Turk, "which is why in the course of a highly controlled clinical study we would want to have test sessions as close to the same time during the day as possible."

Turk insists that this decline in my memory in only a few weeks between the two tests is probably a fluke, although I am not convinced. My consternation at my poor performance, however, is soon mitigated by the next part of our experiment: drinking some inexpensive Bordeaux *vin rouge*. Turk joins me when I ask him, taking just a sip while on duty. I drink a glass and a half and wait about twenty minutes. Then Pat Turk fires up his machine again. I don't feel much different, but I am more relaxed. I am even able to quickly recall four words in the first test, although Turk said earlier that my immediate recall would not necessarily be improved by wine.

As I fear, when I get my test results back, my sober test shows an age score slightly higher than 25, although, other than the word-recognition test, I have similar results to the first time I took the exam. Predictably, my responses were influenced by the Bordeaux. According to Wesnes's report:

> This test was done 20 minutes after a glass and a half of wine. Attention is very sensitive to the effects of alcohol, and David showed a characteristic decline in speed of reaction in the Vigilance Task, which is seen with alcohol. This would have moved him outside of the normal range for his age. Accuracy was not compromised, which is also typical, thus the individual can identify the target properly, but takes longer to do it.
>
> Declines were also seen for simple reaction time and choice reaction time, indicating the overall slowing of responding in tasks measuring attention, but the clearest impairment was with the vigilance task, which requires continuous attention, and as stated above, this is entirely normal. Working memory and episodic memory were not affected at this time by the alcohol. Effects on these latter domains may have occurred later, or may have required more alcohol to be identified.

Wesnes sends me a table (see the next page) that includes my results on all three test sessions, listed under the column "Your Score." (The comments refer to the wine-tasting test.)

For some tests, I performed better under the influence, although, overall, my brain saturated in a small amount of Bordeaux performed accurately but more slowly. Apparently, this ratcheted up my brain age

The Author's Combined Cognitive Tests: Sober and Slightly Intoxicated

Domain	Test	Your Score	OK?	Norm	S Dev	Less or More Is Better	Comment
Attention	Simple reaction time (milliseconds)	236* 242[†] 287[‡]	Yes	261	37	Less	Focused attention normal
	Choice reaction time (milliseconds)	421 454 483	Yes	459	63	Less	Within normal range, normal ability to focus and make rapid decisions
	Vigilance, accuracy	100 100 100	Yes	97	4	More	Within normal range, normal sustained attention
	Vigilance, speed (milliseconds)	421 392 459	Yes	411	43	Less	Speed of responding in normal range, thus when concentrating on task, no problems in decision time

Domain	Test	Your Score	OK?	Norm	S Dev	Less or More Is Better	Comment
Short-term memory (working memory)	Numeric working memory sensitivity range 0–1	0.94 0.83 0.75	Yes	0.83	0.17	More	Within normal range for age, ability to retain information by sub-articulation normal
	Speed (milliseconds)	627 653 609	Yes	914	253	Less	Speed of retrieval from short-term articulatory memory upper end of normal range
	Spatial working memory sensitivity (0–1)	1 0.84 1	Yes	0.79	0.24	More	Within normal range, ability to hold spatial information normal
	Speed (milliseconds)	668 541 536	Yes	1047	289	Less	Speed of retrieval of spatial information from short-term memory upper end of normal range

* First number: original test score
† Second number: "sober" test score
‡ Third number: "vino" test score

by a few years, possibly to my mid-thirties—which is also something I can live with. But it suggests that quaffing more vino than this might push my brain beyond my true age, although at a certain point a moderate to heavy bout of imbibing isn't going to be measured on a scale of how old one is, but how bumbling and incoherent.

At this stage in the Experimental Man project, as I contemplate further brain tests, I feel like a mildly modulating glass or two of wine is in order. Particularly because the next round of tests, on anxiety and fear, is making me, well . . . *anxious.*

High anxiety and the saber-toothed editor

We've all had moments when our hearts are pounding like a Jamaican steel drum and our palms are clammy; when our nerves are fraying just when we need to be sharp, strategic, and thoughtful to save our butts. Evolution, however, has given each of us a giant brain designed for dangers that confronted our great-grandparents many times over: the snarl of a saber-toothed tiger or the crack of lightning on the veldt. Our noggins are not made to deal with the equally vicious snap of a boss on a rampage or the sudden realization that we have been called on to give an impromptu presentation to the zillion-dollar client in front of the executive board of the company.

Deep in the tunnel of another MRI machine, this one at Stanford University, I'm reliving one of the most anxious moments of my professional life. It happened years earlier, when I was a junior correspondent for *Life* magazine. I was in a staff meeting led by managing editor Dan Okrent, a legendary editor of books, magazines, and newspapers. We were discussing one of my first major stories, a possible cover. I had worked hard for weeks and had helped manage a complicated photo shoot. Colleagues were saying I did a great job and that the photos shot by photographer Joe McNally were fantastic. In those days, I was quite shy in such meetings with senior editors, but I was feeling good. Then Okrent, who could be gruff and blunt, blurted out that another reporter would write the story and get the byline.

The suddenness of this put-down felt like one of those ancient lightning bolts flashing close enough to singe me. I couldn't believe it. *Life*, like its sister magazine, *Time*, used to routinely have one journalist report a story and another write it, but this had become rare. As the meeting continued, I felt my heart rate surging and my gut contracting. I felt ashamed, and I'm sure my face was red. I knew I should say something to this editor, who did look remarkably like a saber tooth with glasses, without the long lower fangs. I needed to stick up for myself, but my most overriding desire was to say nothing, to stay quiet and just go along with the managing editor's decree without protest. Part of me was also wondering whether there was any way I could bolt or, better yet, so as not to draw attention to myself by getting up and leaving, to vanish. Poof! And yet I knew this was a pivotal moment for me as a journalist and an adult who wanted to accomplish things. I had to regain control.

In the MRI, the emotions from that long-ago afternoon are flooding back into my brain as I hear the distant whirr of the great magnet wrapped around my head and the clicks and sounds of grating metal. On a monitor, the story of the *Life* magazine meeting is being displayed in reminiscences that I had written at the request of the Stanford scientists conducting this experiment: Philippe Goldin, a research scientist, and Kelly Werner, a postdoc, both working in the lab of Stanford psychologist James Gross. Goldin and Werner asked for stories that were personal and socially embarrassing and had seriously punished my self-esteem. Their idea is to use fMRI scans to investigate the phenomenon of social anxiety in healthy people and in those who suffer from phobias and disorders that make them truly terrified of social situations and interactions with other people.

Fear and anxiety are crucial to human life. They are evolutionary tools developed over the eons to protect us: to alert us to run for the hills when that saber tooth roars and to worry when a son or a daughter is sick so that we will take care of our children. My mitochondrial and Y-chromosome forebearers, however, never had to face a humiliation delivered by the managing editor of *Life* magazine as they slowly migrated out of Africa to the Near East, to Russia, and west into Europe.

And yet this situation might have had antecedents for an ancestor facing a similar loss of face as he came of age in his village in primordial Italy or France. Let's say he had planned a new and improved method for making a spear tip. His tribesmen congratulated him, only to have the top elder assign the actual making of the spear tip to an older, more experienced hunter. I wonder whether he had the same anxiety at speaking up and the same desire to flee.

If so, an MRI, if one had existed, might have shown a similar neural pattern to mine, with the blood flowing strong in the amygdala and the hippocampus, the two most important brain structures pertaining to social anxiety.

The amygdala—Greek for "almond" because of this structure's shape—is part of the limbic system that receives incoming emotional signals and generates a response. The amygdala flashes signals to other brain regions, including the anterior cingulate cortex and the orbitofrontal cortex, to process the emotional input. This triggers other parts of the cerebral cortex that flag our consciousness and thought processes to tell us that something is awry and to consider appropriate behavior and action in response. The amygdala also pings the hypothalamus to begin triggering hormonal changes to ramp up the body's resources to deal with the current threat and future stress. Damage to the amygdala causes emotions to flatten out, and zapping an animal's amygdala with electricity heightens emotions such as anger, aggression, and fear.

The hypothalamus orchestrates a physiological response: an increased heart rate, respiration, and sweating. Sexual arousal uses the same sympathetic nervous pathway. The hypothalamus also activates or shuts down bodily functions that we are unaware of: for instance, your immune responses and digestion radically decrease activity to divert energy from these longer-term functions to systems requiring a boost for a short-term emergency (an attack by a snarling beast or a night of romping). This dual punch by the amygdala and the hypothamus initiates a pattern of behaviors that medical students learn to call the Four "F's": fight, flight, fright, and sex. (I assume "sex" is a substitute for the less family-friendly fourth "f.")

Despite my nervousness about this test on nervousness, I am being tested as a "normal" subject. Yet I am aware researchers have found that imaging normal brains reveals clinically significant findings 8 to 10 percent of the time—"disconcertingly often," says legal and bioethics expert Henry Greely of Stanford. Mild to severe anxiety runs in my family on my mother's side. Mom has gone through periods when she became anxious enough about her career, children, and health that she sometimes had trouble breathing. She still occasionally has panic attacks for no reason, often in the middle of the night. Her mother, my grandmother DeeDee, was sometimes a worrier. She hid this well in front of us grandchildren. Like my mother, she was an accomplished woman and a humorous speaker who actively engaged in her church and various clubs. But she also fussed about ailments real and imagined and drove my mother crazy with calls that her ulcer was acting up or that she had cancer or other maladies that were going to kill her any day now.

The genetics of anxiety is still poorly understood. Recent association studies building on earlier work with animals have found some SNPs that seem to increase one's risk for being anxious. Last year, a study in Boston identified four genetic markers associated with anxiety. "We found that variations in this gene were associated with shy, inhibited behavior in children, introverted personality in adults and the reactivity of brain regions involved in processing fear and anxiety," the study's lead author, Jordan Smoller, a psychiatrist at Massachusetts General Hospital, said in an MGH press release. "Each of these traits appears to be a risk factor for social anxiety disorder, the most common type of anxiety disorder in the U.S."

The SNPs, reported the press release, affect the production of a protein called RGS2, "which mediates the activity of neurotransmitter receptors that are also the targets of many antidepressant and antipsychotic drugs. Mice in whom RGS2 is knocked out [removed] exhibit increased fearful behavior." The scientists genotyped 119 families who were tested for levels of social inhibition and more than 700 college students who filled out questionnaires measuring various personality traits. "Another group of 55 college students had functional MRI brain imaging done after they had completed a standard interview screening for anxiety and mood disorders. While in the scanner, the participants viewed a series of faces expressing various emotions, a test that previously was shown to influence activity in the amygdala. Participants with the inhibition/introversion-associated alleles also had increased activity of the amygdala and the insula, another anxiety-related brain region."

I come out clean for the anxiety genes we have data for; so does the rest of my family, except for my mother, whose SNP variants give her an elevated risk for anxiety. I'm also including SNP data (see the table on the next page) on two genes associated with obsessive-compulsive disorder (OCD), related to social anxiety. Everyone in the family except

				Risk		Risk
Condition	Gene	SNP	Author	Factor	Mother	Factor
Anxiety	RGS2	rs6428136	TT	Average risk	GT	Higher risk (5.54)
		rs1819741	TT	Average risk	CT	Higher risk (3.81)
Obsessive-compulsive		rs4570625	GT	Higher risk	GT	Higher risk
	DRD2	rs1800497	GG	Average risk	GG	Average risk

Gene Markers for Anxiety and Obsessive-Compulsive Disorder: The Author's and Mother's Results

for my father has a slightly increased risk of OCD, according to one SNP; we have a lower risk acording to another SNP. Once again, these results are preliminary and contradictory and may have little or nothing to do with our actual genetic proclivities for these behavioral disorders.

In the scanner, the *Life* magazine story continues to flash in front of my eyes. I'm supposed to rate how anxious the statements make me, on a scale of one to five, by punching the appropriate switch on a button box in my hand.

BACK IN NEW YORK, THE MANAGING EDITOR AT THE WEEKLY STAFF MEETING CONGRATULATED ME—AND THEN ASSIGNED THE STORY TO A VETERAN REPORTER TO WRITE.

I press two, meaning this makes me pretty darn anxious. The screen clears and up pops another sentence in bold and in caps:

I AM A LOSER.

Later, Goldin will tell me that the researchers created these statements based on my story. They are intended to induce "negative self-beliefs" to see what happens in my brain.

NO ONE LIKES ME.

"Some people launch into a cascade of negative self-beliefs," he says. "We want to track how that happens."

I WAS HORRIFIED BY WHAT THE MANAGING EDITOR SAID. I WAS UPSET BUT TOO SHY TO SAY ANYTHING, FEARING EMBARASS- MENT OR THAT I WOULDN'T BE ABLE TO TALK. AS THE MEETING BROKE UP I KNEW I HAD TO SAY SOMETHING, BUT I WANTED TO FLEE.

The test then asks me to make an effort to modulate these nega- tive self-images, to use strategies to tell myself I'm not that bad. This is the second part of the experiment: to see how well people's strategies work to alleviate their anxiety. I am then supposed to press a button for how I feel once I make an effort to tell myself to chill out.

PEOPLE WILL THINK I'M A WIMP.

"Some people are able to modulate their negative self-image. Others have a hard time doing this, and they go into a cascade, and they have a very bad night worrying," says Goldin.

In the MRI, Goldin and Werner run another test on me that shows video clips of actors making comments and using facial expressions.

"Everybody likes you," says one man on the video with an enthu- siastic smile.

"Nobody likes you," says a woman with a sneer.

Again, I am asked to modulate negative feelings with coping strategies. "The amazing thing is that the brain can make changes," says Goldin. "Most of this happens in the connections between the prefrontal cortex and the amygdala. It can be tempered to learn and adapt. People have studied emotional reactivity and regulation before, though this is the first experiment to use autobiographical scripts, negative self-beliefs, and video clips to study how our brains regulate emotional reactions to social feedback in adults with social anxiety disorder." Goldin has published studies on how people mod- ulate behavior when confronted with videos of actors telling them they are worthless or wonderful and what happens in brains dur- ing moments of amusement and sadness. The researchers hope to run the experiment I participated in on 140 people diagnosed with social phobias and also on a group of controls. They plan to finish the project in 2010.

Golden and Werner are among thousands of young neuroscientists unleashing the technology of fMRI to study what happens in the brain during activities as varied as watching a porn video to gazing at fields of tulips. Goldin and Werner, however, are adding a clinical element to their experiment that I did not take part in. Each test subject was asked to participate in one of three therapeutic interventions— psychotherapy, meditation, or exercise—to see if these will cause their brain patterns to change in the MRI, hopefully for the better for the social phobics.

After the fMRI test, the researchers have me answer several questions on a computer about my story (see a sampling of the questions below). I answer "7" for the first question, "8" for the second, and "5" for the third. The questions continue, asking me whether I try to avoid similar situations, whether I've ever talked to anyone about the situation, and so forth.

(S1.1) How vividly can you re-imagine or re-experience that situation right **NOW** in your mind?

1......2......3......4......5......6......7......8......9
Not at all   Slightly   Moderat   A lot   Very much

(S1.2) How much humiliation, embarrass ent or shame did you feel **back in time when you experienced th ituation**?

1......2......3......4......5....   .7......8......9
Not at all   Slightly   Moderately   A lot   Very much

(S1.3) How much humiliation, embarrassm t or shame do you feel right **NOW** as you recall this situation?

1......2......3......4......5......6   ...7......8......9
Not at all   Slightly   Moderately   A lot   Very much

Sample questions on the Stanford Fear and Anxiety

I have no idea how well I did on the test, but I do know what happened on that day long ago at *Life* magazine. I waited until the other staff members had left the room. I stayed, got up, and walked over to Dan Okrent.

"Dan," I said, mustering every nanogram of courage inside me, "can I have a word?"

"Sure," he said, stopping what he was doing to look at me square in the eye. I think I had blurted my words out more aggressively than I had intended, which got his attention.

"About Brad writing the story," I said (Brad being the name of the other reporter). "I worked really hard. I came up with the idea. I don't think it's right that I'm not writing it. I also don't think it was right to announce this in a meeting in front of everyone without telling me about it first."

Okrent listened. When I was finished, he looked me up and down as if he had never really noticed me before.

"You're right about me saying this in the meeting. Next time I'll tell you first." He paused for a minute, then said, "But Brad is still writing the story. The next one, though, will be all yours."

I thanked him, walked down the hall, and ducked into the men's room—where I rushed to the toilet, thinking that I might throw up. I didn't.

This experience was among the first of many where I worked hard to "modulate" my fears, apparently pushing my amygdala to shape up. It took several years, but eventually I became comfortable in public situations. I occasionally have apprehensive moments, but I have learned to beat them back.

This battle to squelch the anxiety beast came out in my results. I showed a definite anxiety response, but I also showed that I could fix it. The first image on page 216 is of my brain reacting to one of the embarrassing stories I sent Philippe Goldin, which included the incident at *Life* magazine. Recalling this story caused my amygdala to ignite on the scan, along with visual-processing centers in the precuneus, the lingual gyrus, and the cuneus. "This means your amygdala was telling your visual areas to pay attention. They might have been visually recalling the incident or putting these areas of your brain on alert to watch out," says Goldin. My fusiform gyrus also lit up, along with areas associated with processing colors, face and body recognition, word and number recognition, and abstraction.

The second image shows my frontal lobe and what happened when I told my brain to settle down. My left ventrolateral PFC is a language area where we think to ourselves and tell ourselves to modify beliefs. "It's the self-talk area," says Goldin. The left middle frontal gyrus lights up when a person is applying a regulating strategy and is trying to change or monitor a negative emotion; the right middle frontal gyrus also ignites when a person is strategizing and analyzing and trying to regulate behavior.

On page 217 are two graphs that compare me to others who took the test: the healthy controls and the social phobics.

Goldin tells me that I was less reactive to the emotional stimuli (harsh facial expressions) than either the healthy controls or the social phobics were. "You seem to be able to control your emotions, which is good. Some people have exaggerated reactions," he says. "Look at the social phobics. To them, these are powerful, uncontrollable emotions."

For my own stories, including the incident in Dan Okrent's office, I also show an ability to mute the emotional impact of the story and to modulate what impact there is.

I tell Goldin I'm surprised at my ability to tamp down my reactions, given my state of high anxiety when I was younger.

The Author's Brain During Anxiety Test

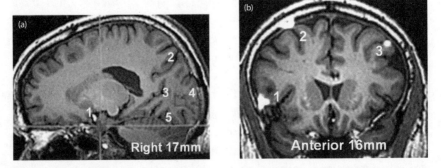

(a) The author's brain: scanned while recalling embarrassing story. 1 = right amygdala; 2 = precuneus; 3 = lingual gyrus; 4 = cuneus; 5 = fusiform gyrus. (b) The author's brain: scanned while trying to modulate an embarrassing story. 1 = left ventrolateral prefrontal cortex; 2 = left middle frontal gyrus; 3 = right middle frontal gyrus. For color versons of these scans, go to www.experimentalman.com.

Anxiety Tests: The Author's Results

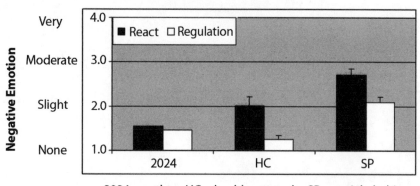

2024 = author; HC = healthy controls; SP = social phobics

Harsh facial expressions: negative emotion ratings. How subjects reacted to negative emotions and attempts to regulate them after seeing harsh facial expressions.

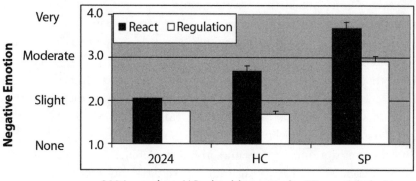

2024 = author; HC = healthy controls; SP = social phobics

Personal stories and negative self-beliefs: negative emotion ratings. How subjects react to their own negative stories and attempts to modulate their reactions.

"You've been alive long enough to learn to use your mind to respond skillfully to anxiety," he says. "That's what people are supposed to do. You came out normal across the board."

This is great, although I can't say it's making me significantly less nervous. As I write this, I'm looking around at people in my favorite café, the Thinker, and wondering, Are some of them even more anxious than I am? Or is my brain just better at covering it up?

Does my brain believe in God?

The question on the screen inside the MRI leaves no room for a nuanced answer:

THERE IS A GOD.

I have a few seconds to answer yes or no on a clicker in my hand, but I have no idea which button to push.

I'm thinking, as blood surges in my head to locales associated with religious belief, that this question, for me, may be unanswerable. I am essentially nonreligious. I seldom go to church, and I find myself agreeing with the likes of Sam Harris and Christopher Hitchins that organized religions such as Christianity are mostly artifacts of premodern cultures that in ancient times created all-powerful deities to explain and cope with the unknown. Overzealous piety has led to horrors such as the Inquisition and to dogma that at times becomes so rigid that it blatantly contradicts scientific proof and sometimes common sense. And yet religion clearly comforts people. Studies show that patients who pray often do better than those who don't. The base philosophies of most long-lasting religions—shorn of dogma and bigotry and the corruption, cruelty, and greed of venal leaders—offer vital moral frameworks for living that are remarkably consistent across cultures, with their admonitions to live a good life, to take care of the weak, to seek enlightenment. Nor can I deny that spiritualism is important: a sense that one's goals can be bigger than just a single person, although I don't frame this as coming from a God who looks like Charlton Heston playing Moses in *The Ten Commandments*, with a white-streaked beard and fiery eyes.

THERE IS A GOD.

These words also conjure memories of my childhood, when my grandfather Duncan was the minister of the Westminster Congregational Church in Kansas City. A deep thinker who wrote books and articles about theology and religious philosophy, Granddad was also the international prelate in one of the highest of the Masons' orders and the grand chaplain of the International Order of DeMolay, the Shriners' youth group. He was an intellectual and theologian who didn't preach much about literal miracles and angels, but he had faith in the morals of his religion and in a powerful deity who has a personal stake in the lives of believers and nonbelievers.

I remember as a boy listening to his sermons, sitting next to my mother in the pews on lazy Sunday mornings, sometimes being fidgety at having to sit so long. Granddad is at the rostrum in his black robes, waving his hands and speaking with a voice that rises and falls in a beautiful cadence as he tells stories about people he knew and from history, blending in biblical references. When I later read some of these sermons, I see that his main themes were frequently about compassion and helping others and that life was hard. In the pews, I remember the smell of slightly musty hymnals and my mother's Chanel No. 5 perfume and the brilliant colors of the stained-glass windows as the sun poured into the sanctuary.

Here is one of Granddad's sermons delivered on January 27, 1952, during a weekly radio address on WDAF-AM, then Kansas City's NBC radio affiliate: "The Christian faith gives positive, firm, and definite help through all tangles, complications and adversities in times of discouragement. . . . Did not the master of Nazareth speak of a heavenly father who cares for his children, so that he makes the 'sun to rise on the evil and on the good, and sendeth rain on the just and the unjust' in such manner as to make it possible for all to follow his command?"

But these Norman Rockwell memories don't convince me to punch yes in the MRI. They can't counter the bottom line for me: that I have no proof that a god exists or that the universe is anything but random, mindless atoms assembling and disassembling without a design or a creator.

As the seconds tick past, and I need to make a decision, my thumb twitches above the "no" button.

And yet, I lack definitive proof that god *doesn't* exist. It's possible that he (or she or it) does exist. Not the man with the beard, but as some force far beyond our brain's comprehension. If there is even a .0001 percent chance that this is so, can I answer no? "Objective evidence and certitude are doubtless very fine ideals to play with," wrote the philosopher William James, "but where on this moonlit and dream-visited planet are they found?"

Time is up and my thumb moves toward yes, and I press it. Almost instantly, my internal skeptic is admonishing me for an answer that could be interpreted as endorsing ignorance, burning supposed heretics at the stake, and vainglorious wars fought in the name of God.

Incredibly, all of these thoughts raced through my brain all at once and in just a few seconds: a seemingly exhausting neuronal exertion that must have lit up my brain like a city at night from thirty-five thousand feet.

This was the point of the test, said neuroscientist Dimitrios Kapogiannis, a postdoctoral student at the National Institute of Neurological Disorders and Stroke (NINDS) in Bethesda. Kapogiannis and a small team of researchers asked forty subjects these same questions about God and religious beliefs while they were inserted in the MRI: twenty who professed to be religiously inclined, and twenty who said they were not. "The purpose of the study has been to discover the underlying cognitive structure of religious beliefs (to find out what cognitive processes take place when religious and non-religious people think about religion)," Kapogiannis wrote me in an e-mail. "Then we wanted to identify brain regions that become active with each such process."

He and his team were specifically testing for three facets of human behavior, including, he wrote,

- Perception of God's level of involvement (i.e., people tend to think spontaneously whether their belief system includes a supernatural agent involved in their lives and how this relates to the issue at hand).

- God's level of anger vs. love involvement (i.e., people tend to think spontaneously whether their belief system includes a loving or angry supernatural agent and how this relates to the issue at hand).

- Dogma vs. practical applied aspects of religion (i.e., people tend to think spontaneously whether the issue at hand relates to a self-evident and unquestionable principle or to a practical aspect, so that they need to think about it under the light of previous experiences).

On this experiment, Kapogiannis is working with senior neuroscientist Jordan Grafman, the chief of the Cognitive Neuroscience Section at NINDS, a researcher with a longtime interest in trying to understand how the brain works when people are identifying with cultural phenomena and beliefs. In 2006, Grafman's lab ran an experiment on political beliefs, scanning the brains of people as they viewed pictures of John F. Kennedy, Hillary Clinton, Ronald Reagan, and John McCain (Barack Obama had not yet risen to prominence). The study found that participants had strong stereotypical associations with specific candidates and political parties—quick emotions and responses—followed by more reflective thoughts when asked whether they would vote for the person for president or whether each person was, or is, a top leader in his or her party. Subjects were also shown the faces with negative and positive words, and their brains tended to react according to their party identification: with Republicans having a negative response to words that dissed their leaders, and vice versa for Democrats. Grafman's team found that patterns of blood flow in specific areas of the brain suggest certain ideological proclivities. "The core of the brain is conservative," Grafman tells me, "the outer shell is liberal; it's playing with things, trying things."

Grafman is an affable, gentle man with large eyes and a willingness to use his scanners to peek into brains as they wrestle with big questions such as politics and religion—and, in another study, why anyone would love the Chicago Cubs. Born in Chicago, Grafman contributed to a recent book called *Your Brain on Cubs: Inside the Heads of Players and Fans*, edited by Dan Gordon. The narrative attempts to explain why anyone would root for a team that has not won a World Series in a hundred years. In one chapter, Grafman wrote about the "paradox" of being in that minority of fans following losers when baseball is supposed to make a person feel good. "There is some evidence that being in the majority [a non-Cubs fan] reduces reflective thinking. . . . The scientific literature suggests that fans of losing teams turn out to be better decision makers and deal better with divergent thought, as opposed to the unreflective fans of winning teams." Grafman says that Cubs worshippers are similar to religious adherents, in the fans' staunch belief that every year is "next year," an activity that he says occurs in the prefrontal cortex, the region of the brain that is responsible for high-level cognitive activities, such as planning, reasoning, establishing context—and, undoubtedly, inventing justifications for perennial losers in baseball.

Other neuroscientists doubt that these tests would tell me much. "They are very impressionistic," says Judy Illes. "I'm not really sure they will enlighten you about your religious beliefs, but maybe they will give a rudimentary idea of what happens in the brain when a person is thinking about or experiencing religion."

Kapogiannis admits that little research has been done along the specific lines of his experiment, which is identifying parts of the brain that activate when a person is asked pointed questions about religious beliefs. Many other studies, however, have delved into the brain's role in belief, God, and religion. For instance, neuroscientist Michael Persinger of Laurentian University in Ontario, Canada, ran studies suggesting that visionary religious experiences in some of his patients were related to seizures. He ran a series of experiments on regions of the brain using magnetic fields to determine whether he could induce visions. In an e-mail, Persinger explained his work:

> Both our experimental and clinical research indicate the visions (or experiences more precisely) are related to electrical lability, sometimes diagnosed as limbic or partial complex seizures, within the temporal lobes. The right temporal lobe is more sensitive than the left. Temporal lobe activity is associated with experiences; the frontal lobe functions are primarily involved with the organization of experience, that is belief. We apply weak, temporally complex magnetic fields across the temporal lobes. The magnetic fields are generated through small solenoids [current-carrying wires or switches that act like magnets when a current passes through them] embedded in a helmet the person wears or a device that is held on the head by Velcro. About 80% of the subjects, including atheists, report a sensed presence along with a variety of other experiences associated with mild activation of temporal lobe functions.[*]

In another experiment, researchers Andrew Newberg and Eugene D'Aquili scanned the brains of meditating Buddhists and praying Franciscan nuns. They saw distinctive brain activity during mystical experiences that "were not the result of emotional mistakes or simple wishful thinking, but were associated instead with a series of observable neurological events," said Newberg. "In other words, mystical experience is biologically, observably, and scientifically real."

David Wulff, a psychologist at Wheaton College in Massachusetts, has concluded that the consistency of mystical experiences around the world, combined with evidence acquired in brain-imaging studies,

---

[*] This experiment was performed under double-blind conditions, meaning that neither the experimenter nor the subject knows who is being exposed to the active test and who is a control receiving a placebo or its equivalent.

"suggest[s] a common core that is likely a reflection of structures and processes in the human brain." One explanation of this phenomenon is that religious belief and the experience of transcendence could be a product of evolution, of an adaptive behavior that helped humans survive. Wulff, however, says he is more inclined to agree with psychologist Lee Kirkpatrick in *Attachment, Evolution, and the Psychology of Religion* that religion is "likely one of those things that is a byproduct of evolutionary processes and not something that is itself necessarily adaptive." (The quote is Wulff summarizing Kirkpatrick's argument.)

British evolutionary biologist Richard Dawkins, who makes no secret of his antipathy for religion in *The God Delusion*, has theorized that genelike entities he calls "memes"—units of cultural inheritance that evolve and compete—are responsible for the persistence and breadth of religious practices and beliefs.

Studies comparing twins raised apart suggest that religion and other behaviors such as traditionalism and conformance to authority are about 50 percent genetic. The rest is a product of environment, culture, and upbringing, which may explain why nearly 90 percent of Americans claim to believe in God, a much larger number than in Europe. In another survey conducted by the Skeptics Society, an astonishing 35 percent of its members said that they, too, believe in God, a finding that surprised the master skeptic and head of the society Michael Shermer. One explanation is that heritability remains a powerful influence, even among doubters.

In 2004, geneticist Dean Hamer of the National Cancer Institute (NCI) published *The God Gene: How Faith Is Hardwired into Our Genes*, about his efforts to isolate a gene or genes that could give people a predisposition to religious revelation. Using data and genotypes from an unrelated study he was conducting at the NCI, Hamer zeroed in on the VMAT2 gene, which is involved in transporting neurotransmitters such as dopamine, norepinephrine, serotonin, and histamine in the brain. Hamer proposes that this God gene changes levels of these neurotransmitters, altering a person's mood and leading to a sense of self-transcendence. People with one version of the gene tended to score higher on a self-transcendence test, he says.

Hamer has been criticized for not publishing his results in a peer-reviewed science journal. Science writer Carl Zimmer was particularly skeptical, writing in a review in *Scientific American* that "The field of behavioral genetics is littered with failed links between particular genes and personality traits. These alleged associations at first seemed very strong. But as other researchers tried to replicate them, they faded away into statistical noise. In 1993, for example, a scientist reported a genetic link to male homosexuality in a region of the X chromosome. The report brought a huge media fanfare, but other scientists who tried to replicate the study failed. The scientist's name was Dean Hamer."

In an e-mail, Hamer told me that the VMAT2 gene and others are being further studied and that they may be part of a block of inherited genes (a linkage disequilibrium block) on chromosome 10.

Needless to say, I checked my results for Hamer's proposed "God SNP," rs33050, but unfortunately I was not tested for this SNP. "It's probably not a functional SNP, but it's in strong LD with multiple other variants," Hamer said, adding that a colleague, neurologist George Uhl at Johns Hopkins, "is still working on the functional effects of all the variants."

Back in the MRI, the questions are still coming. Dimitrios Kapogiannis and his team ask me seventy questions, arranged in subject matter from God as being angry and wrathful to God as being loving, with a variety of topics in between. I answer "no" to the "God is angry" sort of statements:

EVERYONE IS A SINNER.

GOD IS WRATHFUL.

PEOPLE GO TO HELL.

Other ideas popping up on the monitor are particularly irksome:

RELIGION ALLOWS KILLING PEOPLE.

GOD PUNISHES PEOPLE'S MISTAKES.

I also answer "no" to questions about mixing God with politics and social policy:

RELIGION SHOULD GUIDE GOVERNMENTAL POLICY.

RELIGION ANSWERS PROBLEMS OF ABORTION.

RELIGION DICTATES A STANCE ON HOMO-SEXUALITY.

The following statement, however, leaves me arguing with myself again:

GOD IS REMOVED FROM THE WORLD.

If I acknowledge that there might be a God—or that I can't rule this out—then He, She, or It, in my view, must be far removed. But I'm still hung up on the statement that asked whether there is a God.

Finally come statements asking whether God is loving and compassionate:

GOD IS KIND.

GOD CARES ABOUT THE WORLD'S WELFARE.

PEOPLE GO TO HEAVEN.

I would like there to be a good God who is kind and caring and a heaven that virtuous people end up in. But I see no proof of this, perhaps because I lack the appropriate variation in the VMAT2 gene that might allow me to know or sense that God is kind, cruel, or neutral. So I answer no, no, and no.

As the statements roll past, I find the experience stimulating and a little unsettling, given the structure of the "yes or no" statements and my nuanced point of view for some of the statements, but it certainly holds my attention. By now, I have spent several hours taking tests in MRIs. Most of that time, I was fending off boredom and drowsiness, which must have influenced my brain patterns. For this one, however, my vacillations and difficulty in accepting the central premise about the existence of God—and the unexpected intrusion of my childhood memories—must be causing more firestorms of activity in my brain, although in the end, as I say good-bye to Dimitrios Kapogiannis and his team, I find my brain settling back into my usual nonthinking about God or no God.

Weeks later, Kapogiannis sends me my results, along with pictures of my brain and his interpretations. An earnest and thoughtful young scientist from Greece, he warns me that his views are interpretations of one person's results, and that he and his team have not finished tabulating all of the subjects' results to compare me to. "Interpretations are open to questions," he says, "although I think my interpretation is as good as anyone's, since I have been working on this question for some time."

First up is how my brain reacted to the idea that God is involved in my life, and in the world (see the image on the next page).

God's Involvement: The Author's Brain Scan

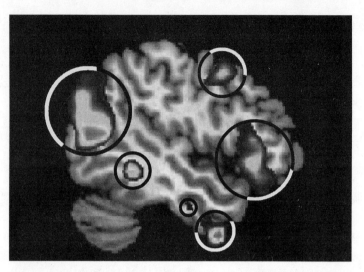

Areas of the author's brain (in light gray surrounded by thick black lines) that are active when asked about the involvement of God in his life and in the world.

Dimitrios Kapogiannis e-mailed me his comments about this image: "You activated a left frontal area . . . that is involved in observation and understanding (or making sense) of the actions of others. You also activated the posterior left middle temporal gyrus, close to a broader area . . . that is involved in 'theory of mind', i.e. the ability to understand the intentions and point of view of others. It seems that when you are thinking of a supernatural force, it comes natural to you to consider its actions and make sense of its intentions as an important part of the thought process."

When I phone him, Kapogiannis explains that this image shows elevated activity in the right side of my brain in action-oriented areas. "God's presence activates your temporal lobe," says Kapogiannis. "You are actively imagining a God in action, visualizing an involved God, imagining a God willing to intervene."

"But I answered no to the statements about God's involvement and willingness to intervene," I say. "I fundamentally don't believe this—unless I'm lying to my brain, which imagines a God in action without me knowing it."

"This is just one interpretation," says Kapogiannis.

I tell him about the barrage of thoughts racing through my head during the "there is a God" question. "Could it be that I'm visualizing my grandfather in action, or that I'm very involved in my thought processes and attempts to decide on taking an action—which is to press the button yes or no?"

"That could be happening, too," he says. "The point is that answering these questions causes a strong response in the action parts of your brain. These are parts of your brain that visualize action in space. You grew up with religion and you do think about it. Maybe that's what gives you a reaction."

He reminds me that "This is all tentative right now; some of it is still very out there"—a notion that I am prone to agree with.

Next are my results for the notion of God getting angry. Kapogiannis e-mails me his thoughts on this scan (see the scan on the next page):* "You activated the left superior parietal gyrus in the medial part of the brain, an area also involved in the so called 'theory of mind', i.e. the ability to understand the emotions and intentions of others. You did not activate classic areas associated with fear (such as the amygdala), implying that you are not really afraid of such a thought (perhaps because you do not believe in it and therefore it does not represent a real threat)."

As for God's love,* Kapogiannis says, "Surprisingly, you activated the primary sensory cortex, an area involved in perception of pain, cold and hot, touch and other bodily sensations . . . almost as if these statements that provoked thoughts of God's love were interpreted as a physical touch. Maybe you can only think of such an abstract and transcending love in very physical terms."

I'm not sure that I buy any of this, although the idea of God's anger or love did not cause much activity compared to other questions, says Kapogiannis. "These inquiries didn't cause a strong activation. This is a pattern seen in other people who are not particularly religious. These pictures can show that you may believe in God, or the idea of God, but He's distant."

Next is my reaction to dogma. "In the questions about religion impacting government policies on abortion and homosexuality," wrote Kapogiannis, "this is again the parietal network, the practical areas. You are imagining how you might take action on a moral issue, an

* For additional brain images (in color) from this experiment, go to www.experimentalman.com.

God's Anger: The Author's Brain Scan

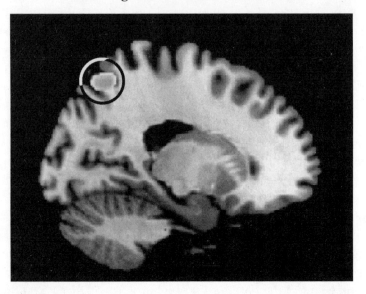

This test did not elicit much activity in the author's brain other than a
small amount of activity on the upper left part of the brain in this image.

action for or against something. You activated bilateral basal ganglia.
Interpretation here is speculative, but these are areas that perform
complex computations, binding cognition with emotions and behavior. It is as if you are trying to find out what is the meaning of an
abstract and puzzling cognitive construct and consider its emotional
significance."

He couldn't tell from reading my brain, however, whether I want
to take an action in favor of or opposed to abortion and other specific
issues. "Your brain is particularly active," he says. "You are wanting
to take action, while others we have tested are only imagining taking
action. This is visual imagery. Some people imagine seeing it as a spec-
tator. You have almost nothing in visual imagining."

So my brain is an activist. My mother, the former environmental
activist, will be pleased to hear this, although we can't tell for sure
from this scan what exactly my brain wants to get active about.

The final test is my "reaction to moral and applied aspects of
religion, which activated my left frontal-parietal network," says
Kapogiannis. "Such a network usually mediates attention and what
we call 'working memory,' keeping many aspects of an issue at hand

while planning and carrying out a course of action. It may imply that you want to combine many different points of view in considering a moral issue (such as abortion), both principles and prior experiences. This is where we look for theological aspects of religion: how a person computes very abstract ideas—God's existence, reincarnation, resurrection. We haven't seen many like this. You appear to be combining complication and emotion. I can't easily read this."

I have been accused before of combining complication and emotion, although not about God.

I find Kapogiannis's work fascinating, and I have no doubt that his research will help us understand the mechanics of how religion works in our brains. But so far, the interpretations seem at best sketchy—which he admits—and at worst little more than a sophisticated version of reading tea leaves. I don't want to be overly critical. I admire Kapogiannis and Grafman for even asking these questions. As Jordan Grafman explained in an e-mail, "The results will help us understand how some people try to process and understand religious statements and beliefs. Clearly you had specific brain areas associated with the different dimensions we identified suggesting different parts of your brain (and most brains) contribute to processing a belief. So we think that these results help us understand how different aspects of thinking are probably necessary to build a personal belief system and they include an ability to understand the intentions and thoughts of others via inference, a translation of emotions in others to bodily sensation, and conceptual organization of belief. All of which you showed. Now you are a specific individual and we are expecting individual differences in activity depending on particular beliefs and individual belief systems."

As I'm sure my grandfather would agree, our beliefs, at a basic level, must offer us a degree of reassurance about our lives: the hardships, successes, and fate of our loved ones; and of our tribes, nations, and species. Perhaps Granddad imbued me with a gene, a meme, or early memories—or all three—to give me a set of beliefs that has, by and large, proved sufficient for me, although, in reading my own neural tea leaves, I clearly see that the meme has not handed me a need to express this through a god or an organized religion.

After taking these tests and pondering what my brain really thinks about God and religion, I decided that if I had this test on belief to do over again, I'd prefer to answer questions about Jordan Grafman's obsession with losing sports teams. Not about the Cubs, but about my hometown's truly flailing professional baseball team: the Kansas City Royals. They haven't even existed for a century, and they won a World Series in 1985. Since then, however, they have fallen on hard times as one of the worst teams in baseball, year after year. Yet I check their box scores and standings every day, often in secret, like a fanatic fearing ridicule and discovery of my true belief—hoping (and praying?) that next year my team at last will go all the way. Not to the World Series, which would be the equivalent of the Second Coming for the Royals, but something that even the faithful can only wish for: a .500 season.

Greed, gambling, and why my brain loves *Dodgeball*, the movie

Ifat Levy is telling me that I am risk-averse. Everyone knows this about me: that I play it safe by writing books, a definitely safe-harbor profession, and by reporting on the occasional violent conflict, and taking long trips by bicycle through Africa and Asia. But this is what New York University (NYU) neuroscientist Levy concluded after studying my brain and behavior: that I shun taking chances, especially ambiguous ones where I don't know the precise odds of success.

Come on, I tell her, my life has been one long plunge into ambiguous and risky situations, which I then write about. Give me a choice between something solid and predictable and something intriguing but indistinct, like running the experiments for this book, and I'll fasten onto the vague and the adventuresome nearly every time. As we know, my DNA tests have also revealed a certain genetic proclivity to risk-taking. I'm a carrier of a higher-risk variant of the DRD3 gene—though, of course, this finding is based on one of those studies that is preliminary and, well, a bit ambiguous.

Levy is thin, intense, and all business, with short-cropped dark hair. She doesn't crack a smile as I try to be humorous about the dichotomy between her findings and my self-perception of living life on the edge. She insists that in the tests I have just taken, which involve gambling inside an MRI, my results place me way below normal for taking risks—an outcome that she will analyze by studying the scans of my brain.

"Okay," I say, realizing that I can't talk my way out of how my brain behaved.

The glaring difference between my perceptions and Levy's results raises a crucial point in all of these investigations: that there seem to be at times two distinct "minds" at work inside my skull, something akin to Descartes' mind/soul and body division, although I agree with modern philosophers that the mind/soul does not float somewhere in the ether as an entity apart from the physical brain. The mind/soul I'm talking about is my consciousness, the "mind" that I'm aware of when I think, believe, or do something, as opposed to what the brain is actually doing when a researcher scans it and interprets the patterns and pathways of activation flashing through my cerebral tissue like heat lightning in a summer thunderhead.

Sometimes the thing I prefer to call the mind/consciousness is in sync with the activity registering in my gray matter; other times, it's not. This raises a number of intriguing questions, including which of my minds I believe when one seems to be contradicting the other. Or perhaps what this dichotomy describes is simply the conscious efforts of the frontal lobe to tell the older parts of the brain to do this or to do that—the process that neuroscientists call top-down modulation. For example, when Philippe Goldin at Stanford asked me to tamp down my anxieties, he could literally see in the fMRI scan a neural discussion going on between my frontal lobe and my amygdala and other centers of anxiety and emotion. In that case, the frontal lobe was persuasive and told the fearful and anxious parts of my brain to settle down—and they did. In many other interbrain arguments, such as trying to resist eating a chocolate chip cookie, my rational brain that wants me to stay healthy and svelte will be handily overruled by the chocolate chip cookie center nestled somewhere deep in the irrational portion of my head that loves sweets.

Trying to read the patterns of activation in my brain is also limited by the state of the science and of MRI scanning. In addition, people have bad and good days, mood swings, and any number of other influences that might alter their neural-prints from moment to moment.

"We will see how you do tomorrow," says Levy. She plans to place me in the MRI at the Center for Neural Science at NYU for a second round of scanning the following morning.

It's January in New York City and brutally cold as the wind cuts like ice through the thick ski jacket I'm wearing, a precaution that I have taken after being caught in Maine with a thin jacket, although the upgrade isn't helping much. Levy and two other researchers and students in the lab of neuroscientist Paul Glimcher are running a battery of fMRI tests—twelve hours' worth over three days—testing my brain on several cutting-edge experiments that study how people make decisions.

Besides Levy's experiments on risk taking, during my three-day electromagnetic barrage, I'll also be tested for what happens in my brain when I win and lose money gambling and, in a third study, the patterns inside my skull when I am considering consumer products I like and don't like.

I originally met Paul Glimcher at the same San Diego meeting where I sat down with James Brewer to hear about my shrinking brain. Glimcher is a leading figure in an emerging subfield of neuroscience called neuroeconomics. This new discipline uses economic principles to represent intricate thought processes in everyday situations marked by competing values and interests. Perhaps more important, it's also searching for a physical basis inside our brains to reveal why people make the choices they do and how we might better predict behaviors. Key to this endeavor is understanding why people make irrational decisions based on emotions such as fear and envy. This proclivity has long frustrated economists who were trying to formulate ever-more-complex economic theories that assume people are essentially rational in decision making. For example, Glimcher asks, "Why do so many traders buy stocks when prices are high and sell when they're low?"

Good question, I think, remembering a certain technology stock I bought in the late 1990s, almost precisely at the top of the bubble. A lot of economists are also wondering about this in the wake of the bank collapse in the autumn of 2008.

According to Glimcher, one explanation is that classic economic theory has no way of testing or truly understanding how different people assess value or how to assign probabilities to outcomes based on these values. He says this is critical in understanding how we make choices and execute decisions. "Bill Gates will value a dollar differently than I will," he says. "What is happening in his brain versus mine when he's deciding what to do with that dollar? What emotions are at work for him, and what are at work for us?"

"I don't know," I answer, thinking that I'd like to be able to value a buck as Bill Gates does.

"Also, how do personal values and decision-making pathways in brain circuitry affect risk taking, and how are repeated experiences with relative values learned by our gray matter?"

Neuroeconomics is an offshoot of behavioral economics and cognitive neuroscience, which runs experiments attempting to quantify how people's behavior affects their decisions. One of the most famous of these experiments is the Ultimatum Game, where two people are offered, say, $100, but for either of them to get the cash, the first one has to decide how it will be split between the two of them. This first person can split the $100 down the middle, keep most of it, or, if he or she is in a generous mood, give most or all of it to the other person. The second person needs to agree to the split, however, or no one gets anything. Rationally, the first person should offer as little as possible to maximize his gain, while the second person should accept whatever the first gives him, since something is better than nothing. But scientists have found that the second person will often refuse amounts that dip too much below 50 percent, believing that it's not fair for the first person to give him or her less. One explanation is that the second person waives away the money to punish the first person for offering such as lowball sum.

Neuroeconomists have asked subjects to play such games and to perform various decision-making tasks in an MRI and when strapped to an EEG and to other brain-reading devices. They have found that rejecting the cash causes high levels of activity in the dorsal striatum, situated in the middle front of the brain—an area involved in reward-and-punishment decisions—which is exactly what should happen if behavioral theorists are correct about the second person wanting to punish the first.

I won't be tested on the Ultimatum Game on this freezing day in New York City. But I will participate in experiments that similarly try to determine why people respond emotionally to different choices, how I differ from others in my decisions, and what this process looks like inside my head.

"Many of us in neuroeconomics believe that we are on the verge of a major new theory using what happens in the brain to explain economic behavior," says Glimcher. He believes that this übertheory could unite neuroscience, economics, and psychiatry, three disciplines that in the nineteenth century were closely connected but in the last century diverged into independent fields that have not talked much to one another.

Glimcher's enthusiasm has not entirely caught on with all economists. Ariel Rubinstein of NYU and Tel Aviv University criticizes the field as overselling itself by relying on experiments that use small numbers of subjects and for the deficiencies of MRI devices that do not accurately capture the speed of decisions made in the brain. The brain usually makes decisions at a speed of 150 to over 1,000 milliseconds, depending on the complexity of the decision, using tiny clusters of neurons on a spatial scale of about 0.1 millimeter. Most MRI machines, however, take images every 2 or 3 seconds and usually don't detect anything smaller than 1.5 to 3 millimeters long. "These MRIs are giant toasters," admits Glimcher. "We have this terrible tool; it's slow, it's a terrible environment and creates noisy images." Another problem is that that the subject pool—usually students—is hardly representative of the wider population.

Neuroeconomists also study greed, altruism, and optimism. Not far from Paul Glimcher's lab just off Washington Square is the lab of psychologist Elizabeth Phelps, another major figure in neuroeconomics. In a 2007 study, she and former graduate student Tali Sharot used functional magnetic resonance imaging to reveal that our brains are far more active when we sugarcoat the future than when we are negative about what's going to happen.

Scientists have long known that people tend to be overly optimistic about future events. Couples getting married believe they will avoid divorce, and contractors insist they will replace the roof on your house faster than often actually happens. Phelps's NYU team scanned the brains of fifteen volunteers who were told to envision various possible life events—positive and negative. Subjects were asked to envisage some events as having happened in the past and some in the future. The question posed by the team: Does your brain interpret bad and good outcomes differently? One way to assess this is to see what areas of the brain light up when a person thinks optimistically or pessimistically. "We wanted to find out whether the same areas of the brain activated for positive and negative events in the past are activated when people anticipate future events," says Phelps.

Vivid memories lit up the amygdala and the rostral anterior cingulate cortex (rACC), says Phelps, which seems to help regulate rational cognitive functions such as reward, decision making, and empathy. Negative reminiscences caused much less energetic activity in these areas of the brain. Curiously, the same thing happened when subjects contemplated future actions, suggesting that the brains in most people may mute the effect of anticipating potentially upsetting events. "Extreme optimism can be harmful as it can promote an underestimation of risk and poor planning," wrote the researchers in *Nature*. "In contrast, a pessimistic view correlated with severity of depression symptoms was more likely to be imagined from an outsider viewing in, than positive future events and all past events."

I won't be tested directly on optimism, but my level of optimism will play a role in the games I play and the decisions I'll be asked to make, since many decisions and choices involve my enthusiasm about the possible results of taking or not taking a risk.

Understanding the brain's joy at winning and its pain at losing is the point of the first test I take on the morning of day two in the NYU imaging center in Greenwich Village. (I'll take the second part of Ifat Levy's test that afternoon.) "We're going to see what your brain does when you lose money," says doctoral candidate Robb Rutledge, who is tucking me into the gurney that will slide into the MRI.

As he finishes locking the monitor device in place above my eyes, I begin to wonder whether all of this whirling magnetic energy running through my brain tissue is dangerous as I spend all of these hours this week in this machine.

"Are you sure these things are safe?" I ask Rutledge.

He smiles in a casual way and says, "It's definitely safe."

He is a dark-haired young man with a three-day beard, a dark T-shirt, and a quiet intelligence. On the Glimcher lab's Web page, all of the researchers have irreverent, yearbook-ish comments about their work and personalities beside their names and contact information. The comment about Rutledge is:

Optimal coding:

Dude, an efficient lifestyle

Why rush or stand straight?

Rutledge starts the test by pulling out his wallet as I'm getting into the MRI. He hands me $100 in cash and says this is mine to gamble with during the session. "That's $100, your own money," he says.

"When you gamble and win money, I'll give you more; when you lose it, I'll take it away."

In the tube, with the computer monitor fitted above my eyes and a button box in my hand, I'm given a choice between two lotteries represented by pie charts, each one offering a chance to make or lose money. One pie might offer a 50/50 percent chance of making or losing $5; the other pie will offer a 75 percent chance of losing $5 and a 25 percent chance of losing $10. (This example is given in the chart below.) Obviously, I pick pie number one, clicking the appropriate button. Then the computer decides whether I win or lose, with the odds of each outcome dependent on the pie I chose. The test is also timed: if I hesitate and don't click a choice in one second, I lose $10.

"Robb is studying how the brain evaluates different values," Glimcher tells me. "This transfers something subjective, like the value of a dollar, to something objective in each person. We are trying to understand and explain this using brain science. We hope to come up with a prediction about how people will act." Part of the explanation may be something called the temporal difference theory of learning, which tracks what the brain does when it expects to win but does not.

Rutledge studies an area of the brain called the striatum, which receives input from dopamine neurons in the midbrain. When you feel a sense of reward, the brain carries extra oxygen into the striatum, which lights up on an fMRI image. "We believe that this activity reflects not just the actual outcomes of the game, but also depends on what your expectations were," says Rutledge. "We hypothesize that your experience is used to update your expectations as you play the game, to learn the value of actions so you make better choices in the future."

For a while, my striatum stays happy as I tend to win, with my earnings reaching $120 at one point: a 20 percent increase. About half-

Lottery Expermient: Sample Test

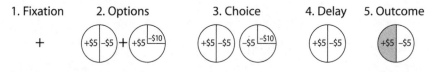

Robb Rutledge's test: 1: shows the subject where to focus; 2 and 3: the subject is given two pie graphs and chooses one; 4: the subject's choice; 5: the computer picks the winning number (+$5 in this example).

way through the session, however, I get drowsy. Lying prone on my back on a comfortable gurney with a blanket over me is a scenario where my brain says, Naptime! I nod off and miss a couple of bets, losing $10 each time, which motivates me to stay awake; I also wiggle my feet and toes. I do this carefully to avoid moving my head and blurring the scans. Then come a disastrous final few rounds of betting, where my luck shifts and my portfolio tumbles in a matter of minutes to a mere $25. I can almost feel the dopamine escaping from my striatum, which is not glad about this sudden loss.

In real life, I would have stopped at about $50 and held on to my money, since the original $100 was not mine, and I would still be ahead unless I hit zero, which anyway would have caused the game to stop. But in the tube, I am forced to continue, a form of subtle torture that is not unlike what happens when a stock one owns is crashing and there is nothing to be done, except to keep it in the hope that it will go back up, or sell at a loss. My midbrain is probably lighting up now even thinking about this hypothetical situation, although I'm not sure whether knowing my neural patterns will make the choice any easier. It might, however, help economists one day understand why I bought that tech company stock back in the 1990s at a ridiculously high price and, of course, sold it for a devastating loss right before the company tanked.

When we break for lunch after the morning session, I brave the arctic air outside and go to a small café nearby on Bleecker Street, only to get a call from Paul Glimcher. He is sitting in his lab with Robb Rutledge and Ifat Levy, and he is chuckling on the phone.

"What's up?" I ask.

"Robb and Ifat are here, and they didn't want me to tell you this— but I told them you are a professional, and you can take it."

"What is it?" I ask, worried that I have done something terribly wrong, although Glimcher's laughter suggests it's not too awful.

"They tell me you're the worst wiggler we've ever had," he says. "The images are blurring because you're moving your feet."

I feel bad, because I thought that my movements were harmless. Glimcher says that any squirm can move a person's head, wasting these sessions that cost hundreds of dollars an hour. All is not lost, however, as long as I promise to stay absolutely still. Later, I read a story by the *New Yorker* writer John Cassidy, who wrote about being tested using a gambling scenario in the same imaging center in 2006 by graduate student Peter Sokol-Hessner:

After about an hour inside the machine, I was more concerned about getting out than I was about making a few dollars. (Sokol-Hessner said that I moved my head around so much that my brain scans were unusable.) "That's the terrible thing about MRIs," Sokol-Hessner conceded. "You are in a long tube, and you might well feel tired or claustrophobic. There's definitely other stuff going on in there besides the experiment. We have to be very careful about how we interpret the evidence."

Lying as still as the dead, that afternoon I am back in the MRI being tested on round two of Ifat Levy's experiment. I'm gambling again, although this time I'm presented with images of rectangular boxes that are meant to simulate glass containers filled with red and blue poker chips. The boxes on the monitor are filled partly with red and partly with blue, with each box having different levels of red and blue. Numbers are attached to the colors, which are amounts of money that would be won if these were real chips and if that color was selected in a random drawing. (The currency is an imaginary one called a "franc.") For instance, one box might pay out 180 francs if the computer "draws" a red chip, but zero francs if it chooses a blue chip. In the game, the player decides between choice number one—the 180 versus zero scenario, or something similar—and a second choice. This is always the same: a "reference lottery" that has 50 percent red chips and 50 percent blue chips, with a blue chip giving the player 50 francs should it be drawn, and a red chip giving zero.

To recap: there are two choices in this game, one that is often risky (though not always) but offers the potential to make big money, and one that is usually less risky and that over time should allow a person to at least break even.

This isn't all. Another twist in the game is that some of the boxes containing the "chips" in option one are partially covered up, obscuring the ratios of red and blue but giving the amounts to be won. This means that the player will know that choosing red, say, will yield 180 francs, but he or she doesn't know how many reds are in the box or the odds of winning. Levy will measure levels of risk-aversion and ambiguity-aversion in my brain by scanning the medial (middle) part of the frontal cortex and the basal ganglia (in the lower middle part of the brain), areas that are known to play a role in decision making. "If activity in your prefrontal cortex is strongly affected by risk, ambiguity, delay of gratification, or loss, then that's the kind of person you are," says Levy—a contention that in my case would make me Mr. Play It Safe, something that I hope in my life doesn't extend beyond a risk aversion when gambling for "francs" in lotteries involving virtual red and blue chips in little boxes.

Players do get real money: $10 for each session, plus a chance to win more money based on their choices while being scanned after the MRI sessions are finished.

I have a bias going into this test because I don't like games of chance such as this, where I have so little control over the outcome. I enjoy gambling but prefer games like poker or blackjack, which take at least a small amount of skill, and where the odds aren't decided by completely random luck or by a computer. Unable to move my feet in the MRI, I also have to fight off my body's growing desire to equate the long tubes in these machines as time to snooze. Like the *New Yorker*'s John Cassidy, I am anxious to finish and get out, as the two hours drag on for what seems infinitely longer.

That afternoon, I meet with Glimcher and his team in their lab's cluttered conference room. That's when Levy tells me I am risk averse, according to my results from the first round of her testing the day before. This result brings a smile to Paul Glimcher's face, especially when I protest that "risk" is my middle name.

The third and final day at NYU, my brain is being scanned more directly to determine how I make decisions as a consumer and what products I like and don't like. It's a realm of testing that is fast catching on with marketing experts who use MRIs and EEGs to assess the desirability of certain products for specific demographics, such as male or female and different ages. A famous experiment in the new science of neuromarketing was a test of Coke versus Pepsi back in 2004. Baylor College of Medicine's Read Montague asked sixty-seven subjects who were given Coke and Pepsi, but didn't know which, to pick the soda that tasted best. Half chose Pepsi, which also caused a stronger neural reaction than Coke in the brain's ventromedial prefrontal cortex, a region that neuroscientists think processes a sense of being rewarded. When the subjects were later told which soda they were drinking, three-quarters said they liked Coke better, as activity increased in memory centers and cognition areas were activated. The take-home on this test is that Pepsi should have half the market share but does not because memory and brand identification with Coke override taste.

One of Paul Glimcher's graduate students, Stephanie Lazzaro, is running a different test on me to see how my brain makes decisions on which consumer goods I like, such as choosing between a DVD of *Dodgeball* and a CD by the hip-hop artist Akon, or a Beethoven CD versus a new planner book. "You're going to be passively viewing images of objects in the scanner," says Lazzaro. "We'll be looking in the scan to see whether you like something or not from the patterns of activity in your brain. We'll be able to give you a ranking of what you liked best and predict your choices based on this ranking."

I spend the next session in the MRI—heroically attempting to keep my feet still—looking at items such as DVDs, CDs, books, and posters, and then outside the scanner, on a computer, I make choices between two products in each frame on the monitor, dutifully choosing one in each match-up.

Several weeks later, the researchers e-mail my results. The first round comes from Robb Rutledge. He shows me what my brain looks like when I'm winning and losing money in his lottery-style game.* "You started with $100 in cash," he wrote in his report. "On each of 200 30-second trials, you made a choice between two lotteries. Then you found out what prize you got." Referring to scans of my brain, he described what happened in the areas that lit up with activity: "These areas responded more (increased blood oxygenation) when you won money than when you lost money. However, the response in these areas *also* depended on what you *expected* to get—if you expected to get a lot, they responded less.

"The striatum and medial prefrontal cortex receive dense inputs from the midbrain dopamine neurons. We think that the signal in the striatum in particular carries a 'reward prediction error' (RPE) signal equal to 'what you got' minus 'what you expected to get.' This signal is positive if you get more than you expected, negative if you get less, and zero if you get exactly what you expected. You could use this signal to learn how much you value items and options. For example, if you get exactly what you expected, your expectation was accurate and your error is zero."

I came out as expected, says Rutledge: my brain gushed with dopamine when I unexpectedly won and gave a sort of neuro-shrug when

---

* To see these and other brain images, go to www.experimentalman.com.

I expected to win and did. Not surprisingly, my brain did not like to lose. Another set of images Rutledge sends me compares my brain scans to those of twelve other subjects tested. I came out within the spectrum of the twelve, he says. "On the right [of the two images] is the area of the striatum that seems to encode a reward prediction error signal in the average of twelve people's brains. You can see that your RPE areas overlap with the group RPE area. . . . At the group level, we took a closer look at the activity in the striatum and the signal does all the things we expect a RPE signal to do: increase signal for higher rewards, decrease signal for higher expected values (when you expect to win), and be the same when you get exactly what you expected (like losing $5 for sure)."

It's hard to tell much for individual subjects, he explains, given the lack of precision of fMRI scans. "One thing we noticed in the group results is that the blood oxygenation changes about twice as much for losses as for the equivalent-sized gains. This may explain 'loss aversion' and now we may be able to determine exactly how much you dislike losses from looking at how your brain responds to different outcomes, which will allow us to better predict how you will learn and make choices between lotteries."

I'm not sure that I see the activation pattern in my striatum as exactly fitting into the group results, but I'll take Rutledge's word for it. It seems rather obvious that most people's brains like to win something of value, especially when it's not expected. This latter reaction, however, might explain why some people—stock traders? entrepreneurs? writers?—repeatedly take risks with unexpected outcomes. It's like taking a drug that gives their brains a huge hit of dopamine if they succeed when they didn't expect to.

Stephanie Lazzaro is next to e-mail her results. She sends me a list of the products I liked best, ranked by how much they activated or didn't activate areas in the medial prefrontal cortex of my brain associated with feeling reward (see the graph on the next page).

Some of Lazzaro's findings:

- I consistently chose the *Dodgeball* DVD over all of the other items.
- *Dodgeball* also had the highest activation of any of the items.
- I consistently chose *The Kite Runner*, by Khaled Hosseini over all of the other items (except *Dodgeball*).

Consumer Products Test: The Author's Likes and Dislikes

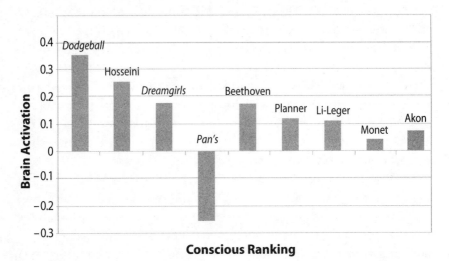

This graph shows two things: the vertical columns display how much the author's brain likes a product (independent of his conscious choices); the arrangement left to right shows how he consciously ranked the products.

In order, the products are: *Dodgeball* (DVD), Khaled Hosseini's book *The Kite Runner, Dreamgirls* (DVD), *Pan's Labyrinth* (DVD), music by Ludwig van Beethoven (CD), planner book, Don Li-Leger print (poster), painting by Claude Monet (poster), and an album by hip-hop artist Akon (CD).

- I chose the *Dreamgirls* DVD once and the Beethoven CD once when given two separate choices between the two items. The activation for these two items is very similar.

Lazzaro found one test result puzzling: my brain's negative reaction to the movie *Pan's Labyrinth*, the dark fantasy-horror film by Guillermo del Toro about a little girl coping with the repression of Francisco Franco's Spain in the bloody aftermath of the Spanish Civil War. "I am not sure why you have negative activation for the *Pan's Labyrinth* DVD," says Lazzaro, "but you did choose it over all of the items to the right of it in the graph. Perhaps you really disliked the way the cover of the DVD looked?"

The truth is, my conscious brain was lying when I said I liked the film. It's the sort of movie that I normally like and, perhaps as important, that my serious film friends think is brilliant. So I claimed to love it, when, as my brain knew all too well, the movie didn't work for me. It was hard to follow, relentlessly dark, and did not, for me, make a plausible bridge between the real world of a little girl having to live with the militarism and horrors of the Franco regime and her fantasy world of strange creatures and a mysterious labyrinth. This was a case where one part of my brain—where I fancied myself an arty intellectual—wasn't able to persuade the like and dislike parts of my brain to love *Pan's Labyrinth*.

This revelation that my brain could "out" me was fascinating and a little disturbing. As I've noted, we all have self-images that we seek to maintain and opinions that we would like to hold that fit those self-images. I tend to support independent film and like the intellectual challenge that the best of them can pose, but my brain would probably prefer an action thriller—or, apparently, a Ben Stiller comedy. I wonder whether a brain scan would reveal other contradictions in a person's head on topics such as politics or, more disturbing, race or ethnicity. Many people have complex feelings and emotions about such issues in their heads, knowing on the one hand what is correct and proper—such as Democrats professing to like their candidate for president. They will vote for him or her, tell their friends they are supporters, but, secretly, do they really like the candidate? Or, more precisely, do the pleasure and reward parts of the brain like him or her, or is this a case of the frontal lobe trying to override more base emotions? This gets even knottier when we realize that the frontal lobe may intellectually like something, but the striatum does not. What are a person's true feelings? Can we say?

When Levy's final results come, I'm pleased to see that my risk-aversion has lessened on the second round of testing, although I still registered in the range of a person who plays it safe with known risks and even more so with ambiguous risks. "Like most subjects, you showed both risk and ambiguity aversion," Levy reports. "In the first session you were very risk averse compared to our population average…, and you showed around-average aversion to ambiguity. In the second session you were more willing to take risks, but, on the other hand, you were less willing to face ambiguous lotteries. Your behavior was pretty 'lawful,' which means you chose the option on the screen more if it paid more or had a higher winning probability, as should be expected from a reasonable subject."

Levy says that neuroeconomists hope to use results on tests like this for large groups of people to predict behaviors, although Levy admits that a person might behave differently in other domains of his or her life. "That raises the question how representative this experiment is of real life behavior," she says. Indeed, I'm not sure whether I'm alone compared to other subjects, but my results suggest that either I have been wrong all these years about my brain wanting to take certain risks, or this test is not the best indicator of my overall behavior. The issues of drowsiness and tedium may also skew a true risk-taker's performance, since people in this group by definition need to be stimulated by challenges they perceive as exciting and real, or they get bored.

My experiences in the MRI at NYU reinforce the notion that our brains are sometimes competing amalgams of thoughts, ideas, and proposed actions that aren't always in complete agreement, with different parts of the brain winning the argument at different times. This seems like it should be a more confusing process, given the competing interests of the reptilian (food, drink, sex), the limbic (love, nurturing, social interactions), and the neocortical (logic, reasoning, decision making). Yet inasmuch as my brain is typical, decisions do seem to be made and lessons learned.

As I leave the NYU imaging center, I find myself wanting to buy the dumbest action movie DVD I can find—undoubtedly driven by my amygdala, or is it my striatum? I also have a new learned trait: to avoid playing lottery games inside an MRI. Much more important, though, as the chilly wind whips around a corner, is a very loud thought coming from the "I'm freezing" part of my brain. It's telling the decision-making part of my noggin that I need to buy a warm scarf and some gloves—now! Preferably, if I have a choice, they should be black and the gloves leather with lining, although any color and texture will do in what is fast becoming a minor emergency, except perhaps bubble-gum pink or phosphorescent orange.

Or does my brain secretly crave these colors?

I'll probably never know, since, for now, I've spent quite enough time in an MRI answering such questions.

Building a new superbrain

It's 2:00 p.m. on a Tuesday in Bethesda, Maryland, and I'm feeling stupid and slightly grumpy. I have lingering jet lag because I took a trip to Europe the previous week and flew in last night from California. Now I'm sitting with two electrodes affixed to my forehead in the Brain Stimulation Unit of NINDS at the NIH, in the same building where Dimitrios Kapogiannis studies the brains of believers and nonbelievers. In a moment, a researcher in the lab of neurologist Eric Wassermann will activate a gizmo the size of a small clock radio, which will send a very slight electric current through my frontal lobe. For the next forty minutes, the flow of electrons will create an electric field that lets neurons having to do with cognition and emotion fire more easily.

For the Experimental Man project, I have established a few simple rules, one of which I am now violating—to do no harm to the subject of these experiments: me. But I couldn't resist this one, and it's unlikely to actually cause me anything but perhaps a mild headache. A magazine editor made the suggestion that I write a review of two mild and safe forms of brain enhancement, one electronic and the other chemical, assessing them as a reviewer might evaluate a book or a new film. Most of my other tests for the book's brain section record what goes on inside my head like a series of snapshots. Yet most healthy people who might one day take these tests will want more than simply a picture of the status quo: they will want to use this information to improve themselves.

We already do this by taking vitamins and vaccines and by washing our hands before we eat. I drink coffee to get a little bump, even if my caffeine fast-metabolizing gene forces me to suck down several cups of joe before I feel even a mild jolt. So is it so awful to try a gadget like the one I'm now hooked up to, which may give my brain a little kick, too?

Wassermann has already told me that his device will not turn me into an Einstein. He is hoping that for people with brain injuries or impairments from disease, it will stimulate the cognitive centers to function better. "We are starting with testing healthy people to get a baseline for how the technique works," he says.

Two days from now, I'm planning to further tweak my mind by taking a brain-boost pill. Called Provigil, it seems to hone in on a section of the brain that helps govern alertness and memory, and it does not diffuse throughout the brain the way that coffee or stronger stimulants such as amphetamines tend to rev up everything. Manufactured by Cephalon of Frazer, Pennsylvania, Provigil has been approved by the FDA only for people who have excessive sleepiness associated with narcolepsy or have disrupted sleep patterns, from switching between shifts at work. In 2006, however, 2.6 million Provigil prescriptions were written, far more than one would think for the relatively rare conditions approved by the FDA. Indeed, more than 80 percent of these prescriptions were reportedly for "off-label" uses—doctors can prescribe pharmaceuticals "off label" for conditions other than those the FDA has approved—such as treating attention deficit disorder and depression. Last year, Cephalon agreed to $425 million in fines and fees to the federal government and several state governments for marketing off-label uses for three drugs, including Provigil.

In the Brain Stimulation Unit, a medical student turns on the juice under the watchful eye of Michael Koenigs, the postdoc running the experiment. I feel a slight tingle and an itch on my scalp as the current rises to 2.5 milliamps: a small amount but enough to give a buzz. A couple of minutes later, I have a metallic taste in my mouth. Koenigs warned me this might happen. Hundreds of people have been safely tested, and this is one of the few side effects they've reported.

In previous experiments on healthy people, Wassermann and others found that this procedure, called transcranial direct current stimulation, improved motor and cognitive performance. In one test, a direct current applied to the left frontal lobe gave a 20 percent boost to a person's ability to name as many words beginning with a certain letter as possible in ninety seconds. Wassermann's team is now testing electric fields with different charges against each other and against a sham, a "control" where subjects get an initial zap and then the machine, unbeknownst to them, is turned off, leaving them to be tested without the electrical stimulation. The team then compares subjects' responses using examinations that measure cognition, memory, and emotions. Direct current applied to the scalp polarizes underlying brain tissue, creating either a positive or a negative charge near the electrode. In vitro (in cells outside of the body) studies have shown that a weak current can substantially change the firing rate of neurons, with an increase or a decrease in the firing rate that depends on the plus or minus charge of the electric field. Evidence suggests that increases in firing enhance local brain function and decreases do the opposite.

Zapping brains is not new. In the 1960s, low-level direct current was used to treat mental disorders, but investigators became more enamored of chemical treatments—until recently, when neuroscientists and clinicians began looking for targeted brain boosters with fewer side effects than pills. Wassermann thinks that one day, we may be able to buy a small device that can be inserted into a hat or attached to a headband and turned on when we need a brain boost.

I feel a slight uptick, like a medium hit of caffeine; it gently lifts the fog of my fatigue, although I don't feel any smarter. I settle down to take some tests of cognition and emotion on a computer. Most telling is a gambling game that presents four virtual decks of facedown cards on a computer screen. When I click on them, cards turn over, and I either win money or lose it, depending on the card. A ticker measures my winnings at the top of the screen. At first the cards seem random, but then patterns develop: I need to figure out which stacks will yield more gains than losses, and vice versa. After a few minutes, my initial mild boost dissipates. I lose at the gambling game, which I find more challenging than the lotteries at NYU, though not by much. The next morning I return for round two of the testing after a good night's sleep, feeling refreshed and awake. Taking the gambling test sans stimulation, I win a modest amount of virtual cash.

Later that second day, I participate in a third experiment. Instead of Koenigs running a negative current through the electrodes attached to my forehead, as he did the first time, he applies a positive current. The effect on my frontal lobe causes a noticeable sense of relaxation and a drop-off in motivation as I play the gambling game. I feel as if I have just taken a relaxing steam bath or a sauna. Oddly enough, I win big anyway. I also experience a strange sensation when I begin speaking to the researchers. I start sentences and then lose my motivation to finish them. Koenigs says this is exactly what his experiment is trying to show: that performance is modulated by different currents. I suspect my results have more to do with yesterday's exhaustion versus today's wakefulness, but the electricity has noticeably messed with my mind, which is the point.

A few days later, I'm in New York City in the office of Steven Lamm, a physician who advocates the prescribing of Provigil for some, but not all, patients with sleep disorders, persistent fatigue, or jet lag. "I would like to prescribe it more than I do," he says, "but because it has only been approved for severe sleep disorders, insurance doesn't cover the cost of the drug for many of my patients." Lamm has used Provigil himself when he is jet-lagged or short on sleep and needs to be sharp. "It is unhealthy to not get enough sleep," he tells me, "but sometimes it can't be helped." Lamm checks my blood pressure and takes a history, tells me about the drug, and scribbles out a prescription for five 200-milligram tabs.

Modafinil, the chemical name of Provigil, has been extensively studied as a treatment for sleep disorders, but data on its capacity for cognitive enhancement is thin. In Cambridge, England, researchers saw a spike in the short-term memory and planning ability of male volunteers who took the drug. Other researchers saw bone-tired subjects who took Provigil stay alert while using helicopter simulators; other tests have confirmed that the drug can improve planning and the ability to remember long strings of numbers.

College students tell me they use Provigil when they are exhausted and need to take exams or to stay awake in class, which is better than popping an Adderall (an amphetamine), the more powerful drug of choice these days for overachieving students who get too little sleep.

I swallow a Provigil at around 2:00 p.m., roughly the same time of day I was first tested in Eric Wassermann's lab. I'm walking down Fifth Avenue in bright spring sunshine and feel nothing. I get a cell phone call and start talking, feeling my usual afternoon dopiness. Later, I board a flight back home to San Francisco, and about three hours after popping the pill, I fall asleep.

In San Francisco, I try Provigil again at 8:00 a.m. the next day, along with my morning cup of French Roast. This time, after fifteen or twenty minutes, I feel an alertness that caffeine alone doesn't give me. The feeling plateaus over the next three hours and resolves into a low-key but constant "up" sensation. I plunge into work and feel highly efficient and bright. For a little while, the sensation is almost too much, as if my brain has been set to fast-forward and can't be turned off.

That morning, while still feeling the effects of the pill, I talk to Jeffry Vaught, the research and development chief at Cephalon. He tells me that the pill is a mild stimulant and does not prevent sleep if people desire it. "For people with narcolepsy," he says, "the impact is not mild; it's life changing." Vaught says the mechanism behind Provigil's effect is not well understood, but scientists know what part of the brain it involves. "It's a pathway involved with wakefulness, with waking you up and keeping you attentive," he explains. "This pathway is activated by Modafinil." Major stimulants such as caffeine and amphetamines act on this part of the brain, too, but they also activate other regions, causing side effects such as jitters, loss of appetite, and that edgy feeling.

As the day wears on and I start writing a story for a magazine, my steady up-ness gets annoying. I'm calm, but I realize that when I write without Provigil, I experience an intricate pattern of short ups interspersed with mild downs, during which I rest my brain. This is quite different from the uniform pharmaceutical lift I'm feeling, which reminds me of what soma, the feel-good drug taken by almost everyone to stay unvaryingly happy in Aldous Huxley's *Brave New World*, might feel like. In Huxley's futuristic society, where everyone is biologically modified and adapted to do exactly what he or she is supposed to do, soma is less of a brain booster than an anesthetic, although the uniformity and constant exposure of this chemical make everyone blithely happy, even if their lives are dull and predictable. I'm not saying that Provigil or TMS—or caffeine—is leading us down the slippery slope of uniform enhancements or mass anesthesia, although, as we understand the brain better and learn more about chemicals that alter it with more precision, such things will become possible.

In fact, I wrote the final section of this chapter after swallowing my final Provigil prescribed by Steve Lamm. You decide—can you tell any difference?

Meta-neuroscience and the elusive whole

As with my tests for gene markers and environmental levels of toxins, I could continue to take tests delving into my brain for years to come and not cover all of them. Neuroscientists are inserting thousands of people's heads into MRIs and hooking them up to EEGs and brain-stimulating electrodes, with hundreds of experiments being run on everything from what our brains look like when we are watching a soap opera or the news to what happens when we are in love. (I wanted to take the latter tests but, alas, was unable to find a researcher currently running scans.) The tests I did take provide a wide range of examples of ongoing research in scanning and analyzing the human brain. I'm not sure, however, how much the testing enlightened me about myself. I am pleased to know that I have no trace of diseases that researchers can identify using scans. As for the rest, I got what researchers promised with this young technology when they told me that it is not yet ready to test individual brains: a number of intriguing images and mostly impressionistic interpretations.

"One problem is that these studies tend to be one-offs," says Judy Illes, noting that there are few follow-up tests or attempts to replicate initial data and to test larger populations, although she expects this to change. "Neuroscience right now needs a meta-approach linking all of this together," she adds, perhaps the neuro-equivalent of the Human Genome Project that sets out to create not only a vast map of the brain, but a schematic of pathways and how different regions connect. Of course, this effort would also need to tie in genetics and the affect of pollutants such as mercury and other environmental influences, although I hate to imagine what this would be called—the Human Neuro-Envirogenomic Project?

At the University of California in Los Angeles, a project led by neurologists John Mazziotta and Arthur Toga is attempting to create what Illes is suggesting: a comprehensive brain atlas that they say will provide a template for what is now known about what brains look like, how they vary, and how they function. Collaborating with researchers from Canada, Europe, and Japan and from UC San Francisco and the University of Texas in the United States, the brain atlas team has scanned 450 "normal" brains and used hundreds of thousands of images taken of seven thousand people around the world to compile color 3-D maps that they say will be able to show everything from relative sizes of anatomical features to differences in brains based on age, race, gender, educational background, genetic composition, and other distinguishing characteristics. Mazziotta and Toga also include cadaver brains, cut into twenty-five hundred microscopically thin slices and mounted on glass slides, where they are stained and digitally photographed. The slides provide information at a much closer range than the scans of living brains do, since MRI scans can resolve down to only 1.5 millimeters. Each slice, called a cryosection, is 60 microns thick: about half the thickness of a human hair. The finished atlas, says Mazziotta, should indeed serve a similar purpose as the Human Genome Project does for geneticists by providing a detailed framework of the brain that researchers can use to perform experiments.

Layered over the anatomical maps will be brain functions such as memory, emotion, language, and speech and how they manifest as complex interactions involving multiple regioins and circuits that shift and change as people experience thoughts and feelings. "The way in which we experience a feeling or recollect a memory is a process that involves a complex circuit that is changing even as we're becoming aware of that sensation," says Toga. "When you scan a subject, all you've got is a picture of that moment in time. The next minute the picture changes."

"You can't just point to an area and say, 'Here's the seat of language,'" says Mazziotta. "For example, the brain handles the challenge of thinking of and initiating a word, and of understanding that word, differently. Execution of these tasks involves complex circuitry throughout the brain. It's incredibly painstaking work."

"What we find out about the brain isn't going to answer all of those ancient and philosophical questions about the nature of the human mind," says Toga. "Still, our work on this project is a way in which we can try to understand complex, hard-to-touch concepts. It is a way to help understand all of those odd combinations of functions that give us our experiential life."

I ask Mazziotta to test my brain by running me through the paces of his "normal's"—assuming I am normal. He tries to arrange it but is unable to clear the project through his Internal Review Board (IRB) because it has approved this study with all results being blind or anonymous, so neither the investigators nor the subjects—nor readers of this book—can know the individual results. I almost participate anyway, but my brain, newly adapted to want to avoid MRIs if possible, strongly objects.

This leaves me at the end of my brain scans pondering not just my own results but also the big picture of what this flurry of testing means. As Paul Glimcher and others have suggested, all of this neuro-activity may one day launch us into a unified theory or perhaps, as E. O. Wilson might say, a consilience of human behavior that includes the work of not only neuroscientists, economists, psychologists, and psychiatrists but also sociologists, anthropologists, and other "ists." I assume that this august grouping will also connect with physicians, geneticists, and environmental scientists for one heckuva Human Everything-You-Ever-Wanted-to-Know Project.

"The idea of the unity of science is not idle," wrote Wilson in *Consilience*, a book devoted to the search for unity among science, religion, and the humanities—and for a few basic natural laws that he believes we are on the verge of unlocking as we link various disciplines together. Wilson admits to having an outsize reverence for the Enlightenment era, which revived in the West an ancient Greek longing for an elegantly simple set of rules that guides the universe. I share this desire, although there remains an immense gap that needs to be traversed before such rules are discovered, much less proved that they are, in fact, both universal and as simple as latter-day denizens of the Enlightenment would like.

From a practical standpoint, my brain, for what it's worth, suspects that science will continue to discover not a simple, clear-cut whole, but a universe that is vastly more complex than we realize. Although there are common laws that great thinkers have already theorized govern most of physics and biology, including relativity and natural selection, the scientific quest continues to yield exceptions to these theories—such as the uncertainty principle—that suggest more, not fewer, complications.

This hardly means that we should abandon the attempt to understand the whole: to link the complexities of the parts to come up with theories and grand designs like the brain atlas and the Human Genome Project. For at least a century, scientists have been parsing, slicing, and dicing the whole into tiny, reductionist pieces. For the human body, this has produced vast troves of data and knowledge about details of genes, organs, cells, pathways, systems, environmental input, and the functions of the brain. This exploration is not even close to being finished. Yet perhaps there is enough material to begin to assemble at least the outline of a total person: to stitch together a model of what we know so far to see what the Experimental Man looks like. I will not pretend that the jigsaw pieces I have gathered will equal a whole or that I can come close in my snapshot of myself to the real person that is me—or, if you took the same tests, you. But it is time, after all of my tests, to make an attempt to see what emerges.

4

BODY

The only true voyage of discovery, the
only really rejuvenating experience,
would be not to visit strange lands but to
possess other eyes, to see the universe
through the eyes of another, of a hundred
others, to see the hundred universes that
each of them sees, that each of them is.

—MARCEL PROUST

Prediction: Heart attack in 2017?

This wasn't supposed to happen.

In a Silicon Valley conference room, I'm being told by a team of biocomputer profilers that if I put on a pound a year—the average weight gain for a man over forty—I have a 60 percent chance of having a heart attack by 2017. I'm studying the line on a PowerPoint chart on the wall that is displaying my future risk according to a sophisticated algorithm that has integrated dozens of my personal risk factors, including genes and environmental inputs. I can see that by 2023, the forecast is even more grim: an astonishing, and frightening, *100 percent* risk.

"So I'll have a heart attack for sure," I say, pausing to absorb this, "when I'm sixty-six?"

"If you follow that trajectory and gain weight, and our model is right," says Tom Paterson, the chief innovation officer and a cofounder of Entelos, Inc. Launched in 1996, Paterson's company develops computer simulations and models for pharmaceutical companies that want to predict the outcomes of drug candidates in humans before and during clinical trials. A fit, compact man with short-cropped hair and wire-rimmed glasses, Paterson purses his lips and smiles anxiously, obviously uncomfortable delivering this news.

I am uncharacteristically speechless, not knowing what to say.

The outcome is even worse if I gain five pounds a year, says Paterson. I will hit the 100 percent risk factor much earlier than age sixty-six. If I become a porker, I'll reach 100 percent in just six years, when I'm fifty-six years old.

The Entelos prognosis is dramatically different from what my internist, Josh Adler, told me: that my heart attack risk is a mere 4 percent in ten years. His number comes from plugging cholesterol levels, weight, age, and other basic information into a formula that medical researchers have developed out of a project called the "Framingham Heart Study." Launched in 1948 in Framingham,

Massachusetts, the study has followed thousands of subjects and their families for six decades in Framingham and elsewhere and has developed models that give heart disease risk factors for individual patients like me based on comparisons with people in the study who closely resemble my profile. Most physicians consider a Framingham risk assessment to be the standard of care, although for me, if the Entelos model is even close to accurate, Framingham is way off.

"I didn't expect this," I finally blurt out.

"Neither did we," says Paterson.

"We thought you would be normal and, frankly, boring, which for a patient is what you want to be," adds Alex Bangs, chief technology officer and cofounder of Entelos. He also looks uneasy delivering this information.

I feel uneasy hearing it: that my first Experimental Man attempt to link multiple tests into a whole picture of a major bodily system have come out like this. If they are right—and they are insisting that the model needs to be refined and further tested—then at least a portion of the fancy tests that I have taken for this book, including genetic tests, are actually useful in informing me of a profound cardiac risk factor that I otherwise would not have known about. Looking at that steep upward curve on the PowerPoint slide is disconcerting, although it's good that I found out about this before having a heart attack, while I still have time to do something about it by adjusting my diet or even taking cholesterol-lowering statins, which Paterson's slide show says would greatly reduce my risk. The next slide presents better news—a graph with a much flatter risk-factor line that tells me if I keep my weight stable, my risk of having a heart attack drops to about 23 percent in ten years. In twenty years, the risk rises to 40 percent. That's still not great, but I'll take it over the other options.

"Guess I'm going on a diet," I say with a smile, trying to exude a calm I don't entirely feel, while promising myself to stay forever slim. In recent years I have watched my weight creep up slightly, from being too skinny at age forty, a slow rise that I vow will now stop.

To develop its predictive model, the Entelos team incorporated dozens of chemical markers from my blood and also CT scans, ultrasounds, and genetic data, including the lead SNP on chromosome 9 that deCode's Kari Stefansson had told me to take seriously, designated as rs10757278. This mass of data was plugged into a computer model that the company has developed from more than a decade of working with pharmaceutical clients such as Pfizer, Johnson & Johnson, and Eli Lilly to simulate how proposed drugs will perform in the cardiovascular systems of test animals and in humans. Entelos researchers have worked in tandem with teams conducting clinical trials on thousands of actual patients, endeavoring to check and refine models that incorporate massive data dumps and analysis for several diseases: heart attack, anemia, type II diabetes, rheumatoid arthritis, asthma, and drug-induced liver damage. The bodily systems they profile include metabolism, cardiovascular, immunological, inflammation, and respiration. They are testing me on cardio only.

The model that Entelos has developed for me is a new version that shifts its algorithms from those used by drug company scientists studying groups of patients to one that Entelos hopes will be useful for individuals. Within a couple of years, Entelos would like to sell this to physicians and consumers as a much more powerful forecasting tool than Framingham and other traditional tests. Entelos plans to price this "PhysioLab personal health simulator" at less than $1,000. This would include the cost of the tests, including the carotid ultrasound. This sounds expensive, though for older people with increased risks of heart disease, if the test proves to be accurate, the cost might be worth it if costly and fatal heart attacks are prevented. Entelos may proffer less detailed versions of the PhysioLab model, with fewer tests taken, for a lower price. "We're still working out the business details," says Alex Bangs. They spent about $50,000 developing my model, he says, on top of millions of dollars invested in their models that were built for Big Pharma.

Paterson and Bangs are engineers and bioinformatics experts who began their careers creating computer models for the aeronautics industry and for the Ronald Reagan–era "Star Wars" program that attempted to build an antiballistic missile shield to defend the United States, with limited success. "We worked on a set of computer simulations that would allow battlefield commanders to understand how to integrate large amounts of information on an incoming ICBM [intercontinental ballistic missile] threat," says Paterson, "and to help them make decisions about what to shoot at them. We got a taste for what it's like to develop very large-scale information systems and simulations to support decision makers in integrating information and evaluating alternatives."

I ask him how he went from missile defenses and planes to human physiology.

"Turns out there are real parallels between aerospace vehicles and people," he says. "Aircraft are self-contained, to some extent they're homeostatic, and you can't change one piece without affecting the other pieces."

The Entelos experiment began with—what else?—more of my blood. This time, my plasma was used to test levels of several dozen naturally occurring chemicals, ranging from triglycerides to plasma glucose and apolipoprotein A-1. I was also tested for a panel of cholesterol tests that only recently have been used by some cardiologists to get considerably more detail than the standard levels of cholesterol: the "total score" and amounts of "bad" cholesterol (LDL) and "good" cholesterol (HDL). My much more detailed analysis also determined levels of different-size particles of cholesterol, including large, small, and intermediate particles of LDL and HDL. Cholesterol, it turns out, is not a static molecule but works in a dynamic system. Created in the liver, cholesterol is part of a package of lipoproteins that are constantly being distributed around the body, delivering fuel in the form of triglycerides (fat) to provide energy to cells from the top of our brain literally to the tips of our toes. As the "packages" dispense their goods, they shrink from large to medium to small. When they become small enough, they are usually taken back up into the liver to be reprocessed into new packages. The numbers of each particle size in the process circulating in our blood at any given time indicate whether too much, too little, or just enough lipoproteins are present. Too much LDL (bad cholesterol) is unhealthy and can cause, among other things, a buildup of plaque in blood vessels. Too little HDL (good cholesterol) is also not great, since these good lipoproteins act to suck up the bad lipids like vacuum cleaners and clean them out of our systems.

My results for almost fifty blood chemistry tests taken for the Entelos test were nearly all safely in the column labeled "In Range" and color-coded a comforting "green."* Four outcome levels were "Out of Range" (see the table on page 259) and color-coded red, although one of these, HDL, is good news, since this is a level one wants to be higher. The other three levels were just over the top of the safe range: total cholesterol, LDL, and an LDL-like molecule called lipoprotein (a) that can build up and also gum up arteries. I am not happy with these three scores—especially since five years earlier, my total cholesterol level was 30 points less—although many people have much higher levels.

* The full results and a description of the Entelos tests are at www.experimentalman.com.

The Author's Key Cardiac Chemistry Results: Higher Than Normal Levels (Out of Range)		
Test	Author's Results	Normal Range
Total cholesterol	220	<200
LDL	134	<130
HDL	65	>40
Lipoprotein (a)	91	<75

My next stop in this heart odyssey is the San Mateo Medical Center south of San Francisco, where I find myself lying back on a gurney with goop lathered on my neck. Just below my left jaw a nurse rubbed against my neck an ultrasound wand, a device the size of an electric toothbrush handle with a cord attached to a computer. The small room was darkened, and I could see a wavy gray image of the inside of my neck on a computer monitor. High-frequency sound waves were showing an image of my tissue and muscle and my carotid artery: a major vessel running directly from my heart to supply blood and oxygen to my brain and, through a separate branch, to my face. The carotids—one is on each side—were scanned for clues as to what was going on in the big arteries of the heart, which are less convenient to scan with an ultrasound. These devices are cheaper to use than a CT scanner and throw off no radiation. Doctors looked for bumps of cholesterol where the carotid splits into two. The nurse showed me on the screen this point of bifurcation in my neck, and I was surprised to see a little gray-white bump, like a very small mountain. "There is some build-up here," said the nurse, who wouldn't comment on what it meant. "That's something the doctor will have to tell you about."

I watched the monitor as she snapped still images of my carotid and that little mound on the computer, and I had one of those rare sensations in this project that this hazy gray image was actually part of me, and that dark substance that seemed to be flowing in there was my blood. More astonishing is that little spur of plaque, sitting there like a nasty parasite. Did this come from one too many burgers in an otherwise healthy diet? At least, I think it's healthy. Seeing that blob left me feeling annoyed at myself, even though later I will find out that my little spur is well within the normal range for my age.

The final Entelos test takes me across the bay to the other side of the Berkeley Hills, to Walnut Creek, a bedroom community where one of those stand-alone CT scanning companies set up shop a few years back to cash in on a sudden demand by the curious—and the well heeled, given the cost—who want to know what might be going on inside their hearts, lungs, abdomens, and other body parts. Computed tomography images were perfected in the mid-1970s, in a marriage between traditional X-rays and computers that allows detailed slices of a person's chest or head to be assembled by programs into 3-D images. We'll get into more detail about CT scans later; for now, you only need to know that I will spend thirty minutes or so inside another steel doughnut, although this time it is okay to sleep, and it is much quieter. They take scans from my neck to my thighs, and after it is over, they are very nice when they wake me up.

The final ingredient for this heart attack investigation is my gene marker results from deCode. When it is added to the mix, Paterson's team goes to work, six people spending hundreds of man-hours to complete a custom analysis for me that was once reserved only for pharmaceutical companies designing drug trials for groups of people.

In the Entelos conference room, Tom Paterson had started his presentation where he delivered my dire forecast with a curious touch: a clip from *Gattaca* that highlights the potential power of genetics and how things might go wrong. I have since spent time with Paterson and Bangs and know they have a sense of humor and are well aware of the potential social impact of having accurate predictive models should anyone want to abuse it as society has in *Gattaca*, but, still, highlighting this dystopic possibility seems odd for a company that's trying to sell a computer-driven crystal ball about people's health that could one day be the engine for the world depicted in the film—although I don't think this will happen.

Paterson explains how his team parsed my data to devise their computer prediction, which he likens to running simulations on complicated new products—aircraft, cars, and factories—before they are actually manufactured. I let it go that my body is being compared to a product, although he makes a case that the same sensibility and algorithms that go into profiling the fuel flow and distribution in a proposed jet engine can and should be used to assess the future performance of something more near and dear: an individual human machine.

The key, he says, is comparing me to data from thousands of participants from large-scale heart studies, including a detailed, long-term study of thousands of people called the Atherosclerosis Risk in Communities Study. ARIC tracks factors such as total cholesterol, HDL, blood pressure, and smoking status as potential contributors to heart attack.[*] The Entelos engineers integrate the ARIC patients and other patients—real and virtual—to create thousands of simulations projecting forward in time that can forecast what might happen based on different scenarios (cholesterol goes up or down; a patient gains weight). The modelers then take this universe of virtual patients and look for a cluster of simulations that have blood chemistry, biomarkers, ultrasound, CT scans, and genetic profiles most closely resembling mine. From all of these possibilities, they create a "Virtual David," as they call it—a "me" on whom they can virtually experiment by running scenarios into the future and assigning probabilities to different outcomes, such as what happens if I have a lousy diet or if I take statins. The Entelos Web site describes the Cardiovascular PhysioLab computer model:

A PhysioLab model is like a flight simulator but for the body. It includes the key physiology involved in a disease and lets scientists try "what if" questions to help understand how disease processes may be unfolding within a patient and how best to treat them. The Cardiovascular PhysioLab model incorporates the physiology involved with cardiovascular disease, including cholesterol metabolism, inflammation, atherosclerotic plaque formation, and how plaque rupture can lead to a heart attack. This model includes thousands of virtual people, each a little bit different in their underlying physiology, in what we call a virtual population.

Paterson uses yet another analogy, saying that the virtual people living in his computers can be like crash-test dummies to test medications. They can be slammed repeatedly against brick walls to test various drugs and doses without anyone getting hurt.

Regaining his composure after his startling announcement about my potential demise, Paterson continues his slide show and concludes that the "Virtual David" in the model—me, more or less—is at high

---

[*] To take an online version of the ARIC test and to find out your heart attack risk over the next ten years, go to http://aricnews.net/riskcalc/html/RC1.html.

risk for having a heart attack because he has an LDL cholesterol level that is unusually sensitive to weight gain. This is because I appear to have an unusual situation with my liver, which does not do a good job of reuptaking small particles of LDL cholesterol once the "package" of lipids has delivered its fuel to the body's cells. According to the team's analysis, if I gain weight, which would cause my LDL levels to rise, plaque will increase rapidly in my cardiac blood vessels. This increases inflammation and could trigger an eventual heart attack.

The other big factor affecting the model, he says, is the deCode genetic marker that Kari Stefansson had called me about from Reykjavik. The modelers used data from studies about the marker and about the two genes that this SNP is close to on chromosome 9—if you remember, this SNP sits on a string of so-called junk DNA outside of an actual gene, but it is close to the CDK2A and CDK2B genes. These genes work to suppress the formation of cancerous tumors but also may play a role in making build-ups of plaque unstable if lipids and inflammation spike upward. When this occurs, it creates conditions that can lead to aneurysms in critical cardiac blood vessels, and, eventually, a heart attack. Paterson cautions, though, that this is just a hypothesis. "We don't really know what this marker does," says Paterson. "Our hypothesis is purely correlative, since no mechanism is known for why this SNP causes heart attack."

"But in your model, this gene marker had a big impact?" I ask.

"Yes, it did," confirms Paterson, who shows me slides of my risk factor if I had been born with the medium- and low-risk variants of this gene, rather than the high risk. Having the lower-risk variant would have lessened my probability of having my ticker stop in the next ten years by more than 20 percent.

Paterson has that pained look again while delivering more unpleasant news, and I take in a breath.

"In effect, you have a double whammy," says Bangs, referring to my problem with uptaking cholesterol into my liver *and* my bum SNP variation.

"But did you factor in family history?" I ask, trying to remain calm by falling back on my old standby of having a mostly heart-healthy family. So far, this history has allowed me to shrug off dire heart attack warnings from Stefansson and others.

"We did," he says.

I gulp, noticing for the first time that in the Entelos conference room, they have laid out next to thermoses of coffee a spread of bagels and cream cheese and doughnuts. I love them, but as we take a break, I suggest that maybe next time I visit, they should put out fruit and nuts, since I now know that eating bagels and doughnuts could kill me. Paterson laughs nervously, and I smile to show him that I'm kidding—sort of. Bangs jumps in and says, "No more doughnuts for you!"

Except for my MRI scans for anxiety, up until this point in the Experimental Man project, I mostly have been the reporter who happened to be covering his own body that was being tested every which way. In a sense, it was an out-of-body experience that only lightly touched my core sense of self and my conviction that I am healthy. But as I leave Entelos and ease my car onto the highway, heading back to San Francisco, I find myself taking these former Star Wars computer modelers seriously, given that most of the tests they have run are traditional ones that have been used in millions of patients—plus a genetic marker that has gotten more validation than most.

I also ponder the diet that I've been more or less following for almost five years: a super-high-protein regimen of mostly meat, nuts, and green vegetables and almost no carbohydrates. A trainer at my gym had suggested it, as well as some books to read about it, including the one written by famed diet doctor Robert Atkins. Not being the diet book type, however, I didn't read the recommended tomes, but I did in a casual way alter my diet based loosely on what the trainer told me—and promptly lost a few pounds. So I kept at it, disdaining most breads, potatoes, desserts, and processed foods. Otherwise, to get protein, I ate meat whenever possible, plus the occasional high-protein bar. It was not an entirely comfortable diet, because I quickly got hungry between meals, probably because my body had no carbs to sustain me even for a short period of time. This led me to eat smaller and more frequent meat-fests and protein bars. But the weight stayed off, so I kept at the diet until it became part of my routine.

Possibly, researchers in Paul Glimcher's lab could provide a neuroeconomic analysis explaining why my brain persisted with a diet that was not entirely rational, one that made me frequently ravenous as I wondered whether all of that fat from the meat was pushing up my cholesterol: a major criticism of the controversial Atkins diet, which encourages low carbs and a higher intake of fat and protein. But I didn't follow the diet religiously. There were occasional lapses with sugar-glazed crullers, pretzels, peanut-butter cookies, and pepperoni pizzas.

Assuming that this diet caused my cholesterol to spike—and now knowing that cholesterol is more of a mortal enemy to me than most others—I waste no time in launching a new diet. I dramatically reduce meat consumption and add in granola, wheat bread, pasta, and other seemingly healthy carbs, changes that I greet with some relief since they eliminate that hungry edge. I continue to exercise for an hour five or six times a week and soon drop about five pounds.

As I am implementing the new diet, the Entelos team asks me to return to Foster City to hear about revisions and corrections of their original results. Back in the conference room, which is now laid out with healthy foods, Tom Paterson and Alex Bangs explain that they have refined their analysis. "You're going to like this better," says Paterson.

He fires up the PowerPoint and shows me an amended risk-curve on their chart, which has dropped to a 28 percent risk factor for heart attack in the next ten years if I gain a pound a year and a 70 percent risk factor over the next twenty years (see the graph on page 265). This is still not great, although the most dramatic news is that the risk line for me if my weight stays stable has dropped down to about a 2 or 3 percent risk factor in ten years, even lower than my Framingham score.

This is a relief, although the core message hasn't changed: if I gain even a little bit of weight, I have a significant chance of having a heart attack.

Paterson apologizes for the earlier, scarier findings, explaining again that this first-time-ever experiment is a work in progress. I remind him that he told me my results would be preliminary, though he and Bangs now believe this is a much more accurate profile that probably won't change a great deal even with further refinement.

I ask what's changed in this version of the model, and Paterson explains that several factors in the model needed to be refined and corrected, including the size of the virtual population they had compared me to. "The population size in the last meeting was too scattered and small," he says. "Last time we said that a difference in stable body weight was huge. Since then, we've created a denser sampling [of virtual patients]." The team also had more closely correlated this population and my profile with real patients. Before, he says, they

Entelos Heart Attack Model:
The Author's Revised Results

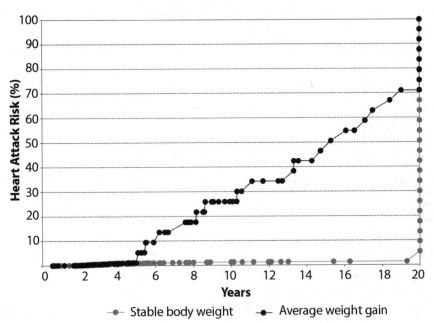

The black line with circles shows a risk factor of 28 percent in ten years and a 70 percent risk factor in twenty years if the author gains one pound a year. The risk if he gains no weight is only about 2 percent.

had found 53 virtual patient profiles, out of thousands of possibilities, that fit closest to me. In the revision, they found 420 virtual and real patients out of tens of thousands of possibilities. "With more information, we tighten up the variables that could be you."

He says that the modelers also didn't factor in the role of the ultrasound results and my lipid levels as much as they should have. "You have good plaque on your ultrasound, so this brings down the risk line," he says. On the other hand, I'm a "hyper-responder" to lipids, meaning that my cholesterol level has this unusual linkage to my weight. Too many cheeseburgers and potato chips will drive my cholesterol up higher and faster than the average person's, hypothetically setting in motion the "double whammy" that can quickly lead to a heart attack.

Paterson ends the presentation by offering a prediction: that if I take another ultrasound test in six months, there is a 90 percent chance that my plaque will rise—that annoying little bump of cholesterol sitting in my neck—although the growth will be very slow. He presents a slide with the two new possible futures: the weight-gain scenario, where my risk flies high, and the one where I don't gain weight, causing the line to rise only slightly.

This time I leave Entelos with an extra banana and a reinforced mission to eat super healthy and to avoid the evil of cholesterol at all costs. Paterson also invites me to join them a few days later in Oakland, where they will present my findings to a well-known cardiologist and heart researcher, Ron Krauss, the director of atherosclerosis research at Children's Hospital Oakland Research Institute.

Krauss's lab is studying how genes and metabolism affect a condition that could be mine if I don't take care of myself, called atherogenic dyslipidemia, where a patient has high triglycerides, low HDLs, and increases in small LDL particles. The Krauss team has shown in studies that patients who go on a low-fat, higher-carb diet can actually trigger a worse lipid profile, which means that cutting out the meat and replacing it with, say, lots of pasta is not necessarily a good idea, especially for someone with my double-whammy profile. His lab and others have also discovered several candidate genetic markers that seem to influence how a person's lipid profile responds to changes in diet. "These results have proven that genetic mechanisms lead to variation in LDL," reports Krauss's Web site.

Reading Krauss's studies and those of other researchers who are trying to untangle the genetics of the lipid system leaves one with a healthy respect for the complexity of this critical bodily system. Dozens and probably hundreds of genes and gene markers are likely to be interacting, with scientists just beginning to sort out a basic outline describing how the pathways work. I peruse the papers, looking to see whether it is possible to assemble a meaningful panel of my own SNPs from those cited in the literature, and I conclude that the data are too preliminary. Yet I do discover a few tantalizing examples of how variations in lipid genes may one day provide people with considerable details about how their personal DNA profile might affect their choice of food: possibly more information than most people will want. Integrating these multiple genetic variables will pose a significant challenge to computer modelers like those at Entelos, who for my assessment included only one gene: the deCode heart attack marker. Imagine trying to integrate dozens or hundreds of genes, along with dozens of other risk factors.

One glimpse into the possibilities offered by lipid genetics is a variation in the APOA5 gene, believed to be important in the metabolism of triglycerides. A SNP variation in this gene seems to allow one to eat a high-fat diet without gaining weight. Wouldn't this be nice! Unfortunately, I am AA—normal—so no gobs of fat for me. Another SNP variation on the APOA5 gene increases the risk of a person experiencing a rise in triglycerides. For this one, I'm CC: also normal, which is good. Other markers in the APOA5 gene are also associated with increased small LDL particles and with changes in lipid patterns based on diet that can impact weight gain and heart attack risk.

"Collectively, these findings have indicated that an understanding of specific interactions of genes with dietary carbohydrate intake may lead to more targeted interventions in susceptible individuals," says Krauss's site. "They also suggest that both reduced carbohydrate intake and weight loss operate to improve dyslipidemia."

A research team combining scientists from Harvard and labs in Sweden and Finland conducted a study that I also am able to apply my genetic results to, called "Polymorphisms Associated with Cholesterol and Risk of Cardiovascular Events." This study tested more than five thousand people and found nine SNPs that collectively convey a risk factor for heart disease based on genetic variations for LDL and HDL production. The researchers came up with a scoring method for determining risk factors. My score is 11, which confers a risk 1.63 times higher than normal that I will have a "cardiovascular event" such as a heart attack (see the table on page 268). This risk factor is similar to the 1.74 times normal originally given to me by Kari Stefansson for carrying that SNP on chromosome 9.[*]

This is another score and risk factor that I'm not sure what to do with, since no one has integrated it with any of my other tests or risk factors. Entelos has not factored it in, making it yet another random, one-off risk factor that probably isn't high enough to really worry about.

We meet Krauss in Oakland in a California Colonial–style stucco building that must have once been a school. I chat with Krauss before the meeting, and it's apparent that he's skeptical of the Entelos model before having seen the details. He says that new models and tests and highly sophisticated measurements can be interesting to a researcher

---

[*] Check the section titled "I'm doomed. Or not."

The Author's Results for Genetic Variations Associated with a Risk for High Cholesterol

SNP	Gene	Author's Results	Score
LDL Cholesterol			
rs693	APOB	CC	0
rs4420638	APOE cluster	AG	1
rs12654264	HMGCR	TA	1
rs1529729	LDLR	GG	2
rs11591147	PCSK9	GG	2
Total			**6**
HDL Cholesterol			
rs3890182	ABCA1	GG	2
rs1800775	CETP	AA	0
rs1800588	LIPC	CT	1
rs328	LPL	CC	2
Total			**5**
Grand Total			**11**

Scoring: low risk (homozygote) = 0; medium risk (heterozygote) = 1; high risk (homozygote) = 2.

but tend to add little to his diagnosis or treatment of a real patient. "When you know about a major variable such as family history or disease in the heart, that's what's important," he says. "It's hard to show that adjusting one small variable really changes much in terms of a diagnosis." He also says that genetic association studies aren't yet clinically useful. "How much do we really know?" he asks, and he is in a position to know. "Not much. In a clinic we have to make judgments in real time that are based on what we know to be true."

On the other hand, he talks about a few years ago when cardiologists were highly skeptical of using the cholesterol measurement as a risk factor for heart attack. "There was a great deal of data, but clinicians resisted this until the proof became overwhelming. That's what's going to happen with genes. They will need to show not just a reduction in risk, but also in improving the disease. This will happen, but the transition is just beginning."

I show him my lipid levels, and he thinks they look pretty good, although he tells me he can't say much more without giving me a thorough checkup. We then settle back and listen to Tom Paterson's presentation. Afterward, the lights come up, and Ron Krauss smiles.

"Well, that was interesting," he says, plunging into several technical questions.

When he finishes, I ask him, "Well, doc, should I take this seriously?"

"I find the presentation impressive," he says, "more than I thought I would. I need to think about it." Later, he tells me he is particularly intrigued by how the model was able to incorporate the risk factor of the deCode SNP. "That could be useful down the road for some of the work we're doing in my research," he says.

I mention my new diet, and he says again that he is reluctant to comment without knowing me a bit better as a patient. Of course, he adds, in general a low-fat diet with reasonable carbs is healthy for most people.

Paterson says that the Entelos team is planning to test me again in a few months to see whether the model's predictions are accurate and whether my new diet will change my results when we retake the tests.

That very night, however, my resolve to eat a better diet is severely challenged when I'm out with my kids, and we walk past a doughnut shop. Of course, they want to go inside. Not that we do this often. It's an occasional special treat, but not tonight! I suggest that we keep on walking. "Come on!" they say. I relent, but I refuse to buy one for myself. Yet I can't help it—I take bites from their doughnuts ("Dad, come on, get your own!")—which makes me wonder whether I can really keep up a super-healthy diet when I feel fine, and I'm at worst just a tad bit over my ideal weight.

To answer this question and perhaps bring some rationality to my diet, I decide that it's time to bring in an expert in this field: someone who can assess the current state of my nutrition, and who might be able to help me understand this intricate swirl of genes, lipids, ultrasounds, and metabolic proclivities.

Raging lipids

I'm not sure what I expected when I decided to visit nutritionist and physician Melina Jampolis. Within minutes of my arrival, though, she is trying to convince me not only to eat plain, nonfat yogurt but also to give up munching cheddar cheese on the quesadillas I typically eat once a week for dinner. Sitting in her exam room in the Presidio Heights area of San Francisco, I protest that at least I eat quesadillas wrapped in *spinach* tortillas, which are supposed to be healthier than corn, but she is firm: spinach tortillas are fine (they are actually no better or worse than corn), she says, but cheese is the issue here. It is loaded with saturated fat and cholesterol!

When I made the appointment with Jampolis, I thought that a visit to a nutrition expert would be a nice amalgam of this book's four sections—genes, environment, brain, and body—by investigating an appropriate diet (which comes from my environment) in the wake of my genetic findings for cholesterol and heart attack and how all of this affects my body. My brain would weigh in by offering its usual back-and-forth argument between the frontal lobes, which would recall the rather frightening Entelos tables of my heart attack risk, and the lower-brain regions that crave cheddar cheese and doughnuts.

But now, as Jampolis launches into the merits of nonfat versus regular cottage cheese, neither of which I like, I find that snappy, unruly part of my brain saying, "You have got to be kidding."

Jampolis is an attractive blond woman who goes by "Dr. Melina" in regular appearances on FitTV. In the office, her face smiles on boxes of her protein bars (Dr. Melina Protein Bars) and on the cover of a recent book on dieting and nutrition. But Melina Jampolis is also a serious physician who counsels overweight and obese people, with a few celebrities thrown in whom she can't mention by name.

Jampolis is part of a growing legion of real and self-proclaimed specialists who are trying to make sense of possibly the most basic need of our bodies: to eat. (Even reproduction and sex can't happen without a steady supply of fuel.) We humans have learned to eat a vast range of foods to fuel bodies that are designed for a hunter-gatherer existence.

Finding food was a precarious challenge for my mitochondrial and Y-chromosome ancestors who moved out of Africa and eventually to Europe. This is why our bodies evolved to eat virtually anything edible; to be able to gorge on meat, grain, and berries; and to store the excess nutrients as fat, since our bodies did not know when the next meal would come. Pigging out was rare for our ancestors, so the mechanisms that processed the sudden influx of fats and sugars went into overdrive for only brief periods of time. Our ancestors also learned to love and even yearn for a wide range of foods, and we have the taste buds and the capacity to sniff out a meal to prove it. We especially home in on the sweetest things, which, for our ancestors, meant that a piece of fruit or a vegetable was safe to eat.

Unfortunately, with about four billion people on our planet now having plenty to eat and perhaps two billion of these having far more than enough, we have had to cope with the downside to our ancestral hunger and our capacity to gorge: heart disease, obesity, diabetes, and the like. (With six billion total people, this still leaves two billion who go hungry.) The rise of nutrition-related maladies has led to vast resources being expended by scientists and policy makers to try to understand nutrition and to suggest how people might make the right choices in what they eat. (For policy makers, establishing ever-shifting food group pyramids has sometimes been influenced as much by food industry priorities as by good nutrition, but that's another story.)

My friend and colleague, the writer Michael Pollan, has popularized the notion of "nutritionism" in his lyrical and informative books, starting with *The Botany of Desire* and continuing with *The Omnivore's Dilemma* and *In Defense of Food: An Eater's Manifesto*. His advice, summed up in a *New York Times Magazine* story: "Eat food. Not too much. Mostly plants." He continues:

> That, more or less, is the short answer to the supposedly incredibly complicated and confusing question of what we humans should eat in order to be maximally healthy. I hate to give away the game right here at the beginning of a long essay, and I confess that I'm tempted to complicate matters. . . . I'll try to resist but will go ahead and add a couple more details to flesh out the advice. Like: A little meat won't kill you, though it's better

approached as a side dish than as a main. And you're much better off eating whole fresh foods than processed food products. That's what I mean by the recommendation to eat "food." Once, food was all you could eat, but today there are lots of other edible foodlike substances in the supermarket. These novel products of food science often come in packages festooned with health claims, which brings me to a related rule of thumb: if you're concerned about your health, you should probably avoid food products that make health claims. Why? Because a health claim on a food product is a good indication that it's not really food, and food is what you want to eat.

For most of my life, I have largely ignored the food debate, preferring to concentrate on what might be called a loosely intuitive routine of eating what seems healthy, with the sporadic lapse. Long ago, I stopped eating fast food and obvious processed foods, after a critical moment when I got terribly sick from food poisoning after eating a fast-food burger near Palm Springs, California. But I have to admit to being driven as much by a fear of getting fat and ugly as I grow older, as being motivated by health concerns. In addition, my body, which once was able to devour anything in any quantity and stay trim, has grown less cooperative. Starting in my twenties, when I was far too skinny, I have put on a few pounds over the last thirty years, which is pretty typical and has gotten me to a weight that fluctuates between average and a few pounds over average for my age and height. Nonetheless, a slight panic that I would become a porker at age forty-three or so was what drove me to try the Atkins-style diet, although my brain played a little game that said, "You are worried about getting fat, but you don't really want to admit this, so you won't read any diet books or acknowledge that you really care about such things."

Proceeding, then, without consulting books or experts to know what I was doing, I ate meat at every meal, and my cholesterol rose, until I got the news from Entelos that something was out of whack with my body. This situation had not yet caused a health crisis for me, but that is the point of predictive modeling: to inform a person using his or her own individual data about what might lie ahead so that he or she can make changes if needed. It was like a dummy warning light going off on my dashboard that my engine needed tending to avoid a breakdown: in this case, a massive coronary. At least, this sort of warning is what Paterson's team is trying to offer up in a way that makes sense to physicians and consumers.

I have also collected some scattered genetic data related to nutrition that I have not yet mentioned. For instance, a Massachusetts-based company called Interleuken Genetics tested me for genes related to how efficiently I metabolize vitamin B. This nutrient is needed for the healthy formation of red blood cells and for the function of nerve tissue, and also for how our cells respond to oxidative stress (this is when oxygen free radicals that are produced during metabolism cause damage to cells). I came out normal for all but one of interleuken's SNPs related to vitamin B. The abnormal one occurs in the TCN2 gene and seems to affect how vitamin B-12 reaches my cells. Too little B-12 can cause anemia and other problems, although the degree of influence or the risk factor is not clear from the Interleukin report that was sent to me after its lab processed my DNA from a buccal swab.

Entelos and other companies are beginning to create models similar to the PhysioLab heart attack program for other systems related to nutrition and diet, such as diabetes, but, as far as I know, no one has yet linked together the hundreds of genetic markers found in association studies—many of which I've been tested for—to come up with a model that correlates the markers with a useful score or a comprehensive predictive model for how our bodies deal with food.

I purposely did not share the Entelos data with Jampolis before I arrived in her office, but I did bring my most recent blood chemistry—the levels of triglycerides and other lipids, proteins, glucose, and the rest—from blood drawn just a week earlier for the second round of the Entelos experiment. My total cholesterol, after about six months on my new diet, had fallen but by only one point, from 220 to 219: a surprise, since my new, self-initiated regimen was supposed to reduce cholesterol (see the table on page 274). My lipoprotein (a) had dropped, too, by a measly 4 points, from 91 to 87. But the shocker: my triglycerides were up 21 percent, from a normal-range 104 to an above-range 126! Egads! And my cholesterol particle pattern (small, medium, and large) had markedly changed—something that the Entelos team would explain to me later when I visit them.

Jampolis isn't overly concerned by my results. "Most of these are fine. I have seen far worse. These are just barely over the normal range. But the cholesterol is still higher than it needs to be." She says that the particle distribution test is expensive and is usually not paid for

	Author's Results		
The Author's Key Cardiac Chemistry Results: Changes after Six Months on a New Diet			
Test	**Round 1**	**Round 2**	**Normal Range**
Cholesterol	220	219	<200
LDL	134	133	<130
HDL	65	61	>40
Lipoprotein (A)	91	87	<75
Triglycerides	104	126	<105

by insurance, so she doesn't use it much for her patients. "I think we do know what to do increase particle size in many people," she says. "Decrease saturated fat and refined carbohydrates and replace them with unsaturated fat and lean protein." She pulls out a chart and asks me to describe an average day of eating, starting when I get out of bed.

Despite my unchanged blood lipid measurements, I came to the meeting convinced that my new diet is healthy. I tell Jampolis that I feel better and that I've lost a few pounds. I am also not hungry all the time, like I was with the old diet—although I'm not happy that my cholesterol hasn't gone down.

I jump into telling her how I eat a banana when I get up and a bowl of healthy granola, and so forth, including the mention of the occasional chicken quesadilla for dinner. She listens and jots notes, saying little until I'm finished. Then she announces her conclusion.

"You are not eating a healthy diet," she says.

"What?" I respond, not sure I heard her right.

"Let's start with your banana and granola in the morning," she says. "The granola may have hidden sugar, while the banana is high in carbohydrates and has a higher glycemic index than most fruits." That is, bananas cause a rapid rise in blood sugar. "This may impact hunger and energy levels. When you combine the banana with granola instead of with lean protein, your breakfast presents a large glycemic (sugar) load to the body, which can cause a subsequent drop in blood sugar and energy level, sending you straight to the coffee cart for a pastry."

"People think they are eating a healthy diet," she adds, "but there can be many places where carbs and sugars can be hiding out." Fruit juice is another food that is unexpectedly loaded with sugars and carbs. She then starts talking about fiber and saturated and monounsaturated fats, and, as hard I try, my mind starts to wander a bit. I'm not sure why, although I suspect that I have a low threshold for this sort of information. Maybe it's because of a secret fear that if I succumb to paying close attention to the nutritional composition of my every bite, I'll become one of those obsessed people who can't move without consulting the tables of dietary facts printed on virtually every product. Possibly, though, it's a good idea to be *mildly* obsessed.

Next, Melina Jampolis has me stand on a device that looks like a typical bathroom scale. It's a "Tanita Body Composition Analyzer," which measures a person's fat percentage by sending a very mild electrical current that enters through one's feet. It sweeps through the water in the body, measuring when the signal is impeded or resisted. Since muscle holds more water than fat does, the current passes quicker through muscle than it does through fat. Jampolis enters my age, height, and weight, and I'm astonished to get a reading of 20 percent fat! I'm within the healthy range for men between the ages of forty and fifty-nine, which is 10.2 to 22.9 percent. But just a few years ago, I had this same test taken during a television story on obesity, and I came out at 15 percent fat.

A printed readout tells me that my target "fat to lose" is 7.2 pounds, which would reduce my fat percentage by 2 percent.

My frontal lobe is humming again with this new information that suggests I may be becoming a fatty, and it demands that I pay attention to Dr. Melina!

"This is within the norm, but it could be less," she says. "We can reduce your fat percentage by two percent and your cholesterol by thirty points if we change your diet." She explains that we're talking about shedding a few pounds, nothing dramatic.

Now that she has the full attention of me *and* my brain, Jampolis delivers her second punch by showing me vials of fat that are conveniently lying around on her desk: test tubes filled with a dense, gooey-looking white substance. "This is the fat in a Big Mac," she says, holding up three thumb-size vials. Other foods, like my beloved cheese, have less fat, but the message is clear: this is the stuff that clogs up blood vessels and causes heart attacks.

Yikes! I'm thinking, she's good. I seriously consider signing up for a whole program, including religious devotion to Jampolis's book *The No Time to Lose Diet: The Busy Person's Guide to Permanent Weight*

Loss. It contains no-time-to-lose meal ideas, exercise strategies, "lose" behavior tips, and even shopping lists.

She pulls out photocopies of pages with titles like "Heart Healthy Nutrition Tips," "Fiber Facts," and others with ideas for what to eat for each meal. For breakfast, I'm supposed to have 0–2 starches, 2–3 proteins or 1 dairy, and 0–1 fruit; 1–2 fats are "optional."

My interest flags ever so slightly as I read this, but I do get one bit of good news. The oatmeal raisin cookie I eat every afternoon is okay, she says, if I follow the rest of the diet. A small piece of chocolate is okay, too. Then comes more bad news: she says to never again eat movie theater popcorn and gives me a handout titled "Movie Madness," which states that popcorn is loaded with fat, including saturated fat from the cooking oil, which I assume is bad.

I promise Dr. Melina I'll give this a try. That afternoon, I go to my neighborhood grocery store and buy cottage cheese and yogurt and one other item that I love and had stopped eating under my old diet: peanut butter. "If you get a healthy brand, a few spoonfuls are good," says Jampolis. But I have to admit that I still have only glanced at her book—old habits, I suppose, die hard.

A few weeks later, I'm back at Entelos, and this time there is no food at all, only nuts at the front desk . . . and candy bars. I grab . . . which one do you think?

I'm here to find out from Tom Paterson and Alex Bangs how I did on the follow-up testing done six months after the first round. I had just given more blood and had a second carotid ultrasound taken, and they had run the numbers through their program.

To my surprise, they have more bad news for me.

My new data suggest that things have gotten slightly worse for my heart attack risk under my new diet.

My main cholesterol numbers have stayed basically flat, says Paterson, when I was sure they would get better, though the real story is a sizable change in my cholesterol particle distribution, following what Paterson calls a classic shift pattern when someone bumps up his carbs. For starters, my blood is swimming in small LDLs (bad cholesterol), which the liver is having trouble uptaking and removing from circulation. If you remember, the liver makes "packages" of lipids—LDL (bad), HDL (good), and triglycerides (can be bad if they stick around)—that deliver energy to cells. These slowly get smaller as they deliver their loads until they are very small particles of mostly cholesterol, which is supposed to be taken back up by the liver. My increase in LDLs is happening in part because the increase of carbs and fats in my system has added a thin layer of triglycerides on the surface of these small LDLs, which makes it harder still for the liver to absorb these particles. It's possible that I have an unidentified genetic trait influencing this uptake problem. This variation might involve a critical enzyme called lipoprotein lipase that breaks down fats and sugars so they can be absorbed as fuel for cells; it is also important in stripping triglycerides from LDL particles so that the particles can be more easily sucked up by the liver.

My carotid ultrasound results are largely unchanged, although the plaque in my left carotid has declined a bit over the previous six months, a small bright spot in my profile.

The bottom line: the Entelos team's model supports what Melina Jampolis says—that the diet I thought was so great is really loaded with extra carbs and hidden sugars and should be scrapped. Tom Paterson tells me that when Ron Krauss reviewed my new data and the model, he commented that given my scores, I was better off with the low-carb, high-protein diet I had been on before.

Paterson's team has not had time to update the table with my risk line on it, but when I later ask Alex Bangs to clarify what my new diet has meant for me in terms of risk, he sends me an e-mail, saying: "Your change in diet doesn't put you on a new curve; your LDL-C (small particle LDL) levels are still sensitive to changes in body weight. Your diet simply shifted where your body is balancing on the same curve. The deleterious effects of a higher-carb diet include certain exacerbations of LDL clearance that would be the case even for individuals on a 'normal' LDL clearance curve. For you, if you were to gain weight on a higher-carb diet, your risk projections are expected to be worse. We don't know yet how much worse until we complete some additional work on our model."

While still at Entelos, I ask Tom and Alex whether working on my profile has caused them to think about their own diets. "The more I work on this, the more I'm trying to incorporate this into my life," says Paterson. "Many of the foods that are in easiest reach are not so good for me, so I got a refrigerator in my office so I can put yogurt in there. It used to be if my stomach acted up, I didn't have the low-fat, high-protein yogurt, so I'd go get a bag of potato chips from the vending machine."

"I love cereal," says Bangs, "but now I eat high-protein cereal. That would be my adjustment. I'm also trying to decide what kind of breakfast gets me through the day. You can learn so much through self-experimentation; then you have to follow it. I'm close to starting a food log so I can keep track."

Here we are, three guys talking about yogurt and food logs—not a scene I expected to be writing for this book.

"What are you motivated most by?" I ask Paterson and Bangs. "Avoiding a heart attack and staying healthy, or looking good? I'm not talking about obsessive vanity, just the desire to not look fat and unattractive. I'll admit that in my case, I like to look good. Who doesn't? As a healthy person, this is probably more of a motivation than my fear of a heart attack."

"Me, too," says Paterson. "But it's not just looking good. I want to be around for a while. I want to be a part of my son's life. He's five years old, and it's such a privilege to be around for him."

"Are we being too compulsive about this?"

"I want to make it easy for myself to do the right thing," says Paterson, "to have the right foods around, a gym nearby that's nice, my workout stuff in the car."

"Habit is good, but our bodies are not designed to do what we do," says Bangs, "sitting behind a desk, eating fast food—that's bad. Giving a little extra attention to this is a good thing. If you're farming, you're closer to evolution."

"But won't some people really obsess and get worried when they have the Entelos profile done?" I ask. "If I was prone to worry, I sure could be going crazy over my results."

"If people are now hypochondriacs, they will become well-instrumented hypochondriacs," says Bangs.

Not long afterward, I speak with Ron Krauss over the phone, and he confirms his conclusion that my old diet was better. "You changed to a lower-fat, higher-carb diet, and carbs are known to raise triglycerides, and they are known to affect small-particle LDLs," he says. "You had an apparent worsening; in your case, it's subtle. You are not on an unhealthy diet. We have notebooks full of data that fit this profile."

"How bad is this?"

"You have to be careful; there is an overall phenotype that is higher LDL and lower HDL; it's called the metabolic syndrome. We would have to test you further to know whether you are at high risk for this, but you have some of the indications." He adds that my levels are still close to the normal range, although these ranges may not be entirely accurate for everyone. Different people may have different ranges of what's safe and what's not. That's something Krauss and other national leaders in cardiology are discussing now, he says, to see whether lipid levels need to be revised.

I say to Krauss that I didn't much like my previous diet. Eating all of that meat didn't seem healthy, and I was starved half the time. "The point is to eat more protein and less carbs. You don't have to get protein from meat. Eat vegetables and nuts and peanut butter," he says, echoing Melina Jampolis. "And some carbs are okay."

Since the first presentation by Entelos, Krauss has been impressed enough with the validity of the model to sign on to the company as a paid consultant, in part, he says, because its model integrates the complex data he has been collecting and trying to make sense of better than his more simple statistical models do. "They are much more comprehensive as a predictive model. It's the same process we're using but with far more brainpower. This is what I need to take this information from the lab to the clinic."

Over the weeks since I talked to the nutritionist, the computer modelers, and the cardiologist, I have tried to eat more nonfat cottage cheese and yogurt, and I'm amazed to report that I'm okay with it. I don't love it, but it's not awful. Using nonfat milk in my coffee is tougher: it tastes like water. But I am enjoying a few scoops of peanut butter and, of course, my cookie each day.

And, okay, the intermittent doughnut—which I may regret one day as I sink to the ground clutching my chest. But hey, I can't be a complete diet saint here. My brain won't let me. Also, I've decided that I'm willing to live dangerously (did you hear that, Ifat Levy?) in exchange for a sugar-glaze hit a couple of times a month.

Bumps on my kidneys . . . oh no!

Radiologist Judy Yee has agreed to decipher an X-ray scan of my body, but to reach her office, I first need to pass down several hallways and go up an elevator filled with aging veterans of U.S. wars. By the looks of it, most are from the Vietnam era or earlier as I stroll through the Veterans Affairs Medical Center in San Francisco, a pleasant cluster of stucco and brick buildings on a hill with palm trees overlooking the Pacific Ocean. These men are mostly ailing, with some in wheelchairs and pushing little tanks of oxygen on wheels, with plastic breathing tubes snaking down from their noses. They wear hats with names of ships and battalions. I feel a wave of reverence for these former soldiers I hadn't expected to run into. I am also reminded of the inexorable progression of aging. In these ex-soldiers' faces, I can read flashes of what they looked like when they were young and in uniform, with their lives ahead of them, many facing war and perhaps the horrors of combat that some may still struggle with.

This random dip into the world of the infirm makes me feel ridiculous as a healthy person carrying a CD burned with scans of the inner me for a physician who obviously has better things to do than spend an hour with me. This raises a crucial question about the Experimental Man world to come: who will have the time to analyze the data and to indulge the healthy in possible scenarios of future ailments? Machines can do much of it: scanning for anomalies, running raw results through ever-more-complex algorithms. But it will be a very long time, I suspect, before machines will be able to replace the knowledge, judgment, and art of a physician and other highly trained experts to make the final determination about the severity of a diagnosis or possible treatments or about distant maladies that might be prepared for or prevented. True prevention based on these tests will be good for society and for cutting medical costs, but it will also require a rethinking of resources and of how and how many physicians and health-care workers are trained. In my own borderline case with my heart, I would not want to choose a course of taking statins or any other treatment without first showing my data to Josh Adler and Ron Krauss and requesting their advice.

I had asked Josh to help me find a radiologist to read my full-body CT scans that were taken at HeartScan in Walnut Creek at the same time that the company's technician shot the heart scans used for the Entelos project. Josh recommended Judy Yee. She has a joint appointment at UC San Francisco, although most of her time is spent here at the VA Medical Center. Yee is a well-known expert in virtual colonoscopy—a scan and an analysis of the colon that looks for cancerous polyps by using a CT scan. She is one of those A-list physicians who chooses to practice at the less glamorous VA, where she says she has great research opportunities and loves working with the vets.

The week that I visit, Yee and a team of researchers from around the United States have just released a seminal study on virtual colonoscopies that proves they can be used to locate suspect polyps with almost the same accuracy as a traditional colonoscopy. People fifty years and older dread colonoscopies because this invasive procedure involves sticking a tube and a scope up one's—well, let's just say that despite evidence suggesting that early detection of polyps almost always prevents colorectal cancer, many fifty-pluses do not get colonoscopies.

In the study, published in the *New England Journal of Medicine,* almost twenty-six hundred participants age fifty and older were given virtual and traditional colonoscopies at fifteen research centers, including Yale, the Mayo Clinic in Arizona, and the San Francisco VA Medical Center. The scans identified 90 percent of subjects who had adenomas (benign tumors) or cancers measuring 10 mm or more in diameter. The scans did better than traditional colonoscopies at detecting smaller lesions. Colorectal cancer is the third-most-common cancer and the second leading cause of death from cancer in the United States, says Yee. "Seventy-five percent of the cases have no known risk factor, other than age, so it can come out of nowhere for people who have no colon cancer in their family." This number is rising, she adds. I ask whether anyone has made a connection between this rise and chemical toxins, and she says not directly that she knows of. We both agree that without an obvious genetic component, the increase in cases suggests there may be an unidentified environmental toxin at work.

"We need to be able to better make these links between environment and disease," she says.

She walks me over to a computer screen and launches excitedly into a demonstration video of the inside of a colon, showing me how one can "fly" virtually through the narrow tubes that are colored orange-pink by the computer. The virtual colon's walls look like they are made out of plastic, although the demo is fascinating: a glimpse of the Experimental Man in the future that has now arrived.

"The computer simulates what you would see [using a traditional scope up someone's bum], with a three-dimensional map based on the density of the tissue and of any polyp that appears." She sees a polyp on the sample image and zooms in, clicking on detection-and-analysis options that provide data on density and size. "This could revolutionize the detection and prevention of colon cancer," she says. Programs are also available for a similar super-scan of muscles and the spine, she says. At $800 or more, the virtual colon test isn't cheap, and it is not yet covered by Medicare or by many insurance companies, although Yee expects that to change in the wake of the recent study.

One test that I, um, haven't gotten around to for the Experimental Man project—or for my own routine health maintenance—is a traditional colonoscopy. My genetic profile suggests that I am at low risk for this form of cancer, with almost all of my SNP results falling under a 1.0 risk factor (reminder: odds ratio scores greater than 1.0 mean that one is in the higher-risk group) (see the table on page 283). Many of the SNPs associated with colorectal cancer come from solid studies with large populations that have been replicated, although the usual caveats apply, such as the need for clinical validation. Also, given what Yee has told me about how many cases are not hereditary, genes may play a lesser role with this cancer than others. Or, perhaps, there are genes as yet undiscovered that make one susceptible to a chemical or other environmental influence that increases the risk of acquiring this form of cancer.

At one point, I had talked to physicians at Stanford University about getting a virtual colonoscopy, but the issue of cost came up, and we never did it—a rare incidence of the Experimental Man not following through on a test.

"This may sound strange, but I wish we were looking at *my* virtual colon," I say to Yee, as I peer at these remarkable images. She mildly chastises me for being age fifty and not having had the test. I promise her that I'll get it, although the $800 price tag means that I may have to wait for my insurance to decide to cover it.

As a marriage of traditional X-rays and computers, computed tomography technologies have been getting ever more sophisticated since they were first developed in the 1970s. Newer machines cut

Sampling of the Author's Genetic Results for Colorectal Cancer			
Gene	**SNP**	**Author's Risk**	**Results Factor**
CRAC1	rs4779584	CC	0.91
EIF3H	rs16892766	AA	0.97
POU5F1P1	rs6983267	GT	0.99
SMAD7	rs4939827	CC	0.84
SMAD7	rs10795668	AG	0.96
SMAD7	rs3802842	AA	0.94
ADIROQ	rs266729	CG	0.95
ADIROQ	rs822395	AA	1.09
ADIROQ	rs822396	AA	0.14

ever-finer slices of images, down to 1 mm apart for the newest high-resolution scanners. My scan diced up my insides into ultra-thin slices every 5 mm, which is a space about half the size of my little fingernail.

One downside to CT scans, of course, is radiation. Most likely, I was exposed to about 6 to 8 millisieverts (mSv), a measurement of the physiological effects of a dose of radiation.[*] Yee says that we are exposed to a background level of about 3 mSv of radiation each year from the sun and other natural sources. CT radiation from her virtual colonoscopy is about 5 mSv for a man and 7 mSv for a woman.

I ask her whether 6 to 8 mSv is dangerous, and she surprises me by saying there is little long-term data on this, despite X-rays being around for more than a century and more than sixty-nine million CT scans being performed each year in the United States. "We don't have the studies to know the effects of long-term exposure," she says. Data on radiation exposure exist from survivors of the nuclear bombs dropped on Japan in 1945 and from exposure studies of employees working in nuclear power plants. The Japanese bomb victims had an average exposure of about 40 mSv of radiation (some had much higher levels), the plant workers about 20 mSv. Both groups have suffered

---

[*] Sieverts are named after the Swedish radiation expert Rolf Sievert (1896–1966).

from high rates of cancer. A recent paper by David Brenner at the Columbia University Medical Center in New York City makes a case for conducting fewer CT scans because of radiation, although Yee says that the paucity of information means that no one knows for sure whether the CT levels are safe.

She says, however, that the risk-benefit of radiation exposure versus getting colon cancer tips heavily in favor of getting a virtual colonoscopy, although I'm not entirely sure how she knows this, if the radiation risks of sixty-nine million scans a year are poorly understood and may themselves cause cancer. Yet the traditional procedure can cause other complications, such as an accidental perforation, which happens in about one out of a thousand procedures. "There is a much lower risk of a perforation with a virtual colonoscopy," says Yee.

Yee starts to pull up my CT images on her monitor and notices immediately where they came from: HeartScan.

"Is this a stand-alone CT scanning place?" she asks.

Yes, I answer, explaining how Entelos sent me there to get the heart scan for its experiment. The HeartScan people were nice enough to examine pro bono the rest of my body from my ears to my pelvis, a scan that usually costs nearly $1,000.

"Normally, I wouldn't look at these," she says. She has nothing against HeartScan, but, like many physicians, she does not approve of retail scanning facilities that take images on demand from consumers, often without the involvement of a patient's physician. These facilities began popping up in the 1980s, soon after the CT and, later, MRI scanners became widely available. Lately, however, the demand for scans outside of medical centers has dropped, apparently due to cost, the radiation issue, and the fact that the scans often don't detect anything untoward; nor is there good data on the risk-benefits of scanning seemingly healthy people. (Does this sound a little bit like some of the criticisms one hears about genetic tests for consumers, minus the radiation?)

"Do people bring these in to you?"

"Not to me, I work for the VA. We don't do drive-by readings here." She's doing this one for me because she was asked by my physician, Josh Adler, and to help out with my project. "A lot of these retail CT centers are shutting down," she says. "The vogue is over."

Josh Adler had been adamant about his opposition to healthy people getting scans for no reason. "There are dangers in taking these scans," he told me. Besides the radiation, if a lump appeared on a liver or a lung, even if it probably was nothing, a physician might feel obligated to order a biopsy or another invasive procedure that has its own risk factors, such as infection or the accidental puncture of an organ—although these risks were low. "But why expose yourself to this when these scans usually turn up nothing?" he had asked during my very first exam for this project.

I ask Yee whether scans like mine for healthy people help find maladies that might otherwise be missed. She agrees with Josh that there is no evidence that random scanning of healthy people is useful or worth the risk. She does say, however, that during the virtual colonoscopy study, she and her fellow radiologists, when they had taken the colon images of patients, did find suspect polyps and other growths and abnormalities on organs in the torso that were part of the scan. "We are obligated to look at other things that fall within the scan," she says, "and we find that about eleven percent of the patients scanned for the colon have something else going on that we see: a lung mass or a polyp that could be cancerous." She notes, however, that people getting the colon screening are not getting a random CT scan. "This scan is indicated for people fifty and older to prevent colon cancer."

As part of their service, HeartScan had already sent a report on my full-body scan, signed off on by a physician whom I never met. From the head down, I had a clean slate, according to HeartScan's radiologist, except for a slight calcification in my pericardium, the membrane sac that surrounds the heart and holds in fluids. The only other finding was two "small, right renal cysts." The report was signed by a Thomas Atlas, MD. When I did an online search on him, I found out that he is a board-certified radiologist working in Tustin, California, who went to medical school at the University of Southern California and did his residency at Baylor College of Medicine—both good programs. Dr. Atlas suggested several times in the HeartScan report that I follow up with my primary physician.

So I first took the scans and reports to Josh Adler, and he said that these sorts of findings are exactly what is wrong with these scans. "You are healthy and there are no indications for you to get scanned," he said. "For some people, getting back a report like this, which mentions the word 'cyst,' might make them very anxious for no reason. Also, if these cysts had been a bit more ambiguous, it might have put you in a position to investigate further, even though the chance they would be something significant is very low."

At the VA Medical Center, I'm looking over Yee's shoulder as she brings up my images on the screen. I want to know about the kidney cysts, which we'll get to, but I first ask Yee whether we can start with my head, at the top of my body, or near the top, and then proceed downward. The scans were taken in three parts—neck: from ears to shoulders; chest: from shoulders to diaphragm; and abdomen: from diaphragm to pelvis. Beyond this, I've had the top of my head (along with the rest of my noggin) scanned in the MRI images, and my body from the pelvis on down in a separate set of MRI images taken by an orthopedics friend of mine. Cobbling together these images gives me the full-body scan from head to toe.

Looking at the series of my sliced self, now displayed on the monitor, I recall as a boy seeing a gruesome but fascinating exhibit at the Field Museum in Chicago. Someone had diced up a human cadaver in what looked like half-inch or so slices, with each section pressed between glass and preserved. Like a grotesque display of family portraits hanging on the wall along a stairway, the slivers hung in a progression up one of the wide flights, going up several floors. The specimens were old and a bit brown and crumbly when I saw them. I remember thinking that this person had been someone's son or father. While alive, he had friends, had worked, and maybe liked baseball and Babe Ruth candy bars (my favorite snack in those innocent, pre–Melina bar days). Now my human slices have been taken live and virtual, and here they are for me to look at.

Judy Yee first inspects my lower head. The most remarkable thing is the metal in my mouth, she says. My gold and silver fillings reflect the X-rays back as glimmer rays, looking like a bright light shining off a metal surface. I explain that I didn't take great care of my teeth when I was a teenager. "That obscures some things," she says.

"There are your nostrils; they're clear," says Yee. My inner ear is fine, "no density or growths"; so are my neck and tongue. The bottom of my brain also looks okay, confirming the MRI scans analyzed by Jim Brewer. She toggles through the images using various contrasts available on the computer program: grays, whites, and blacks. She then shows me how the program can assemble all of the images into a movie, and we watch the slices become a continuous, downward scan from my nostrils to my upper shoulders, and back up again using a setting called a "yo-yo": an odd appellation for going up and down through my body. It reminds me of another childhood moment, when other kids in elementary school called me "Yo-Yo," after Duncan Yo-Yos.

Next come my upper chest and lungs. "There's something right there," she says, pointing with a cursor and expertly zooming in on what looks like a fleck of white paint. She says it's a small scar about 6 mm long on the right middle lobe, probably from a past infection. "That's nothing to worry about."

I ask what my lungs would look like if I smoked. "We would see little black holes," she says. "You don't want that."

Moving to my heart, she tells me it looks healthy, with no coronary calcification. "Since you're a young man, you shouldn't have any," she says, something I'm very happy to hear and will share with the Entelos fellows. "I'll scroll down below the diaphragm," she says, and tells me that my liver, stomach, colon, and spleen look fine. "Your gallbladder has no stones."

Continuing downward, we pull up the series of slices that show my lower abdomen, which includes that pesky kidney (see the illustration on page 288). On the scan this organ looks like a fetus, or a shrimp with a giant head. Yee sees one of the bumps right away: a protrusion. She uses a tool in the program to check its density. It's less dense than the tissue of the organ, she tells me, which means it's not a tumor. "If it was a tumor, it would be much denser."

But the image abruptly stops below my bladder. "The prostate is missing and the rectum," she says. "They didn't do a complete scan."

This means there is a break in the full-body scan—which I'll have to live with. Either that or I'm rectumless.

I ask Yee about the future of scanning. She believes that the sophistication of the virtual colonoscopy will be available for most of the body in the next few years. "You'll be able to fly through many organs and vessels," she says, "to see if there are any anomalies."

But my full-body scan is not yet finished. Besides the MRI images of my head, I had one other set of scans run of my lower body, which literally got me down to my toes—at least, on one foot.

The final scans were ordered by San Francisco orthopedic surgeon Kevin Stone. He ran them in 2007 after I injured my knee surfing. A wave whipped me hard against the sand, tearing a meniscus on my left knee. Stone ran an MRI of that knee and also, for the sake of the Experimental Man project, of the rest of my leg. This MRI showed the small tear in the meniscus, which eventually was surgically repaired, and also revealed a small divot in my ankle bone that Stone said must have been a bad sprain years ago. Actually, it was a series of repeated sprains that happened when I ran cross-country in college and kept twisting my ankle. I couldn't believe that this long-forgotten wound

The Author's Abdominal CT Scan

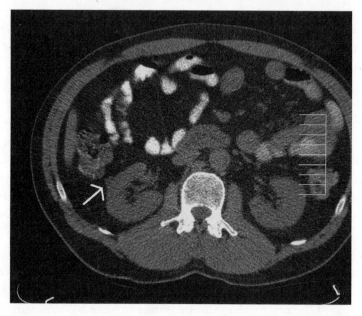

The arrow points to a protrusion in the kidney caused by a benign cyst.

was still there, an artifact of my life at the age of twenty. "Your bones tell a story of your life," remarked Stone, a tall, angular expert on knees and the meniscus who had the classic intensity and passion of a surgeon who loved what he did and was used to being in command in the operating room and in his office. He showed me another "artifact": a spot in my knee where the femur attached to the joint. At the base of the femur was a patchwork of worn spots in the outer layer of the bone that looked like peeling plaster on a wall. "That's from running too much on pavement," said Stone. He told me it would be a good idea to avoid much running in the future, especially on pavement. "You've banged that up, and the damage is done," he said, making me again feel older than I liked to feel.

I wanted to tell him that I love running—it is a time when I can think and relax my mind. But looking at that battered femur, I knew he was right.

"Bike all you want, and swim," he said with that surgeonlike authority, "but if you keep running, you will be getting new knees sooner rather than later."

Like my diet, this would be a major change initiated by my testing: an end to most jogging. I did try running again anyway, once my meniscus was healed, but I promptly injured my other knee by bruising the bone. I looked again at the image of my battered femur and realized I couldn't fight this.

The bone bruise healed in a few weeks, but that was all I needed to convince me that my lifelong habit of running is now, sadly, an occasional indulgence, although I've replaced it with other exercises.

Stone reviewed the rest of my leg, which was fine, right down to the tip of my big toe. Looking at the ghostly image of my white foot bones, I remembered getting what was probably my first-ever X-ray as a small child in Steve's Shoes in Kansas City. For some reason, in the mid-1960s, in possibly the first (and hopefully the last) retail shoe craze with X-rays, for a while shoe stores were offering to X-ray people's feet, I suppose to help determine the best shoe size. It was a gimmick that must have exposed thousands of people to radiation, at a time when the implications were perhaps less understood than today—or people didn't care as much. Shoe salesmen would have absorbed the most radiation. I wondered what the cancer rate was among salesmen at Steve's and other stores from that era as the workers aged. I certainly didn't remember them wearing lead aprons or any protective gear.

Once again, I marvel that anyone my age has survived with all of the chemicals and foolishness we were exposed to, in part because we didn't know better, but also because even when we did, we ignored the risks. It makes me wonder what we are exposing ourselves to now, out of ignorance or willfulness.

After I completed my head-to-toe scanning, minus one prostate and rectum, I was left feeling unhappy about the running but otherwise pleased that I came out with just a few bumps and bruises inside me. Most of the images, however, are still crude and fuzzy black-and-white pictures and "yo-yo" videos. As Judy Yee says, coming soon will be a full-body virtual image of the inside of a human, probably color-coded by organ: orange-pink for the colon, maybe lime-green for the stomach, and fire-engine red for the heart? One day, when radiation issues are hopefully resolved, we might all be able to fly through our own innards on a laptop as if we're in a flight simulator or playing a game of Halo without the violence. Possibly, there will be online, direct-to-consumer sites that will facilitate this for a price: "VirtualBodyandMe," or some such thing.

If this sounds strange or even macabre to be cruising one's insides looking for polyps, it wasn't long ago that the idea of mucking around inside our DNA—or our brains—seemed intrusive and frightening. For some people, it still does. For me, if I remain healthy, I will probably find better things to do than spend time flying through my virtual spleen, though perhaps one day this journey through our insides will be as easy as getting a simple X-ray is today, without the radiation.

Life at age 122 (the gene that regulates forever)

Now that you've gotten this far in the book, consider this: there may be a class of supergenes that could trump almost everything I've been tested for in the Experimental Man project. These are regulator genes that command whole systems of other genes, launching cascades of reactions that give a huge hereditary health advantage to a person who carries certain variations. These genes may protect carriers from a range of environmental assaults and even explain why certain people's brains remain stubbornly, even obnoxiously, convinced that despite evidence to the contrary, they *will* be healthy for a very long time.

Put another way: What are the chances, despite the Entelos model and a few deleterious, high-risk genes nestled in my genome, that I'll be alive in 2080, on my 122nd birthday? Or that this will never happen—that I'll disdain nonfat, plain yogurt, gorge on cheddar cheese and sugary glazes, and have the predicted heart attack in 2017, which will kill me just shy of my turning sixty years old?

To find out, I dispatch my very last batch of blood for this book to Cambridge, Massachusetts, to Sirtris Pharmaceuticals, a five-year-old company named after the family of übergenes in question. Called sirtuins, there are seven of them inside humans: SIRT1 through SIRT7. In mice, these genes and the enzymes they create appear to control a cornucopia of desirable functions in cells that lead to improvements in diseases such as obesity, diabetes, Alzheimer's disease, and cancer—effects that contribute to mice and other organisms living longer and healthier lives.

Out of the seven sirtuins, SIRT1 has been most intensively studied. When its enzyme is activated by the body, researchers have found that it improves insulin sensitivity, decreases glucose levels, potentially lowers fat stores and body weight, acts as an antioxidant, repairs DNA, and increases the number of mitochondria—the structure that powers a cell—which augments the energy output of the cell and increases endurance during exercise. (Mitochondrial capacity decreases as we age and may play a role in increased insulin resistance and age-related cancers.) In rodents, boosting the activity of SIRT1 increases life span by up to 24 percent; in flies and fish, up to 59 percent.

Perhaps more interesting is that SIRT1 activation slows down aging, meaning that old age comes more gradually, as do the diseases of aging, such as diabetes and heart disease. Given our long life span, humans have not been tested, since the experiment would take more than a century. Yet many scientists are convinced that this gene family plays a fundamental role in regulating life span in numerous animals, including us.

Recently, researchers have learned more about the other six SIRT genes. Much needs to be done, but it appears, for instance, that SIRT2 is beneficial for mitigating neurological and metabolic disorders and cancer and that SIRT3 and SIRT4 may improve metabolic and mitochondrial disorders. SIRT5 is still poorly understood, and genes 6 and 7 affect cancer and cardiovascular disease, respectively. Sirtris is developing drugs for other sirtuins as it creates a whole family of potential fountain of youth pills.

The company plans to test a biomarker in my blood that will reveal how much SIRT1 is circulating inside me. High levels would suggest that I am lucky enough to be experiencing at least some of the beneficial influences of this enzyme—although, as we shall see, the enzymes need to be turned up by environmental input to work at maximum effect. Low levels would mean that I lack these cell-boosting chemicals that might help people live a longer time. So one afternoon, in a cubicle at UCSF, I have a small amount of blood drawn directly into a tube that was sent by Sirtris with a special chemical to preserve it for the test. I take that tube, pack it on ice, and ship it overnight to Cambridge.

Scientists have discovered two primary mechanisms that are used by nature to stimulate a rise in SIRT1 activity. The first is caloric restriction. Since the 1930s, researchers have known that a severe restriction in calories triggers a battery of cumulative cellular defenses in many organisms that works to increase life span. This "CR" effect on the SIRT1 pathway undoubtedly evolved to allow organisms to survive major stresses in their environment, such as drought or famine. Animals subjected to a CR diet not only live longer, but they also show reduced rates of cancer, diabetes, inflammation, and cardiovascular diseases. A few hearty people have tried prolonged CR diets, with some of them cutting calories by 10 to 25 percent for up to six years. These diets have shown some benefits when scientists tested them for blood pressure, glucose, insulin, cholesterol and triglyceride levels, and the thickness of heart walls. Other tests have measured biomarkers and levels of oxidative stress. But no one has conducted systematic long-term human studies. Nor do I plan anytime soon to cut back my calories by 10 to 25 percent to mimic starvation. Even if I did and managed to eke out a few extra years, what kind of life is it to be constantly ravenous?

The second method is far more pleasant. It was discovered in 2003 by Harvard scientist David Sinclair and involves resveratrol, a molecule found in red wine that appears to activate the SIRT1 system. Grapes and other plants make resveratrol and other polyphenols when they face environmental stress. Or, in the case of grapes for making wine, when winemakers intentionally stress them as part of the winemaking process. For the grape, resveratrol stokes its cells to increase the plant's chances of survival. These chemicals do the same for animals that eat or imbibe them. As Sinclair theorizes, plants and animals have developed this symbiotic process as a tool for plants to trigger a response similar to caloric restriction, which seems to be controlled by the SIRT1 gene.

Sinclair is a slight, soft-spoken Australian with a sometimes devilish smile. A cofounder of Sirtris, he began working on sirtuins as a postdoctoral student at MIT in the lab of pioneering longevity researcher Leonard Guarente. While there, in the late 1990s Sinclair discovered that the SIRT1 equivalent in yeast, called SIR2, prevents aging by slowing down the accumulation of ERCs, circular strands of DNA that build up in yeast organisms as they age, eventually killing them. During this time, Guarente's lab made another crucial discovery: that a link may exist between SIR2 and caloric restriction, and that this link could mean that SIR2 in some way regulates aging in yeast.

By 2003, Sinclair had acquired his own small lab at Harvard and was searching for a chemical activator for SIR2, still working with yeast—a simple and cheaply maintained organism that lives only a few days, which is useful for longevity research. In February of that year, Sinclair learned that scientists at Biomol Research Laboratories, based in Plymouth Meeting, Pennsylvania, had observed that sirtuins are activated by certain polyphenols, including resveratrol, which occur naturally in red wine. Sinclair and Konrad Howitz, Biomol's director of molecular biology, collaborated to isolate resveratrol and test it in yeast and fruit flies. "Never in my wildest dreams did I think we would find an activator of SIR2," says Sinclair.

The discovery brought attention to the formerly obscure Sinclair, who told anyone who would listen that he was sure that sirtuins would work in more complex organisms—including, possibly, humans. In a 2004 interview in *Science*, Sinclair called resveratrol "as close to a miraculous molecule as you can find. . . . One hundred years from now, people will maybe be taking these molecules on a daily basis to prevent heart disease, stroke, and cancer."

Few people believed him, with critics suggesting that aging in complex organisms, including humans, is too multifaceted to be controlled by a single gene or family of genes. This attitude prevailed, despite a decade of research by Guarente and another longevity-science pioneer, UCSF geneticist Cynthia Kenyon. In 1993, she coauthored a paper that introduced the notion of a super-regulator gene for life span. Her work was conducted on tiny roundworms called *C. elegans*, which are about a millimeter long, and a worm gene called DAF2. When mutant, DAF2 can double a worm's life span; when combined with another life-extending technique developed by Kenyon, it increases the life span up to six times normal. Naysayers, however, doubted that a single gene would have such an effect on humans or other mammals—never mind the notion that a molecule found in a beverage could trigger this effect.

I first visited Sinclair's lab in 2005, shortly before he published a paper providing strong proof that resveratrol does work in mammals. With little white mice crawling on his arms in his newly expanded lab at Harvard, he told me that his resveratrol-fed mice were healthier than the controls, and their cells were aging more slowly, despite the mice being fed an unhealthy, fatty diet. When the paper on these experiments came out in 2006 in *Nature*, the results showed that mice on a high-fat diet that were fed large doses of resveratrol were as healthy as mice on a regular diet. Resveratrol also improved the mice's insulin sensitivity and increased their energy production.

This study grabbed headlines around the world, given the connection to wine and to longevity. But there was a catch: the mice were given a dose of resveratrol that was the equivalent of a 150-pound man needing to drink roughly 1,500 bottles of wine a day. (Subsequent studies have reduced this intake to just a few hundred bottles.)

Sinclair's paper came out within days of another paper in *Cell* from the lab of Johan Auwerx of the Institute of Genetics and Molecular and Cellular Biology in Illkirch, France. Auwerx's team had given their animals even higher doses of resveratrol. These mice stayed slender and strong on a high-fat diet, exhibiting the energy-charged muscles and reduced heart rates of rodent athletes, if there were such a thing. The number of mitochondria in their cells increased, which supercharged the cells' energy output.

Sinclair's and Auwerx's success in extending the life span and improving the health of mice silenced many doubts about resveratrol working in mammals, although some people wanted to see whether this molecule also worked for mice on a normal diet. Skeptics continued to insist that resveratrol would not work in humans. Sinclair disagreed. "The system at work in the mice and other organisms is evolutionarily very old, so I suspect that what works in mice will work in humans," he told me at the time.

In 2004, Sinclair cofounded Sirtris with serial entrepreneur and venture capitalist Christoph Westphal, a cofounder of two bio-tech successes, Alnylam Pharmaceuticals and Momenta Pharmaceuticals. Sirtris quickly raised millions of dollars and set out to develop an improved version of resveratrol that could be taken as a pharmaceutical pill. They developed a new formulation dubbed SRT-501, which is much more readily assimilated by the body than the natural form of resveratrol. SRT-501 delivers enough of a punch that it replaces the need to drink hundreds of bottles of wine. Sirtris also spent some $20 million on screening compound libraries and investigating the chemical structures discovered, and came up with compounds with no structural resemblance to resveratrol that also activate SIRT1.

In 2008, Sirtris began testing STR-501 in human trials to see whether it improved the outcomes of type II diabetic patients. The FDA does not consider "longevity" an acceptable endpoint, nor is aging a disease like cancer or Alzheimer's, which is why Sirtris is targeting diseases of aging. Preliminary results showed that the drug not

only was safe but lowered blood glucose and increased insulin sensitivity for the eighty-five patients tested—which is what it was supposed to do. Besides diabetes, Sirtris is conducting very early human
trials for victims of MELAS, a rare neuromuscular disease caused by
a mutation in the DNA of mitochondria. CEO Westphal says that the
company believes that STR-501 may act in MELAS victims' cells to
ramp up mitochondria production.

At least one major player in Big Pharma is willing to bet on the
notion that a single family of genes and compounds will revolutionize
medicine by slowing or eliminating the diseases of aging. This would
be GlaxoSmithKline (GSK), the British pharma giant that acquired
Sirtris last year for $720 million—a deal that Westphal says will free
him up from having to constantly raise money to more quickly develop
the drug. He and Sinclair also made a little profit from the transaction,
I suspect.

Despite GSK's faith in STR-501 and other SIRT1 activators, it
still has years of testing to go and faces many hurdles. Ultimately,
it may fail. But if it works for diabetes and a wide range of age-related
diseases, then the world will have its first true longevity elixir. Medical
science will also have to reconfigure itself from focusing on studying and developing new treatments for specific diseases to delving
more deeply into targets such as SIRT1 that cause system-wide, or
organism-wide changes.

The idea of emphasizing systems and pathways may also be has-
tened by an increased understanding of the role of other types of environmental input. Resveratrol and STR-501 are beneficial chemicals
that come from the environment, but, as we know, there are also
chemicals that cause harm, such as mercury and PBDE flame retardants. They perturb entire systems of genes and proteins. We had a
peek into this new world in the environment section of this book when
Carolyn Mattingly in Maine showed me sample pathways that were
impacted by toxins such as mercury, which may cause certain genes in,
say, the respiratory pathway to increase or decrease activity and lead to
a cascade of changes in other genes in the system.

A few days after I shipped my blood to Sirtris, I receive results
back from the company's associate director of biology, Olivier Boss.
He had searched my blood for a specific biomarker that indicates how
much SIRT1 enzyme my body is churning out. This marker is called
a messenger RNA (mRNA), a molecule that is produced by cells to

be an exact match for a stretch of DNA in a gene. The RNA attaches itself to this DNA and creates a copy of its code, which it then takes as a messenger to another part of the cell that translates the code into the construction of enzymes and other proteins. In this case, the mRNA is carrying as the "message" the specific code to make SIRT1 enzymes.

Boss calculated my level of SIRT1 mRNA and compared it in an informal test to the levels of three other people at Sirtris: David Sinclair, CEO Christoph Westphal, and the company's senior vice president for development, Peter Elliott (see the graph below).

Incredibly, I came up with almost a thousand times the levels of Sinclair and Westphal and seven hundred times the level of Elliott. Oliver Boss summarized these results in an e-mail: "You have *very high levels of SIRT1*. Or, more accurately, there are very high levels of SIRT1 mRNA in the blood sample you sent me."

This astonishingly high level of mRNA prompts a round of e-mails that has Sinclair declaring in his characteristically short and playful e-mail style that "David is going to live to 122."

This is why I started this chapter wondering whether I will still be around in the year 2080, when I turn 122 years old.

I ask Boss what these levels mean for me. He writes in an e-mail:

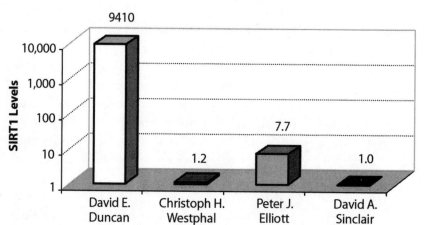

The Author's Level of SIRT1 (mRNA) Compared to Others' Results

From left to right: Results for the author; Christoph Westphal, CEO of Sirtris; Peter Elliott, senior vice president for development of Sirtris; and David Sinclair, Harvard longevity scientist and cofounder of Sirtris.

We do think that having high levels of SIRT1 (in any tissue) is "good for health." SIRT1 seems to help cells adapt to many different kinds of stresses (e.g., low calorie diet, cold environment, (mild) oxidative stress.

I ask him what would happen if I took their new drug, STR-501, which apparently acts to increase the activity of whatever SIRT1 enzymes are floating around inside a person. He answers,

Our SIRT1 activators increase the activity of SIRT1. Not 100% of SIRT1 is all the time maximally active in cells. Our activators necessitate the presence of at least a little SIRT1, and whatever is there will become more active in the presence of a SIRT1 activator.

Thus, even with your high levels of SIRT1, David, you can still benefit from a SIRT1 activator (should you ever need one, and if we ever "make it on the market"). You can live to . . . I don't know . . . is there a limit?

I ask whether there are fluctuations in the levels of mRNA—in other words, do I normally have lower levels, but, for some reason, I had a super-spike of these biomarkers when I had my blood drawn? Boss answers,

Fluctuations: There are no detailed studies reporting on the possible natural fluctuations in SIRT1 (mRNA or protein) levels. We do know that the level of expression of SIRT1 (i.e., the quantity of SIRT1 mRNA and protein) can increase in most cells/tissues studied upon environmental stressors (as mentioned before).

For you, David, with a one-time measurement we cannot tell whether the level measured is constant-stable or not. However, with this very high level, I would think that most of the time your levels are on the high side. (Natural increases/variations of SIRT1 expression are usually of a few fold on amplitude, very rarely 100-fold. Yet, your level is more like 1,000-fold higher than the average in this study.)

Boss cautions me, however, that this is a test of only four people, although it seems as if I do have high levels. Either that, or the other three test subjects have extraordinarily low levels. More people will need to be tested to determine a true average.

So what *does* this mean for me? Besides all of the potentially beneficial effects throughout my body, I notice in a paper that was recently
coauthored by Olivier Boss that increased activity of the SIRT1
enzyme is beneficial to the heart, in both mice and humans. It miti-
gates oxidative stress and inflammation and improves the health of
heart cells. In another study in the Netherlands, researchers found a
SNP in SIRT1 that seems to reduce the risk of heart attack. I am heterozygote for this SNP, meaning that I have a higher chance of having
this beneficial effect, although the lead researcher, Maris Kuningas of
Leiden University, e-mails me that when they repeated their study, the
power of this SNP to reduce heart disease lessened. The SNP, though,
does appear to provide beneficial effects during a cardiac "acute crisis,"
thereby reducing mortality from heart disease. Another study suggests that taking resveratrol with a statin improves "cardioprotection"
against heart attacks in mice.

I ask Boss if there are more SNPs or genetic markers that have been
identified for SIRT1:

> A few polymorphisms/genetic variants in the SIRT1 gene have
> been reported recently, but there is no data yet on the possi-
> ble impact of these polymorphisms on the expression level of
> SIRT1.

"Wow!" is my main response after this fevered afternoon of e-mailing
back and forth with Sinclair and Boss. Despite my abundance of SIRT1
mRNA, I feel depleted by the excitement even as I'm wondering if it
can possibly be true. Reading study after study about the miraculous
properties of SIRT1 and resveratrol, my supergene eyes begin to get
sore, even as my SIRT1-charged neurons refuse to shake loose a core
of disbelief that anything can be this wonderful.

And yet . . . what if it *is* true?

It certainly will reinforce my notion that I'm healthy, a core faith
that has been shaken some by my visits to Entelos and to orthopedic
surgeon Kevin Stone, who revealed my battered knee.

On the other hand, I'm not sure that anyone can yet explain the
significance of my results. In a way, this is another random piece of
information, added to all of the other indiscriminate bits and pieces
of SNPs, biomarkers, risk factors, and the rest that have yet to be

linked up, analyzed, and weighted to give me a holistic assessment or a total score for what lies ahead in my health future. The Entelos model has come closest to giving me a systemic score, although I wonder how its model would integrate and interpret my SIRT1 levels. Would this new information cause that risk-factor line in the table to go down?

Since I can't answer these questions, my thoughts turn to how I feel about living to be 122 years old. This was a joke coming from Sinclair—he has no idea whether I will live another day, let alone another 72 years—but I find myself pondering whether I want to live to perhaps a more reasonable enhanced age: say, 100 years old. The number of centenarians has steadily risen in recent years to about eighty-one thousand in the United States today and is expected to reach over six hundred thousand by my ninety-second birthday in 2050, according to the U.S. Census Bureau. Will I be one of them? Will this number go up dramatically if STR-501 works to slow aging and is approved, and people start popping longevity pills?

I can think of many people I wish would live or would have lived to be 122 years old: Jonas Salk, Leonard Bernstein, Nelson Mandela, Charles Dickens, Voltaire, and my grandmother, to name a few. Others I would not want to live so long: Joseph Stalin, Vlad the Impaler, and the bully who lived across the street from me growing up in Kansas City. One person, a supercentenarian French woman named Jeanne Calment (1875–1997), apparently did live to be age 122 (and 164 days). But imagine if we had tens of millions of people living to be 122—or if 122 years old became as common as 82-year-olds are today. In the year 2000, more than nine million people were age 80 or older in the United States, a number that had nearly quadrupled in forty years. Even without resveratrol or STR-501, this number will reach over thirty million by 2040 at the current pace.

What will we do with these extra people, who will be aging more slowly and will be healthier longer?

We already have had a preview of what would happen. In 1900, life expectancy in the United States was 47.3 years. Then came better hygiene, medicine, nourishment, and safer streets, and voilá, life expectancy increased in 2005 to 77.8 years—the age of my father now. In 1970, when I was 12 years old, life expectancy was 70.8 years. By 2015, this will increase to almost 80 years. By 2050, the Census Bureau says that average life span will top out at over 90 years.

Recent increases in the rates of obesity, diabetes, and heart disease in the United States and other countries might derail the steady
rise in life span, although the average total years continue to tick
upward, despite expanding waistlines.

So far, the West has more or less absorbed this unprecedented
surge of older people, although we routinely hear that Social Security
is threatened in the near future and that health-care costs for hordes
of seniors will one day bankrupt the country. No one knows what
will happen, but subtle shifts are already occurring in our perceptions
of old age. When I was a boy, anyone over sixty-five was considered
old. Now the age for being old has drifted upward to seventy and
beyond, depending on the person. Also, more people are working until
an older age. The recent economic crisis may also affect the elderly as
their investment portfolios and retirement funds whither, increasing
stress and forcing some retired seniors to go back to work.

Technologies other than miracle pills may also help us to live longer
and healthier—machine augmentations, growing replacement parts
using stem cells, nanobots (tiny robots unleashed inside us to make
repairs and deliver treatments), and possibly mergers between organic
minds and supercomputers. Some prophets of ultra-long life and
immortality talk excitedly of a day coming when humans will shed
our frail organic shells in their current state and merge with technologies that will allow us to endure for ages to come—and, presumably,
to be smarter, faster, and free of disease and disorders. I also assume
that people craving immortality will want to be beautiful, tall, and fit,
although I wonder whether these concepts will be relevant in a post-
human world where minds are living inside computers. I also wonder
who will pay for all of this and what happens to people who live a
thousand years on a modest fixed income. I seldom see issues such as
beauty and finance discussed by transhumanists and others who seek
an ultralife based on technological improvements. Nor have I heard
how postpeople will procreate and otherwise have sex. I assume they
will, although I'm not sure how one does this mind-to-mind inside a
computer. Maybe with avatars?

The quest for immortality is nothing new. In ancient Greek mythology, the goddess Eos fell in love with Tithonus, the son of a Trojan king
and a water nymph, and asked Zeus to make her lover immortal. But
she forgot to ask for eternal youth, leaving poor Tithonus to wither

away and eventually turn into an insect longing for death. Alfred Lord Tennyson wrote a poem about Tithonus, excerpted here:

> The woods decay, the woods decay and fall,
> The vapors weep their burthen to the ground,
> Man comes and tills the field and lies beneath,
> And after many a summer dies the swan.
> Me only cruel immortality
> Consumes; I wither slowly in thine arms,
> Here at the quiet limit of the world,
> A white-hair'd shadow roaming like a dream.

More recent is the story of Juan Ponce de León and the Fountain of Youth that he was supposedly looking for in Florida in the early sixteenth century—although, most likely, the story of a magical spring with rejuvenating waters was told by Native Americans who were trying to get rid of the Spanish invaders by telling them this spring was located in someone else's village far beyond their borders. During the Enlightenment, with science on the rise, immortalists such as the eighteenth-century French philosopher and mathematician Marquis de Condorcet suggested that advances in medicine might one day greatly extend the human life span. Mary Shelley's *Frankenstein* is in some ways a Gothic horror story about a young scientist's efforts not only to create life, but to prolong it by using electricity to reanimate dead tissue. In 1923, geneticist J. B. S. Haldane published *Daedalus: Science and the Future*, predicting that genetics might hold the key to improving human health and also that early critics would view these developments as "indecent and unnatural."

Transhumanists have traced the first use of their term to biologist Julian Huxley, who defined transhumanism in 1957 as "man remaining man, but transcending himself, by realizing new possibilities of and for his 'human nature.'" More recently, transhumanists such as the computer scientist Marvin Minsky and the inventor and futurist Ray Kurzweil have expanded this view to encompass the possibility that humans may evolve into a state beyond what is currently considered human, as advances in computers and artificial intelligence create what Kurzweil believes will be a singularity event where humans merge with machines or where superintelligent machines simply overwhelm organics. Kurzweil believes that humans will be able to upload the contents of their minds into a computer by the 2040s (this is an amendment: he used to say the 2030s). He has set a date when he thinks the singularity will occur: 2045, the year I would turn eighty-seven years old. That is the year when Kurzweil predicts that artificial intelligence will surpass humans as the most intelligent entities on Earth.

I have attended transhumanist meetings and find some of their postulating fascinating as an exercise in hypothetical possibilities. But parts of the movement are spiraling into what at times seems like a pseudo religion or a philosophical school in the style of the pseudo-religious Pythagoreans in ancient Greece. Over the years, transhumanists have developed subcategories of adherents following creeds such as Abolitionism, which advocates using transhumanist technologies to alleviate the suffering of sentient beings everywhere. (The emphasis is on sentient, I suppose, instead of human, so as not to exclude artificial intelligence, assuming such beings will suffer.) Other derivations include democratic transhumanism, immortalism, libertarian transhumanism, postgenderism, and, of course, singularitarianism.

Doubters, from the conservative bioethicist Leon Kaas (formerly the chairman of the President's Special Advisory Committee on Bioethics, appointed by George W. Bush) to *Wired* maverick editor Kevin Kelly, have criticized transhumanists for everything from violating basic tenets of humanity (Kaas) to simply being foolish. Kelly has suggested that they are overly optimistic about technological breakthroughs because they hope to save themselves from their own deaths.

My friend Gregory Stock, a bioethicist from UCLA and the author of *Redesigning Humans: Our Inevitable Genetic Future*, also questions whether technology will allow us to transcend death or whether humans will want to become machines. Stock has told me that if Ray Kurzweil is right about the singularity, then humans should be doing everything possible to keep artificial intelligence from taking over, because one of the first things it is likely to do is snuff us out—at least, those of us who have not had our minds uploaded, in what sounds like a film by the Wachowski brothers.

Another problem that I have shared with Kurzweil is why humans would want to create a singularity with hardware when it has already occurred in evolution. Arguably, the singularity happened approximately four to five million years ago when humanoids first became sentient. No machine that I'm aware of has yet come close to duplicating everything the human brain can do. Computers can process and analyze complex masses of data at blinding speeds and operate complicated systems far beyond the capacity of the human brain. But even at this moment, my gray matter is performing a range of functions, including thinking about what to write in this paragraph, which is far beyond the capacity of any machine. I suspect that if a "second" singularity occurs, the new intelligence will look more like an organic brain than like a silicon-based computer. As Greg Stock has suggested, humans of the future are likely to live longer and to have an array of enhancements, both through chemicals such as drugs and through hardware. We have long had simple augmentations, such as eyeglasses. In the last fifty years, innovations have included heart pacemakers and devices that feed sound waves into the brain so that the deaf can hear. "I think we will get much more of this," says Stock.

I also wonder who will get the miracle drugs and devices, and how many people will want to live to be 122 years old and beyond. In my travels in Africa and Asia and also in less affluent places in the West, I have seen countless people whose lives are hard and often miserable. Would they have access to longevity fixes—and, if they do, would they want them? Even for the well-off, life can be difficult. I have watched my parents gradually shift from wanting to take on the world and accomplish great things—and achieving a great deal—only to pull back in their late seventies into a very quiet life on the coast of Maine, writing, painting, and enjoying each other's company. If I follow their lead, I'll have almost fifty years of this sort of quiet living beyond their current age should I live be 122 years old. Can I handle that much quiet? Or will my body on STR-501 compel me in my enhanced youthfulness to engage in a more ambitious lifestyle longer than I would have as a "natural"?

During a visit to Cambridge, I ask David Sinclair when his company's superpill might be ready, if it works.

"This will impact humans within a decade," he says. "That's why I don't think there is anything more important than this quest."

Perhaps, though, the importance may have more to do with how profoundly it affects society and who we are as humans as it does in fending off old age and disease.

EPILOGUE: ETERNITY

Everybody's a mad scientist, and life is their lab. We're all trying to experiment to find a way to live, to solve problems, to fend off madness and chaos.

—DAVID CRONENBERG

Experimental children

For the cover of this book, photographer Art Streiber created an image of the Experimental Man done in the style of a David Hockney collage. It's lovely as a work of art but optimistic as an interpretation of science. In Streiber's work, the Man is a pastework of parts, but he is easily recognizable as a whole human being—minus some crucial features digitized out to keep the book cover rated PG-13. A more accurate image of what I've learned about the Experimental Man might be a Pablo Picasso or Georges Braque painting during high Cubism: a figure that looks vaguely human but is incomplete, misshapen, and impressionistic, with perhaps a touch of the grotesque. It's a grab bag of genes, cells, proteins, compounds, synapses, and body parts joined together haphazardly, without a complete understanding of how the pieces fit.

Keeping the fine art metaphor going for another few lines, we could also add to Streiber's photograph bits and pieces of my family and their results: my parents, my brother, and my daughter and perhaps hints of the generations of humans and other creatures that preceded me during the last three billion years. For my brother, there would be a blank patch as we search for the cause of his malady. Many question marks would be inserted, along with symbols for toxic chemicals detected inside of me; a scan or two of my brain; an enlarged image of my heart, which might seize up if I get chubby; and also perhaps an outsize image of the SIRT1 gene that may or may not be pumping a life-extending enzyme throughout my body.

There are many more gaps in the image because either an experiment isn't yet available, or I didn't take a test that is offered somewhere in the world. At the beginning of this project, I set out to take every assessment that made sense, but of course I didn't take them all. Some were too expensive, too obscure, or, like the cell assay in Chris Austin's lab, too fraught with bureaucratic hurdles and ethical concerns—although Austin and I still hope to run that test at some point in the future.

I feel as if I have taken enough tests to get a good idea of what is possible at this moment and in the near future—and to appreciate that further tests right now would simply add to the heaps of data I have accumulated, most of it preliminary that has yet to be integrated into a true picture of my quantified self. Yet I did take a number of additional tests that I will briefly mention.

Out of my five senses, I had assessments run on three: sight, hearing, and taste. My eyes were examined in Hyderabad, India, at the L V Prasad Eye Institute, where I was working on a story about physician-researchers in India who were using stem cells to help restore eyesight. In between surgeries at this state-of-the-art eye hospital, eye and stem cell surgeon Virendar Sangwan had an ophthalmologist on the institute's staff peer into my eyes with a light and pronounced them healthy.

For taste, I met with Dennis Drayna of the NIH, whose lab has isolated a number of genes that account for a person's ability to detect a bitter taste. While in San Francisco for a conference, Drayna ran a very simple investigation to verify the genetic proclivity revealed in my genetic results for not being able to taste bitter things. Sitting across from me in a downtown hotel restaurant, he handed me a strip of paper about the size of a stick of gum but shorter, coated with a bitter substance called PTC. "This is a simple test they run in high school biology class," he said, "but it's effective."

First, he had me taste a plain piece of paper. Nothing. Then he gave me the PTC paper.

"How bitter is this?" I asked, reluctant to pop something awful-tasting in my mouth.

"Really bitter to me," said Drayna, "because I have the bitter taste version of the gene."

"Will you try it, too?" I asked, still holding the paper.

"I'd rather not."

"That bitter, huh?"

I slowly placed the PTC paper on my tongue, preparing to scrunch up my nose, but—nothing. Just like the plain paper.

"That's amazing," I said.

"You and about twenty-five percent of humans are unable to taste bitterness," he said.

He explained that scientists think the ability to taste bitterness was important to our ancestors so that they would avoid food that was toxic to them. This is the opposite of why we like to eat sweet things—because these foods are usually safe to eat. I asked what possible evolutionary advantage there would be for 25 percent of people not to be able to taste bitterness.

"We don't know," he said, telling me that his research and other studies on taste can help people with taste and smell disorders. How people taste might even offer clues to diseases such as obesity.

For hearing, I took an online exam devised by Dennis Drayna on his Web site. The test has twenty-six very short, familiar tunes such as "Yankee Doodle" and "Silent Night" that one listens to and then answers yes or no to whether they were played correctly.* This simple test is part of Drayna's research into identifying genes that are responsible for communication disorders, such as problems with auditory pitch recognition. My results:

Hearing Test Results
You correctly identified 24 tunes (out of 26) on the Distorted Tunes Test. Congratulations! You have a fine sense of pitch.

In another test I took in La Jolla, California, near San Diego, researchers at a start-up company called NeuroVigil investigated what my brain was doing during the one-third of my life when I am asleep. A young neuroscientist at the Salk Institute and the founder of NeuroVigil, Philip Low, tested me while I was snoozing, using the prototype of an invention called the iBrain, which used a single electrode attached to the forehead that measured brain activity while I slept. Most sleep-measurement devices are multiple electrode caps that need to be run in a controlled setting, such as a hospital. The iBrain, when it is finished, will be used in a person's home—or, in my case, it was used in the bedroom of a suite in a hotel next door to the Salk. The finished iBrain will be wireless, said Low, and will report data during the night to NeuroVigil's data centers via the Internet. NeuroVigil, which plans to sell the iBrain to researchers, physicians, hospitals, businesses, and consumers, will use a patented algorithm to analyze the brain patterns. The program, said Low, uses complex grids to detect

* Try the test yourself at www.nidcd.nih.gov/tunetest.

structural changes that occur in a sleeping person's brain that reveal neurological disorders. "We are using sleep as a microscope to study brain activity," said Low.

With long, wavy, black hair, a penchant for European-cut suits, and a charismatic intensity, Low came to the Salk in 2001. In 2007, he finished his Ph.D., which he said was just one page in length, with a very long appendix titled, "A New Way to Look at Sleep: Separation & Convergence." Still in his twenties, Low is now a fellow at the Crick-Jacobs Center for Theoretical and Computational Biology at Salk. When I met him, Low was getting attention from major media for using his iBrain technology to test brain waves in songbirds. Using some of his new algorithms, Low analyzed EEGs of zebra finches and discovered in these bird brains a hereto unknown similarity to mammalian brains. They show periods of rapid-eye-movement (REM) sleep, slow-wave sleep (SWS), transition stages, and quick EEG transitions. "No one thought that birds, which lack a neocortex, would show these patterns," he said. He also was able to record patterns in the sleeping finches' brains that were identical to when they were awake and singing, possibly suggesting that they were dreaming about singing, findings that may offer clues to how humans dream.

Low talked fast and excitedly, with a slight European accent from his upbringing in Paris, Monaco, and Switzerland. He told me over dinner at the resort in La Jolla that the iBrain technology and his algorithms could be used to detect neuro-abnormalities that can cause shifts in how the brain performs in EEGs much in advance of cognitive symptoms. Low believes the device will be added to the growing panoply of diagnostic tools that range from cholesterol levels to genetic profiles and that one day scanning brain waves with his portable device will be routine for everyone from transportation workers to soldiers. He even sees applications for detecting pandemics and bioterrorism threats, although he wouldn't elaborate, he said coyly, for proprietary reasons.

I suggested to him that this product could face some of the issues that confront genetic testing if it is used in the wrong way by employers, the government, or insurers to screen for abnormalities and behavioral quirks—or, possibly, dating services such as Match.com as part of the vetting for your next date.

"Like any test of this sort, we would need to protect people from abusing this technology," he said.

After dinner, Low and his team from NeuroVigil, plus friends and observers from Salk and MIT, led me to my suite, where they had set up machines to monitor my sleep and to make sure the iBrain was functioning properly. This was the first time they had tested the device outside of a hospital or a clinic, so they brought in traditional sleep-monitoring devices to make sure they were getting a strong signal from my snoozing brain.

After I got ready for bed and was wearing a soft terry-cloth robe provided by the hotel, Low attached the single electrode to my fore-head and connected it to the device: a bundle of electronics and a battery crammed into a stuffed otter. Low smiled. "We didn't know how to protect the electronics, so we went to Ralphs grocery store and bought this otter."

"It was just the right size," said one of the engineers.

"You want me to sleep with the otter?" I asked.

"Yes," said Low as everyone laughed.

Low was worried that I would have trouble falling asleep with an electrode stuck on my head and the wires—which will be eliminated in the upcoming wireless version—leading to the fluffy otter. But he didn't know how easily I could snooze, especially where brain tests were concerned. Soon after they turned out the light and closed the door to my separate bedroom, I quickly fell asleep while the team gathered around the equipment in the anteroom.

The next morning, I woke up to find the otter on the floor and Low and his team gone. Later that morning, I caught up with them, and they said that they had stayed until about 4 a.m., when they got what they needed and left. Several weeks later, I received my results. (See the graph on page 312.) I was normal except for a funny alpha wave that sometimes suggested alcoholism or a debilitating muscle disorder, but Low said that the wave pattern was more likely caused by the placement of the electrode. Here is NeuroVigil's report on my brain while sleeping:

Patient had a good Sleep Efficiency (92.7%). Sleep Onset Time was normal. Distribution of sleep stages was typical in that he displayed greater SWS [slow-wave sleep] early in the night and a greater tendency for REM [rapid eye movement] later in the night. REM latency and Total REM time were normal within this recording period. The patient is not likely to suffer from Depression or Sleep Apnea. The patient also displayed

NeunVigil Sleep Test: The Author's Results

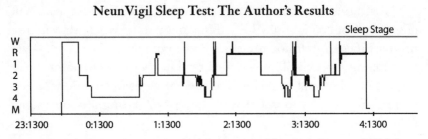

The author's brain waves registered while sleeping. The thick lines are REM sleep.

Alpha rhythms superimposed on the EEG throughout the night, which is most likely due to the frontal positioning of the sensor rather than fibromyalgia [a debilitating syndrome involving pain and fatigue in muscles, joints and bones] or alcoholism. Total wake time for the bedtime period was short. In-depth analysis of sleep stages using SPEARS [Low's algorithm] revealed no sign of pathologic brain rhythm generation.

One other test I took provided me with a tiny peek into a realm that will dominate medical research in coming years: the proteome. A proteome consists of the thousands of enzymes, hormones, and other strings of amino acids that conduct most of the business and activity in our bodies. Proteomes provide everything from superstructure materials, such as collagen in skin and bones, to enzymes, such as the SIRT1, that activate processes large and small in cells, tissue, and organs, body-wide, in a dynamic, ever-changing process. Humans produce about a million different proteins, which are made according to instructions from genes in the genome, with different sub-proteomes working in individual organs and systems such as the cardiovascular and the respiratory—or, for this test, the proteome contained in my blood.

Oddly enough, this test began at the launch party for a genomics company—Navigenics—in New York City, where I met one of that company's cofounders, David Agus. A very busy man, Agus helped develop Navigenics as a sideline to his day job: a cancer doctor and a researcher at the Cedars-Sinai Medical Center in Los Angeles. One of Agus's positions there is director of the Spielberg Family Center for Applied Proteomics, which has been working to use detectable changes in patients' proteomes to diagnose and to better treat cancer.

This research would form the scientific basis for my proteomic assessment, although this was only part of the technology that would be deployed to see what was mucking around in my blood. The other crucial bit would be supplied as the result of a series of phone calls to Agus in 2006 from a former vice president of the United States, a billionaire venture capitalist, and a famed entrepreneur—these were Al Gore, John Doerr, and Bill Berkman. All of them phoned Agus with the same suggestion: that he link up with the quintessential California-tech mind of inventor Danny Hillis.

For years, Hillis was a major force at the Walt Disney Company's legendary Imagineering Unit, designing computer systems for theme-park rides and other highly complex projects that involved integrating visionary ideas with unseemly mounds of data. In 2000, he left Disney and cofounded Applied Minds, a secretive firm that helps companies and governments solve big problems in aerospace, entertainment, electronics, and biotechnology, among others. Applied Minds' known clients include General Motors, Sony, NASA, and aerospace and defense contractor Northrop Grumman. Hillis and Agus both laughed when they told me about the calls they got from Gore, Doerr, and Berkman.[*] According to Agus, Gore said, "You gotta meet this guy from Disney." They did meet and ended up cofounding Applied Proteomics, combining Agus's biology and Hillis's computer savvy to try to make sense of the hyper-complicated realm of the human proteome, where even now inside you billions of interactions are happening, everything from processing that glass of orange juice you just drank (loaded with hidden sugars) to fending off a virus you inhaled while riding to work next to that sniffling man on the bus.

Once more, I had blood drawn and shipped to Applied Proteomics headquarters in Glendale, California, north of Los Angeles—to a couple of low, nondescript buildings across the street from the sprawling campus of the Walt Disney Company. There's not much else I can say about what Agus and Hillis are doing at Applied Proteomics. They asked me to sign a nondisclosure agreement before running my test and told me that the company would be in stealth mode for an indeterminate amount of time as they develop their technologies.

---

[*] Doerr is a major investor in Navigenics, and Gore is now a partner at Doerr's venture firm, Kleiner, Perkins, Caufield and Byers.

When my results were ready, I traveled to visit the top-secret Applied Proteomics headquarters, where Hollywood, Washington, and Silicon Valley triangulate. I wondered what I would find. I had read a 2005 *Wired* story by Xeni Jardin that described Danny Hillis taking her down a hallway at Applied Minds—Applied Proteomics' parent company—that dead-ended into an old-fashioned red phone booth. "The phone rings," Jardin wrote. "He places receiver to ear. 'The blue moon jumps over the purple sky,' he says, and hangs up. Suddenly, the booth becomes a door, swinging out to reveal a vast, open room filled with engineers, gadgets and big ideas. . . . 'This is where the secret laboratories are,' Hillis says."

I visited Applied Proteomics on a hot June day in Glendale and quickly entered an alternate reality where bookshelves and wall lamps were tilted and false skylights glowed. I looked for the red phone booth but didn't see it, though it might have been in another building. But I was given a tour of a facility filled with computers, electronics, and a central conference center overlooking yet more computer banks that reminded me of a high-tech villain's lair in a James Bond movie, with touches of the hideouts of the Joker and the Riddler in the 1960s *Batman* series, but smaller.

One room we visited was where proteins in my blood had been smashed into fragments of amino acid chains in the company's mass spectrometers—the same sort of instruments used at AXYS in Canada to detect levels of environmental toxins in my blood. This time the machine was measuring the mass of the protein fragments in my sample and others. "We compare these results to databases of known or predicted [protein] fingerprints in order to identify the original, parent protein," said Applied Proteomics scientist Bruce Wilcox, who, along with fellow scientist Dan Ruderman, was conducting the tour. In a few minutes, Ruderman would show me my results. But first, I sat down with David Agus and Danny Hillis in the company conference room.

In his early forties, Agus looked the dapper physician in a dress shirt with cuff links but no tie, while Hillis's graying hair was pulled back into a ponytail. His round face made him look like a big kid; so did his flannel shirt and tennis shoes. Right away, he plopped his feet up on a chair as the two men explained what they were doing in this fledgling field.

"We're the first settlers in the West," said Agus, sitting eagerly on the edge of his chair, a contrast to Hillis's laid-back style, "and we don't yet know how the kids will turn out in this new land of proteomics."

"Fundamentally, we're using the mass spec to study the dynamism of disease, how it really works at the proteome level. It's a physics and computational problem. That's why Applied Minds did it, because it encompasses physics, chemistry, and biology—"

"And it helps mankind," added Agus, finishing Hillis's sentence, something these two did a lot.

"That's right," said Hillis. "At Applied Minds we look for three things in new projects. First, they need to cross-fertilize among different disciplines. We also want to make money. And we want to change the world. We often turn down projects when they meet only two out of the three."

"As a physician, I'm also looking for better ways to get answers for patients—" said Agus, who is cut off by Hillis.

"What he means is he can look at complicated things and come up with what's broken and how to fix it."

"We need new ways to measure complicated things," said Agus, "to better understand interactions of bodily systems, of proteins and the environment."

"You can sequence the genes and measure the environmental, but it doesn't tell you what's going on unless you combine it all," said Hillis.

"It's superficial," agreed Agus.

Hillis told me they took a picture of my proteome using the mass spectrometer—or, at least, a picture of the proteins swirling around in my blood on a certain day in April 2008, when I had my blood drawn. But it was a picture that included measurements at the atomic level of such complexity that it took about 24 gigabytes of storage space to hold all of the sample data (the picture I saw represents only about 1/24th of the total data from the sample) —that was fourteen hundred times the amount of digital space it took to store this entire book.

"It's a high-res picture of your whole proteome," said Ruderman.

We got up and walked over to an enormous flat-screen monitor on a wall of the lair, a TouchTable device invented by Hillis that models complex three-dimensional shapes on a flat screen—aircraft, buildings, cars, and proteomes. Ruderman clicked on my proteome file using his fingertips and pulled up functions that zoomed up and down the screen like an iPhone—although the touch-table technology had come first. Up popped a field of yellow dots that looked like a 3-D star field from outer space.

"This is your plasma," said Agus. "This is you."

It looked beautiful, I thought, not expecting something inside my blood to appear in the guise of a starry, starry night. "The star field you see is a multidimensional representation of the data from your sample," explained Ruderman. "Horizontal X and vertical Y positions reflect the physical properties of the proteins." The colors are used to indicate the abundance of a protein at a given position in the three-dimensional space, with the most abundant appearing to be "closer" in the star field. "The color scale we use is arbitrary and can be adjusted at will, given the analysis that is being undertaken. For example, the spectrum of color might represent the range of abundance of proteins in a single sample. We also use color to compare two samples."

Ruderman and Agus showed me the identity of some of the proteins by pulling up selected "stars" on the screen: there was plasma albumin, an abundant protein that maintains pressure in the blood system and transports fatty acids and other molecules; and fibrinogens that help coagulate blood. The initial processing of my blood had identified several hundred proteins out of thousands of possibilities. "This was more than we usually get in a quick look," said Ruderman. "We got a rich sample from you." He said that if they were running a sample from a patient, they would be more focused in their search. "To a significant extent, we can 'dial in' what part of the proteome we want to examine, depending on the focus of the experiment."

"This is like Google Earth," said Agus, "but much deeper."

Ruderman zoomed in on a yellow dot and it said "Bos Taurus," a cow protein. "That could be from lunch," he said. "Did you eat beef?"

Later he told me he had been kidding, that this protein was probably a human protein that looked similar enough to a cow protein that the analysis program made the connection. "That is part of the power of having genome information from related species," he said, "they can help each other fill in the blanks."

So what did the "stars" forecast for me?

Agus says I appear to be normal, although he adds that the technology is not yet ready to ascertain normal versus abnormal.

Ruderman explained that they are comparing patient results to online databases from the NIH and elsewhere and to samples that had been designated as "normal," looking for both structure and quantity.

Their database will be created and developed for a specific disease to test patients in one clinical trial at a time, Agus explained.

"We can hopefully detect anything strange or abnormal in the profiles that we identify from the clinical trial specimens," said Agus. "But we need to run lots more clinical trials over time to understand what all of this data means. Right now, we don't know much, it's so early, but we're on the verge of getting there."

"So, from looking at this, you can tell I don't have cancer."

"You don't have cancer from what we can tell here, though right now we haven't run enough data sets to answer that question. But our hope and belief is that over time we will have those answers, and we can not only identify whether someone has cancer but also help guide the treatment of the disease."

I left Glendale, delighted to be cancer-free in my blood, inasmuch as Dan Ruderman and David Agus could tell from smashing up my proteins in their mass spectrometer. Though, of course, their proteo-glimpse of only 320 proteins covered a minuscule fraction of all of my proteins. Nor did this investigation—or any other I had taken—cover the other "omes" that exist in our bodies.

Take the microbiome: the trillions of microorganisms such as bacteria and fungi that live inside each of us, mostly harmoniously. I spoke with geneticist Craig Venter about scanning my microbiome, and he agreed, but even at a steep discount his institute needed to charge me $40,000, which meant putting off this test for now. Another "–omics" I want to test is epigenomics, a field that is studying how our DNA can be repro-grammed by environmental exposures and other factors. Until recently, geneticists didn't realize this happened. They assumed that DNA stayed essentially the same throughout one's lifetime, which turns out to be untrue. I'm not sure how one could be tested for epigenetic activity, but perhaps one day there will be a simple test available.

I could go on, but I think, for now, that this Experimental Man has taken quite enough tests. It's time to draw some conclusions about what I've learned.

First, I'm incredibly lucky that nothing serious was found, dooming me to grow ill and die anytime soon—although apparently I need to watch my weight more than most people do. Scans of my genes, blood, brain, and body revealed no hidden horrors, although it's possible they missed something. For my family, who took only the genetic tests, we found nothing immediately alarming, although I remain frustrated that science is unable to nail down the exact cause of my brother's brittle bone disease or to find an effective treatment.

One criterion for the success of a medical test is whether it changes a person's treatment or lifestyle. By that measure, I did make two substantial changes: to my diet and in my exercise preferences. I have described both alterations to my lifestyle in detail, so I won't revisit them here, except to say that I have stuck to both changes, more or less: to the healthy diet suggested by Dr. Melina to avoid activating the heart-attack scenario provided by the Entelos model. I also avoid running on asphalt and other hard surfaces to preserve what's left of my battered knee joint, as shown on Kevin Stone's MRI. Living in a city, this has effectively meant an end to the running I have always enjoyed, although I've picked up the slack by bicycling like a fiend.

Beyond this, I have to admit that little of the testing has had much effect on me and how I live or how I view my future health and well-being, other than to teach me a great deal about the current state of the science and what experts who are developing these experiments expect to be available in the near future. My generally good results have also reinforced my almost hardwired, dare I say religious, faith that I'm healthy, although I was at least temporarily rattled by a few of the brain tests that suggested my gray matter is aging. Yet I still can cling to my brain-age score of twenty-five years old, even if I did not do quite this well the second time around.

But I have acquired a deeper understanding of the potential disruptive power of these tests that deign to forecast my future and yours—or at least to suggest a *possible* future. A troubling example is how my family would have reacted to a genetic test for my brother's condition if it had been available when he was born. Without question, they would have treated Don differently—any parent would, with a child at high risk for brittle bone disease. Being ignorant of his genetic secret, they didn't think twice about allowing him to participate in sports, roughhouse, and take risks.

Don's case is extreme. Yet buried in my gigabytes of data, should anyone else be mad enough to attempt to replicate the Experimental Man project, is the potential revelation of equally disturbing and dangerous information about individuals and their families. Take breast cancer. I don't dwell overly on the possibility that my daughter, Danielle—the experimental daughter—has some high-risk genetic variations. Nor, apparently, does she. I remain concerned, however, and will encourage her to be extra vigilant about testing herself.

We remain years, perhaps decades, away from an iHealth-like device that will collect and integrate massive amounts of data, from genetic proclivities to proteomic dynamics and daily exposures to toxic chemicals. But the day will come when such a device will be available, possibly in several choices of colors and accessories.

Meanwhile, huge chasms in our knowledge need to be filled before the Experimental Man will be complete down to every SNP, copy number variation, and synapse. Perhaps the biggest gap is the affect of the environment on our DNA, cells, organs, and bodies. Few of the tests I've taken for the Experimental Man project provide much useful information about how the environment interacts with genes, neurons, and proteomic systems and pathways. As cardiologist Eric Topol of Scripps told me, "You could almost say that giving genetic results without environmental data is inaccurate." The same is true about any system in the body, since the whole point of evolution has been to create defenses inside organisms to fend off the daily onslaught of the environment, from natural challenges such as UV rays and flu viruses to the thousands of toxic chemicals that we humans have unleashed into the air, the water, and the earth—and into the food we eat, such as that halibut I caught off the coast of Bolinas.

Of course, my last test for this book offers me a convenient excuse to ignore the Entelos predictions and even the disturbing news delivered by neuroscientist Jim Brewer that my brain is shrinking. This is the Sirtris test for SIRT1 mRNA. This is an ace in the hole for a man who is inclined to believe he is healthy, no matter what a bunch of other tests say.

And yet I remain ambivalent about the prospect of living a very long life—to age 122 or beyond, if I can acquire a steady supply of SIRT-501, Sirtris's super-resveratrol pill. What I would prefer, rather than growing steadily older for another seventy-two years, would be to run a different sort of experiment: preserving my body in some sort of stasis while I am still relatively young, to be awakened periodically to see how things are going. This technology does not exist, but perhaps one day it will.

I'd like to wake up in twenty-five or fifty years to see how the Experimental Man revolution is going: whether all of the self-knowledge that our children in the 2030s or the 2050s will have access to will be beneficial. I'd like to know what toxic chemicals were banned and what replaced them—and also whether humans came to their senses in time about global warming. And, of course, I'd want to know whether the Kansas City Royals broke .500—or, for the sake of Jordan Grafman, whether the Chicago Cubs went another twenty-five years without winning the World Series. Most of all, I'd like to see how my children are doing and, hopefully, my grandchildren—although, come to think of it, I would never want to miss the years when they were moving from teenagers to young adults by putting myself in stasis.

With all of the tests run and the results in, I return for one last visit with Josh Adler, nearly two years after our first meeting. He looks the same, although his glasses have become more stylish, even as his hair is slightly longer and unkempt, in that same "too busy to get a haircut" look. He does not conduct a full exam this time. "It doesn't seem necessary after all you've been through," he says. He does check my blood pressure and my weight and listens to my heart. I weigh eight pounds less than when I visited two years earlier, and my blood pressure, which had been slightly high, was a normal 134/82 and my pulse a healthy 70.

"I'm glad that blood pressure is down," he says. "It may be the new diet."

"You think so?" I say. "If it is, then we were both wrong about how this project would turn out. We both thought that I would not learn anything dramatically different about myself."

"To my surprise, we did learn something with those tests on your heart. That was unexpected, that the results were so different from your Framingham score.* And yet I'd have to have more clinical data on the test before I really trusted it as a predictor for your future heart disease. Until then, I think the genes from your parents are still the best predictor."

"What about the rest of the tests?"

"You took a risk in taking these tests," he says. "You could have found a lot of false positives. Every test has the ability to be wrong, which requires further tests and causes anxiety and concern. Let's say the radiologist had used the word *mass* on your kidney CT scan instead of *cyst*. We might have needed to investigate that further, when most likely it was nothing."

"What about the genetics?"

---

* To recap, the Entelos model gave me a 28 percent chance of having a heart attack in ten years, compared to a 4 percent chance according to the Framingham heart study, which is the standard of care for physicians.

"At the moment, we are getting lots of new information from these tests, but physicians like me have few strategies about how to use them for preventive care."

I ask him how many patients have brought in their own test results from the online genetic testing sites or CT scans or other self-tests. He answers that only a handful of people have brought in results from 23andMe and the other sites. "I have sent a couple of people to DNA Direct to take the BRCA tests, because they were worried about their insurers finding out." He has had maybe fifty patients bring him body scans that they have had taken at retail centers on their own. Only one had anything noteworthy—"a person with a nonspecific nodule in the left lung," he says. "This caused the person great anxiety. With further testing, turns out he had nothing dangerous."

Josh reiterates that wearing seat belts and not smoking are still going to help a person stay healthy more than a small risk factor for a disease caused by a genetic variation. "People are worried about bisphenyl A right now, and they're smoking," he says.

I ask Josh where he thinks all of this testing is heading. "Do people want to live forever, with no disease at all?"

Josh doesn't answer for a moment and then sighs, looking suddenly weary—the friendly, smart doc who always appears sympathetic and full of energy for his patients is experiencing a momentary lapse. "Maybe. But I doubt it. Life is hard. And anyway, I can tell you that day is such a long way off; it's not something I would give much thought to. I am around sick people all the time. Some are dying. It would be great to help them more than we can, but I have to deal in the here and now."

A couple of days after my exam with Josh Adler, I call my daughter on Skype and catch her just returning from a day's kayaking on a frigid river in Scotland. She is still bundled up in rain gear, her blond hair messed up and wild as she hovers in the wavy, ghostlike image. I ask her whether she has a minute to talk about the project, and she says she has just a couple of minutes, that she needs to change and meet some friends for dinner. I ask her whether she has learned anything from her testing.

"Not really," she says with her usual bluntness. "But I think I'll be finding out a lot more in a few years. Then I'll pay more attention, I guess." She pulls off her rain jacket and says, "Dad, I've gotta go."

The screen goes dark, and I wonder again whether I did the right thing involving her—and then I put this thought to rest. Of course,

I did the right thing. Really, this whole experiment has been about her. She is the product of my genes and her mother's genes. We have provided the basic programming for who she is and how she behaves, and we have provided her with whatever defenses she has against environmental assaults that might make her get sick or die—just as I am in part the product of my parents' genes. Yet there is a huge difference between my generation and hers: they will know the details of those similarities and differences and a whole lot more.

This makes my daughter the true experimental woman, and her generation the experimental generation. And it will be up to them to decide to take this new torrent of data in stride or to reject it—or to let it overwhelm them with worry.

I may live to see what happens to her and to society, or not. But it's my daughter, my two sons, and their peers who will decide whether there is a limit to what they want to take away from this vast human experiment to pass on to their own experimental children.

ACKNOWLEDGMENTS

I am grateful to the many people, companies, universities, and labs that took time out of their busy schedules to help with the Experimental Man project. This was a labor-intensive book that made demands on some of the smartest and most innovative people in the world. My family was also understanding and supportive in ways large and small.

First, thanks to my parents, my brother, and my daughter for taking the courageous step of joining me in getting genetic tests and in submitting to interviews and requests to do everything from giving blood to vetting the manuscript. I am grateful to my father and mother for their encouragement and help with edits. Also, my love and appreciation to Alex, Sander, Bobby, Maddy, and Tessa—and to my beloved, Lisa—they all put up with my spending many hours holed up writing on weekends and on vacations. My love goes out to my brother, Don, and his family and to Helen and Tony Conte.

I appreciate the patience and encouragement of my editor, Stephen S. Power, and my fabulous agents, Mel Berger and Eugenie Furniss— and special thanks to Ellen Wright and Kimberly Monroe-Hill at Wiley.

Special thanks to Andrew Robertson, Nadia Mustafa, Nathaniel David, Gregory Stock, and Kevin Kelly.

Appreciation to my media colleagues and friends: Moira Gunn of "Biotech Nation"; Hilary Stout, Mark Stein, Dan Colarusso, Blaise Zerega, Jacob Lewis, and Joanne Lipman of *Portfolio*; Tim Appenzeller and the editors of *National Geographic*; Adam Fisher and Chris Anderson of *Wired*; Stephen Petranek, Corey Powell, and Bob Keating of *Discover*; Bruce Auster of NPR's *Morning Edition*; Richard Harris; and Michael Caruso.

Thanks to Martin Reese and Omicia, Po Bronson, Olivia Judson, Jane Ganahl, Cathryn Ramin, Julian Guthrie, Jen Itzenson, Rodes Fishburne, Caroline Paul, and the San Francisco Writer's Grotto.

Much thanks for special help, in some cases for many years, from Josh Adler and Wendy Frago of the University of California at San Francisco; Leo Trasande of the Mount Sinai Medical Center; Francis Collins, formerly of the National Human Genome Research Institute; Craig Venter of the J. Craig Venter Institute; Arthur Caplan of the University of Pennsylvania; Eric Topol of the Scripps School of Medicine; Judy Illes of the University of British Columbia; David Agus of Cedars-Sinai Medical Center, Navigenics, and Applied Proteomics; Topher Sharp and Ralph Horwitz of the Stanford School of Medicine; Henry Greely of Stanford Law School; and fisherman Josh Churchman.

Genes: Thanks to Kari Stefansson, Edward Farmer, Berglind Ólafsdóttir, and deCode Genetics; Jonathan Rothberg of Raindance Technologies, 454 Life Sciences; Charles Cantor and Andi Braun of Sequenom; Jay Flatley and Maurissa Bornstein of Illumina; Steve Fodor and Kat George of Affymetrix; Ann Walker of the University of California at Irvine; Randy Scott of Genomic Health; Linda Avey and Andro Hsu of 23andMe; Mari Baker, Dietrich Stephan, Amy DuRoss, Michele Cargill, and Kari Kaplan of Navigenics; Martin Reese and Andrew MacBride of Omicia; David Altshuler at Harvard University; Peter Byers of the University of Washington; Ryan Phelan and Trish Brown of DNA Direct; Michel and Yves DuBosc; Robert DuBose; Bennett Greenspan of Family Tree; Kathy Duncan Cawley; Bryan Sykes of Oxford University; Jack Horner of the Museum of the Rockies; Michael Christman and Courtney Sill of the Coriell Institute; James Lupski and John Belmont of Baylor College of Medicine; Greg Feero of the NIH; Ellen Barron and Doug Fambrough of Oxford

BioSciences Partners; Barb Short, Joy Redman, Charles Strom, and Raj Pandian of Quest Diagnostics; Dariush Mozaffarian of Harvard Medical School; Cathy Fomous of the NIH; Steven Murphy and Jonathan Freed of Helix Health; amd Michael Cariaso of SNPedia.

Environment: Thanks to Karin Broberg of Lund University; Robert Taylor of Texas A&M; Mathew Rand of the University of Vermont; Åke Bergman of Stockholm University; James Pirkle and Antonia Calafat of the Centers for Disease Control; Karl Rozman and John Doull of the University of Kansas; Linda Birnbaum of the Environmental Protection Agency; Ron Leon at Mount Sinai Medical Center; Dale Hoover and Laurie Phillips of AXYS Analytical Services; Denise Jordan-Izaguirre of the federal Agency for Toxic Substances and Disease Registry; Shelley Brodie and Leo Rosales of the EPA; Dennis Prevost; Ed Fitzgerald of the State University of New York at Albany; Carolyn Mattingly of Mount Desert Island Biological Laboratory; Chris Austin, Ruili Huang, and Menghang Xia of the National Institutes of Health's Chemical Genomics Center; George Gellert and Pradeep Babu of Napo Pharmaceuticals; Brenda Weis of the Broad Institute; and the Environmental Working Group.

Brain: Thanks to James Brewer of the University of California at San Diego; Eric Wassermann of the National Institute of Neurological Disorders and Stroke; Adam Gazzaley and lab at the University of California at San Francisco; Pat Turk and Keith Wesnes of Cognitive Drug Research; Philippe Goldin, Kelley Werner, and James Gross of Stanford University; Dimitrios Kapogiannis and Jordan Grafman of the National Institute of Neurological Disorders and Stroke; Ifat Levy, Robb Rutledge, Stephanie Lazzaro, Elizabeth Phelps, and Paul Glimcher of New York University; Michael Koenigs and Eric Wassermann of the NIH; Steven Lamm of New York City; and John Mazziotta of the University of California at Los Angeles.

Body: Thanks to Tom Paterson, Alex Bangs, Michael Gishizky, and James Karis of Entelos, Inc.; Ron Krauss of the Children's Hospital Oakland Research Institute; Melina Jampolis of San Francisco; Interleuken Genetics; Judy Yee and Marilyn Pique of the Veteran's Administration Hospital in San Francisco; Heartscan; David Sinclair, Christoph Westphal, and Olivier Boss of Sirtris Pharmaceuticals; and Leonard Guarente of MIT.

Epilogue: Thanks to Dennis Drayna of the NIH; Philippe Low and his teams at the Salk Institute and NeuroVigil; David Agus, Danny Hillis, and Dan Ruderman of Applied Proteomics; and Lee Hood of the Institute for Systems Biology.

Also, thanks (and apologies) to anyone I have forgotten!

Passages in the book originally appeared as articles in *Wired, National Geographic, Portfolio, Portfolio.com, Discover, Fortune,* and *Technology Review*.

NOTES

For more information on sources and a guide for finding information on genes, environment, brain, and body, go to www.experimentalman .com

Introduction: The Fish and Me

The epigraph to this section is taken from *The Journals and Miscellaneous Notebooks of Ralph Waldo Emerson*, volume VII, 1838–1842 (Cambridge: Belknap Press, 1969).

3   *At the right dose*   U.S. Department of Health and Human Services Agency for Toxic Substances and Disease Registry, "Toxicological Profile for Mercury" (March 1999), pp. 58–66.

4   *In my "before" test*   U.S. Environmental Protection Agency, *Draft Report on the Environment Technical Document* (June 2003), pp. 4–56.

4   *Other chemicals, once on board*   Christopher Lau et al., "Exposure to perfluorooctane sulfonate during pregnancy in rat and mouse. II: Postnatal evaluation," *Toxicological Sciences* 74 (2003), pp. 382–392.

5   *Some people take more*   Hipolito M. Custodio et al., "Polymorphisms in glutathione-related genes affect methylmercury retention," *Archives of Environmental Health* 59 (November 2004), pp. 588–595.

7   *I do have a DNA marker*   M. R. Munafò et al., "Association of the dopamine D4 receptor (DRD4) gene and approach-related personality traits: Meta-analysis and new data," *Biology Psychiatry* 63 (January 2008), pp. 197–206.

8   *These results were even more*   David Ewing Duncan, "The Pollution Within," *National Geographic* (October 2006).

9   *The fishes' exposure levels*   U.S. Environmental Protection Agency, "An Assessment of Exposure to Mercury in the United States," *Mercury Study Report to Congress* (December 1997).

9   *In 2004, her team*   Custodio, "Polymorphisms in glutathione-related genes."

10   *Even if my genes*   U.S. Food and Drug Administration and Environmental Protection Agency, "What You Need to Know about Mercury in Fish and Shellfish" (2004), www.epa.gov/waterscience/fish/advice.

11   *"There are . . . things"*   Fyodor Dostoyevsky, *Notes from Underground* (1864), www.kiosek.com/dostoevsky/library/underground.txt.

14   *"Illness is the night-side"*   Susan Sontag, *Illness as Metaphor and AIDS and Its Metaphors* (New York: Anchor, 1989), p. 3.

16   *Take the BRCA tests*   National Cancer Institute, "Genetic Testing for BRCA1 and BRCA2: It's Your Choice" (February 2002), www.cancer.gov/cancertopics/factsheet/risk/brca.

21   *I conducted no self-experiments*   Karl Grandin, ed., *Les Prix Nobel: The Nobel Prizes 2005* (Stockholm: Nobel Foundation, 2006).

21   *Nor am I trying*   Douglas Martin, "John Paul Stapp, 89, Is Dead; 'The Fastest Man on Earth,'" *New York Times*, November 16, 1999.

1. Genes

25   *I was working*   David Ewing Duncan, "DNA as Destiny," *Wired* 10, no. 11 (November 2002).

33   *The quest to know*   David Ewing Duncan, *Calendar: Humanity's Epic Struggle to Determine a True and Accurate Year* (New York: Avon, 1998), p. 18.

35   *One of the more telling*   Patrick Sulem et al., "Genetic determinants of hair, eye and skin pigmentation in Europeans," *Nature Genetics* 39 (October 2007), pp. 1443–1452.

35   *Like most SNP markers*   H. Elberg et al., "Blue eye color in humans may be caused by a perfectly associated founder mutation in a regulatory element located within the HERC2 gene inhibiting OCA2 expression," *Human Genetics* 123, no. 2 (March 2008), pp. 177–187; R. A. Sturm, "A single SNP in an evolutionary conserved region within intron 86 of the HERC2 gene determines human blue-brown eye color," *American Journal of Human Genetics* 82, no. 2 (February 2008), pp. 424–431.

35   *About one-third of Caucasians*   Sulem, "Genetic determinants of hair, eye and skin pigmentation in Europeans."

35   *In a noncarrier*   F. O. Walker, "Huntington's disease," *Lancet* 369, no. 9557 (January 20, 2007), pp. 218–228.

37   *It causes the degeneration*   National Eye Institute, "Age-Related Macular Degeneration" (April 2006), www.nei.nih.gov/health/maculardegen/armd_facts.asp.

37   *AMD is the leading*   Macular Degeneration Partnership at the Discovery Eye Foundation, www.AMD.org.

38   *Also, gene chips*   James R. Lupski, "Structural variation in the human genome," *New England Journal of Medicine* 356 (March 2007), pp. 1169–1171.

39 *In 2007, Venter*   J. Craig Venter, *A Life Decoded: My Genome, My Life* (New York: Viking, 2007); S. Levy et al., "The diploid genome sequence of an individual human," *Public Library of Science Biology* 5, no. 10 (2007), www .jcvi.org/cms/research/projects/huref/overview.

39 *In 2007, the NIH*   National Human Genome Research Institute, "Study to Probe How Healthy Younger Adults Make Use of Genetic Tests" (May 2007), www.genome.gov/25521052.

41 *This is bad news*   Anna Helgadottir et al., "A common variant on chromosome 9p21 affects the risk of myocardial infarction," *Science* 316, no. 5830 (June 8, 2007), pp. 1491–1493; N. J. Samani, "Genomewide association analysis of coronary heart disease," *New England Journal of Medicine* 357 (August 2, 2007), pp. 443–453.

41 *"Because they are usually safe"*   The side effects of statins haven't been well studied. Reported side effects are rare, they include muscle pain, cognitive problems, and impotence.

42 *But there are two nearby genes*   Anna Helgadottir et al., "A common variant on chromosome 9p21 affects the risk of myocardial infarction."

42 *No actual link*   Cardiologist Eric Topol of the Scripps Research Institute, in conversation with the author, July 9, 2008.

42 *DeCode, however, was kind*   DeCODEme can be found at www .decodeme.com.

43 *In the deCode study*   Anna Helgadottir et al., "A common variant on chromosome 9p21 affects the risk of myocardial infarction."

44 *Not long after my*   Navigenics can be found at www.navigenics.com.

44 *Nearly 158,000 of these people*   Navigenics Web page on heart attack, www .navigenics.com/member/healthcompass/Summary/d/MI.

44 *Soon after this, I add*   23andMe can be found at www.23andme.com.

46 *Differences in calculating*   Navigenics, "The Science behind the Navigenics Health Compass Service," www.navigenics.com/science/WhitePaper.

46 *The U.S. Department of Health*   Helen Palmer, "Genetic Test Standards," American Public Media (December 2005), http://marketplace.publicradio. org/shows/2005/12/01/AM200512019.html.

49 *I lack a high-risk variation*   C. Kissling et al., "A polymorphism at the 3'-untranslated region of the CLOCK gene is associated with adult attention-deficit hyperactivity disorder," *American Journal of Medical Genetics*, Part B, *Neuropsychiatric Genetics* 147, no. 3 (April 5, 2008), pp. 333–338.

50 *I also have a SNP*   S. Seki et al., "A functional SNP in CILP, encoding cartilage intermediate layer protein, is associated with susceptibility to lumbar disc disease," *Nature Genetics* 37, no. 6 (June 2005), pp. 607–612.

52 *My seventy-seven-year-old father's hair*   P. Sulem et al., "Genetic determinants of hair, eye and skin pigmentation in Europeans," *Nature Genetics* 39 (October 2007), pp. 1443–1452.

55 *About one person in*   P. E. Andersen Jr. and M. Hauge, "Congenital generalised bone dysplasias: A clinical, radiological, and epidemiological survey," *Journal of Medical Genetics* 26 (January 1989), pp. 37–44.

55 *Others have only mild*   P. H. Byers, "Osteogenesis Imperfecta," in P. M. Royce and B. Steinmann, *Connective Tissue and Its Heritable Disorders: Molecular, Genetic, and Medical Aspects* (New York: Wiley-Liss, 1993), pp. 317–350. Also see McKusick-Nathans Institute of Genetic Medicine at Johns Hopkins University School of Medicine, "#166200 OSTEOGENESIS IMPERFECTA, TYPE I," Online Mendelian Inheritance in Man (OMIM), www.ncbi.nlm.nih.gov/entrez/dispomim.cgi?id=166200.

55 *Most people with this problem*   For Michael Petrucciani, see Mike Zwerin, "Michael Petrucciani: A Triumph of Spirit," *International Herald Tribune*, January 12, 1999; for Julia Fernandez, see Web site of the Disability Foundation, www.the-disability-foundation.org.uk; for more information on osteogenesis imperfecta, go to the Web site of the Osteogenesis Imperfecta Foundation, www.oif.org.

61 *"Word of First Grandchild"*   "Word of First Grandchild Is Awaited with Eagerness," *Kansas City Star*, March 10, 1958.

62. *It's likely that my*   National Cancer Institute, "Genetic Testing for BRCA1 and BRCA2: It's Your Choice."

62 *The BRCA genes*   Ashok R. Venkitaraman et al., "Chromosome stability, DNA recombination and the BRCA2 tumour suppressor," *Current Opinion in Cell Biology* 13 (June 2001), pp. 338–343.

62 *Mutations in these genes*   Tuya Pal et al., "BRCA1 and BRCA2 mutations account for a large proportion of ovarian carcinoma cases," *Cancer* 104 (December 2005), pp. 2807–2816.

62 *Female carriers have*   David Ewing Duncan, "Frontiers of Science," *Discover Magazine* (October 2005).

63 *For a more extensive*   SNPedia Web site, page on "Breast Cancer," www.snpedia.com/index.php/Breast_cancer.

63 *The first SNP associated with*   Nichola Johnson et al., "Counting potentially functional variants in BRCA1, BRCA2 and ATM predicts breast cancer susceptibility," *Human Molecular Genetics* 16 (May 2007), pp. 1051–1057.

64 *Myriad's breast cancer test*   William A. Hockett, Myriad Genetics, via e-mail, August 27, 2008.

64 *In 1994, Myriad*   Ibid.

64 *Called DNA Direct*   DNA Direct, "Breast & Ovarian Cancer Risk," www.dnadirect.com/patients/tests/breast_cancer/index.jsp.

65 *According to the site*   DNA Direct Web site, page titled "Should I Take This Test?" www.dnadirect.com/patients/tests/breast_cancer/should.jsp.

68 *He is an AG*   H. Li et al., "Candidate single-nucleotide polymorphisms from a genomewide association study of Alzheimer disease," *Archives of Neurology* 65, no. 1 (January 2008), pp. 45–53.

75 *He first got interested*   I have been unable to locate a full reference for this book; my source is Michel DuBosc.

76 *Now armed with buccal swabs*   Family Tree DNA can be found at www.familytreedna.com.

76 *Greenspan's site partners*   Bennett Greenspan, Family Tree DNA, via phone, August 2008.

76 *In 1997, Hammer*   Michael Hammer, the Hammer Lab at the University of Arizona, via e-mail, August 29, 2008.

77 *In another study, Hammer's team*   Ibid.

77 *Greenspan's company also*   Bennett Greenspan, Family Tree DNA, via e-mail, August 29, 2008.

78 *Uncle Bob turns out*   "Geno Project—U5a1a—R1b1c—Our Sicilian Family DNA Project European—Norman Sicilians or Native Italians, Sicilians?" www.genoproject.com.

78 *A sixty-seven-marker panel*   Greenspan, Family Tree DNA, via phone, August 2008.

79 *This progression was verified*   deCODEme, "Frequently Asked Questions: How Does the Ancestry Tracing Work?" www.decodeme.com/information/faq.

80 *Through Family Tree*   Clan Donnachaidh DNA Project, www.familytreedna.com/public/clandonnachaidh&fixed_columns=on.

80 *One year he traveled*   Herbert Ewing Duncan, *Logbook: An Architect's Tale*, unpublished manuscript (2008), p. 33.

81 *I strolled over*   Katherine Duncan Smith, *The Story of Thomas Duncan and His Six Sons* (New York: Tobias A. Wright, 1928).

81 *In about five minutes*   Ibid.

83 *I had a near-perfect*   Results provided to the author by Family Tree DNA. (For complete results, consult www.experimentalman.com.) The primary scientific study on this topic is Bruce Walsh, "Estimating the time to the MRCA for the Y chromosome or mtDNA for a pair of individuals," *Genetics* 158 (2001), pp. 897–912. Also consult an excellent series of tutorials by Bruce Walsh posted on his Web site at the University of Arizona in Tucson, cosponsored by Family Tree DNA (Walsh is a scientific adviser to the company), http://nitro.biosci.arizona.edu/ftdna/models.html.

84 *He had agreed to test*   Oxford Ancestors can be found at www.oxfordancestors.com.

84 *At the time this stunning genetic connection*   Bryan Sykes, *The Seven Daughters of Eve: The Science That Reveals Our Genetic Ancestry* (New York: W. W. Norton, 2001).

84 *In 1995, he confirmed*   Ibid.

84 *Sykes disproved explorer Thor Heyerdahl's*   Ibid.

85 *According to Sykes*   Dan Mishmaram et al., "Natural selection shaped regional mtDNA variations in humans," *Proceedings of the National Academy of Sciences* 100, no. 1 (January 7, 2003), pp. 171–176; also see Spencer Wells, "Deep Ancestry: Inside the Genographic Project," *National Geographic* (reprint edition, 2007).

86 *Some famous H people*   See the Web site deCODEme, http://demo.decodeme.com/ancestry/get_mitochondria.

88 *As a scientist, Horner*   John R. Horner, Montana State University, via e-mail, September 1, 2008.

88 *In 2000–2001, Horner's teams*   Ibid.

89 *This analysis was organized* J. M. Asara et al., "Interpreting sequences from mastodon and T. rex," *Science* 316, no. 5822 (April 13, 2007), pp. 280–285; also see Mary Higby Schweitzer et al., "Analyses of soft tissue from Tyrannosaurus Rex suggest the presence of protein," *Science* 316, no. 5822 (April 13, 2007), pp. 277–280.

90 *This system is present* J. William Schopf, "Fossil evidence of Archaean life," *Philosophical Transactions of the Royal Society Biological Sciences* 361 (June 2006), pp. 869–885.

90 *Scientists think that* Anthony M. Poole et al., "Evaluating hypotheses for the origin of eukaryotes," *BioEssays* 29 (January 2007), pp. 74–84.

90 *For instance, we share* Anna Salleh, "Rice Genome: Very Early Days," ABC Science (April 2002), www.abc.net.au/science/articles/2002/04/05/521385.htm.

90 *Jumping ahead a few* Samuel Aparicio et al., "Whole-genome shotgun assembly and analysis of the genome of *Fugu Rubripes*," *Science* 297 (August 2002), pp. 1301–1310.

91 *We share some 98.7 percent* Konstantinos T. Konstantinidis et al., "The bacterial species definition in the genomic era," *Philosophical Transactions of the Royal Society Biological Sciences* 361 (November 2006), pp. 1929–1940.

91 *When Horner and his crew* Horner, via e-mail, September 1, 2008.

91 *Following is a comparison* Nathaniel David, via e-mail, April 1, 2008. The SWISSPROT database can be found at www.ebi.ac.uk/swissprot.

94 *The main exception is a disease* Gudmar Thorleifsson et al., "Common sequence variants in the LOXL1 gene confer susceptibility to exfoliation glaucoma," *Science* 317 (September 2007), pp. 1397–1400.

95 *"In the brain"* G. Bart et al., "Substantial attributable risk related to a functional mu-opioid receptor gene polymorphism in association with heroin addiction in central Sweden," *Molecular Psychiatry* 9 (June 2004), pp. 547–549.

96 *For this reason, 23andMe* For information on 23andMe's system of designating studies as preliminary or established, go to www.23andMe.com/more/science.

97 *Then there is one* 23andMe, "Caffeine Metabolism," www.23andMe.com/health/pre_caffeine_metabolism.

98 *Most geneticists consider* Having too few people analyzed means that random outliers who have or don't have a gene or a disease can overly influence the results by causing the risk factors to be too low or too high. Imagine polling a hundred people in a neighborhood that, unbeknownst to you, is Republican about their choice for president. You would get a skewed result, compared to a test in a larger or more statistically relevant population. 23andMe's rating criteria:

Established Research

These topics meet our criteria for findings that are very likely to reflect real effects.

Four-stars: At least two studies that examined more than a thousand people with the trait/condition, or smaller studies where there is a consensus that the effect is real.

Preliminary Research

Includes results of studies that still need to be confirmed by the scientific community.

Three-stars: More than a thousand people with the trait/condition were studied; however, the effect has not yet been confirmed in a second independent study of similar size.

Two-stars: Fewer than a thousand people with the trait/condition were studied.

One-star: Fewer than a hundred people with the trait/condition were studied.

99   *In one scene, a flashback*   "Gattaca Script—Screenplay," www.script-o-rama .com/movie_scripts/g/gattaca-script-screenplay.html.

100   *Last year, Congress*   "H.R.493," http://thomas.loc.gov/cgi-bin/bdquery/ z?d110:h.r.00493.

100   *Already, law enforcement*   Ellen Nakashima et al., "U.S. to Expand Collection of Crime Suspects' DNA," *Washington Post*, April 2008, pp. A01.

103   *Baker, a small, feisty woman*   Navigenics, "Celiac Disease," www .navigenics.com/healthcompass/ConditionDetails/d/CelD.

104   *Some geneticists have criticized*   From www.23andMe.com: "23andMe Research Reports are based on high-quality but limited scientific evidence. Because these results have not yet been demonstrated through large, replicated studies, we do not perform complete quantitative analyses of their effects. We do, however, explain how they may—if confirmed—affect your odds of having or developing a trait, condition or disease."

105   *Many of the gene markers*   See the media section of the deCode Web site for details on association studies conducted by deCode: www.decodedi-agnostics.com.

106   *In 2006, the Federal Trade Commission*   Federal Trade Commission, "Facts for Consumers: At-Home Genetic Tests: A Healthy Dose of Skepticism May Be the Best Prescription" (July 2006), www.ftc.gov/bcp/edu/pubs/ consumer/health/hea02.shtm.

106   *And over the years*   FDA, "FDA Approves New Genetic Test for Patients with Breast Cancer" (July 2008), www.fda.gov/bbs/topics/NEWS/2008/ NEW01857.html.

106   *His institution and several*   David Harvey, Feinstein Kean Healthcare, via e-mail, September 2, 2008.

106   *I suspect that the*   GenomeWeb, "Rep. Kennedy Revives Obama's Personalized Medicine Bill for Next Congress; Adds Incentives" (September 2008), www.genomeweb.com/issues/news/149316–1.html.

106   *Within weeks of the company's*   U.S. Department of Health and Human Services, "U.S. System of Oversight of Genetic Testing: A Response to the Charge of the Secretary of Health and Human Services Report of the Secretary's Advisory Committee on Genetics, Health, and Society" (April 2008). A summary of the findings in the 192-page report and appendices can be found in the cover letter to former secretary of Health and Human Services Michael Leavitt,

dated April 30, 2008. Also see a letter dated August 18, 2008, to Secretary Leavitt from the SACGHS about a meeting held in July 2008 that included testimony from major online testing sites and other interested parties, www4 .od.nih.gov/oba/sacghs/reports/letter_to_Sec_08–18–08.pdf.

106   *State governments, too*   Steve Johns, "Two Bay Area Gene Testing Companies Get State OK to Resume Business," *San Jose Mercury News*, August 8, 2008.

108   *In Camden, New Jersey*   The Coriell Institute for Medical Research can be found at www.coriell.org.

109   *Quest, for instance*   Barb Short, Quest Diagnostics, via e-mail, September 2008.

109   *Quest runs five hundred thousand*   Ibid.

2. Environment

The epigraphs to this section are from *The Poetical Works of George Herbert* (Boston: Little, Brown and Co., 1863) p. 46; and Judith S. Stern, University of California Davis, via phone, October 20, 2008.

117   *I had asked him to help*   These tests were run in 2005–2006 as part of my *National Geographic* story "The Pollution Within" (October 2006).

117   *They save hundreds of lives*   Lisa Stiffler, "PBDEs: They Are Everywhere, They Accumulate and They Spread," *Seattle Post-Intelligencer*, March 28, 2007.

118   *Since 2001, CDC's Environmental Health Laboratory*   Dagny E. P. Olivares, Centers for Disease Control and Prevention, via e-mail, September 24, 2008.

118   *In Bhopal, India, in 1984*   Dan Kurzman, *A Killing Wind: Inside Union Carbide and the Bhopal Catastrophe* (New York: McGraw-Hill, 1987).

118   *Another incident happened*   Timothy S. George, *Minamata: Pollution and the Struggle for Democracy in Postwar Japan* (Cambridge, MA: Harvard University Asia Center, 2001).

119   *More than three thousand people*   Ivan Valiela, *Global Coastal Change* (Hoboken, NJ: Wiley-Blackwell, 2006), p. 203.

119   *From 1987 through 2002*   California Department of Developmental Services, *Autistic Spectrum Disorders, Changes in the California Caseload, An Update: 1999–2002* (April 2003).

119   *From the early 1970s*   L. L. Robison et al., "Assessment of environmental and genetic factors in the etiology of childhood cancers: The Children's Cancer Group Epidemiology Program," *Environmental Health Perspectives* 103, supplement 6 (1995), pp. 111–116; L. J. Paulozzi et al., "Hypospadias trends in two US surveillance systems," *Pediatrics* 100, no. 5 (1997), pp. 831–834.

119   *These diseases buck*   "US Mortality Data 1960 to 2004," in *US Mortality Volumes 1930 to 1959*, National Center for Health Statistics, Centers for Disease Control and Prevention (2006).

119   *In 1971, the U.S. Surgeon General*   Jesse L. Steinfeld, "The Surgeon General's Policy Statement on Medical Aspects of Childhood Lead Poisoning" (August 1971).

119   *It's now known*   R. L. Canfield et al., "Intellectual impairment in children with blood lead concentrations below 10 microg per deciliter," *New England Journal of Medicine* 348, no. 16 (April 17, 2003), pp. 1517–1526.

120   *In mice and rats*   Linda S. Birnbaum et al., "Brominated flame retardants: Cause for concern?" *Environmental Health Perspectives* 112 (January 2004), pp. 9–17.

120   *In 2001, investigators in Sweden*   Deborah C. Rice et al., "Developmental delays and locomotor activity in the C57BL6/J mouse following neonatal exposure to the fully-brominated PBDE, decabromodiphenyl ether," *Neurotoxicology and Teratology* 29 (July–August 2007), pp. 511–520.

120   *In 2005, scientists in Berlin*   Sergio N. Kuriyama et al., "Developmental exposure to low dose PBDE 99: Effects on male fertility and neurobehavior in rat offspring," *Environmental Health Perspectives* 113 (February 2005), pp. 149–154.

120   *My blood level*   Andreas Sjödin, *Serum Concentrations of Polybrominated Diphenyl Ethers (PBDEs) and Polybrominated Biphenyl (PBB) in the United States Population: 2003–2004,* Centers for Disease Control and Prevention, National Center for Environmental Health, p. 21.

120   *The news about another*   This is PBDE-99.

120   *Decas tend to be*   Linda S. Birnbaum, U.S. Environmental Protection Agency, via e-mail, September 5, 2008.

120   *Decas have been shown*   Ibid.

121   *Scientists have found*   Suzanne M. Snedeker, "BCERF Briefs: PBDEs" (February 2007), http://envirocancer.cornell.edu/pbde/brief.cfm.

121   *In 1999, Bergman*   D. Mieronyté et al., "Analysis of polybrominated diphenyl ethers in Swedish human milk. A time-related trend study, 1972–1997," *Journal of Toxicology and Environmental Health* 58 (November 1999), pp. 329–341.

121   *In Nicaragua, children*   Maria Athanasiadou et al., "Polybrominated diphenyl ethers (PBDEs) and bioaccumulative hydroxylated PBDE metabolites in young humans from Managua, Nicaragua," *Environmental Health Perspectives* 116 (March 2008), pp. 400–408.

121   *In Oakland, across the bay*   Douglas Fischer, "What's in You?" *Inside Bay Area*, March 27, 2006, www.insidebayarea.com/bodyburden/ci_2600879.

121   *In 2004, the European Union*   Birnbaum, via e-mail, September 5, 2008.

122   *One curious possibility*   Agency for Toxic Substances and Disease Registry, *Public Health Statement for Polybrominated Diphenyl Ethers (PBDEs)* (September 2004), www.atsdr.cdc.gov/toxprofiles/phs68-pbde.html.

122   *Since 2004, Boeing*   Terrance Scott, Boeing Commercial Airplanes, via e-mail, October 14, 2008.

122   *To find out the exposure*   Anna Christiansson et al., "Polybrominated diphenyl ethers in aircraft cabins—A source of human exposure?" *Chemosphere* (September 2008).

124   *This may be why*   Ronald A. Hites, Indiana University, via e-mail, September 2008.

124   *Bergman's breast-milk study*   Mieronyté, "Analysis of polybrominated diphenyl ethers in Swedish human milk."

126   *The EPA has listed*   Birnbaum, via e-mail, September 5, 2008.

126   *Those that I did have*   This was for paid by *National Geographic* for my story "The Pollution Within."

129   *These undoubtedly included*   National Research Council, *Toxicological Effects of Methylmercury* (Washington, D.C.: National Academies Press, 2000), p. 325.

129   *A 2005 study*   Leonardo Trasande et al., "Public health and economic consequences of methyl mercury toxicity to the developing brain," *Environmental Health Perspectives* 113 (May 2005), pp. 590–596.

131   *I grew up reading*   Donella H. Meadows et al., *The Limits to Growth*: (New York: Signet, 1972).

132   *In an article responding*   *Kansas City Star* (1967).

132   *Three years later*   *Kansas City Star* (1970).

132   *According to the* Star   *Kansas City Star* (January 12, 1970).

132   *In 1971, the* Star *reported*   *Kansas City Star* (September 23, 1971).

132   *Sulfur dioxide levels*   *Kansas City Star* (June 8, 1972). The companies were cited for producing sulfur dioxide levels as high as 150 to 175 micrograms per cubic foot; the highest level allowed according to regulations at the time was 60 micrograms per cubic foot for SO-2. KC Power and Light was specifically cited, according to *the Kansas City Star* (December 16, 1971).

132   *Even one of my favorite*   *Kansas City Star* (1975).

132   *Worse was my discovery*   EPA, "National Priority List Sites in the Midwest," www.epa.gov/region7/cleanup/npl_files/index.htm#Kansas.

133   *In 2005, the second*   EPA, "Second Five-Year Review Completed—Doepke-Holliday Superfund Site, Johnson County, Kansas" (November 2005), http://epa.gov/Region7/factsheets/2005/fs_2nd_5yr_rev_doepke-holliday_sprfnd_johnson_co_ks1105.htm.

133   *Incredibly, there was*   The Chemical Commodities Inc. (CCI) site can be found at www.epa.gov/region7/cleanup/npl_files/ksd031349624.pdf.

136   *Doull was the coeditor*   Curtis D. Klaassen, ed., *Casarett & Doull's Toxicology: The Basic Science of Poisons* (New York: McGraw-Hill, 2007).

136   *The long-term effects*   R. Scott Frey et al., "Cancer morbidity in Kansas farmers," *Transactions of the Kansas Academy of Science* 99 (1996), pp. 167–170.

136   *Another more recent study*   M. P. Montgomery et al., "Incident diabetes and pesticide exposure among licensed pesticide applicators: Agricultural Health Study, 1993–2003," *American Journal of Epidemiology* 167 (May 2008), pp. 1235–1246.

137   *It was used in seeds for crops*   EPA, "Hexachlorobenzene" (January 2000), www.epa.gov/ttn/atw/hlthef/hexa-ben.html#ref5.

139   *This is probably because*   National Center for Health Statistics, "Health, United States, 2007," www.cdc.gov/nchs/fastats/lifexpec.htm.

139 *But some of the largest*   EPA, "Hudson River PCBs," www.epa.gov/hudson; also EPA, Hudson River PCBs: Frequently Asked Questions, www.epa.gov/hudson/faqs.htm#27.

140 *In animals, they damage*   EPA, "Polychlorinated Biphenyls (PCBs): Basic Information," www.epa.gov/epawaste/hazard/tsd/pcbs/pubs/about.htm.

140 *In separate accidents*   M. Kuratsune et al., "PCB poisoning in Japan and Taiwan," *Progress in Clinical and Biological Research* 137 (1984), p. 155.

140 *GE has spent*   Kristen Skopeck, EPA, via e-mail, September 2008.

140 *PCBs collect in a fish's fatty tissues*   EPA, "Hudson River PCBs: Actions Prior to EPA's February 2002 Record of Decision (ROD)," www.epa.gov/hudson/actions.htm.

140 *For many of the cogeners*   National Center for Health Statistics, "National Health and Nutrition Examination Survey," www.cdc.gov/nchs/nhanes.htm.

142 *Home owners are upset*   "Pollution Upsets Homeowners," MasonryConstruction.com, www.masonryconstruction.com/industry-news.asp?articleID=550135&sectionID=0.

142 *The original plant*   EPA, "General Electric Company—Fort Edward," www.epa.gov/Region2/waste/fsgefort.htm.

142 *In 1982, PCBs*   "Pollution Upsets Homeowners," MasonryConstruction.com; also Leo Rosales from the EPA, author's interview, January 2006. (Note: Leo Rosales is no longer with the EPA.)

144 *In Texas, New York's problem*   Valhi, Inc., representative, via phone, September 2008.

145 *Sipping bottled water*   Britt E. Erickson, "Bisphenol A under scrutiny: Congress, media call into question safety of widely used plastics chemical," *Chemical & Engineering News* 86 (June 2008), pp. 36–39.

145 *Credit it to those*   Food and Drug Administration (FDA), "Phthalates and Cosmetic Products" (February 2008), www.cfsan.fda.gov/~dms/cos-phth.html.

145 *An expert panel*   National Toxicology Program, "NTP-CERHR Monograph on the Potential Human Reproductive and Developmental Effects of Di(2-Ethylhexyl) Phthalate (DEHP)" (November 2006), http://cerhr.niehs.nih.gov/news/index.html.

146 *My inventory of*   EPA, "Perfluorooctanoic Acid (PFOA)," www.epa.gov/oppt/pfoa.

146 *More recently, Penta and Octa PBDEs*   Lisa Stiffler, "Limited Ban Placed on Flame Retardants," *Seattle Post-Intelligencer*, April 3, 2007.

146 *The report concluded*   Gary Stevens et al., "Risks and Benefits in the Use of Flame Retardants in Consumer Products," UK Department for Trade and Industry/University of Surrey Polymer Research Centre (January 1999).

146 *Swiss chemist Paul Hermann Müller*   Nobel Foundation, "The Nobel Prize in Physiology or Medicine 1948," http://nobelprize.org/nobel_prizes/medicine/laureates/1948/index.html.

146 *DDT was considered* Rachel Carson, *Silent Spring* (Boston: Houghton Mifflin, 1962).

147 *No one celebrates chemicals* Marjorie McNinch, Hagley Museum and Library, via e-mail, September 10, 2008.

147 *In 2007, U.S., European* American Chemical Society, "Demand and costs rise in tandem," *Chemical & Engineering News* (July 2008).

149 *"There are no laws"* EPA, "Food Quality Protection Act (FQPA) of 1996," www.epa.gov/pesticides/regulating/laws/fqpa/176.

149 *But even if we* CDC, "National Report on Human Exposure to Environmental Chemicals: Spotlight on Bisphenol A," www.cdc.gov/exposurereport/pdf/factsheet_bisphenol.pdf.

150 *The process, they say* Faraz Alam, internal medicine physician, author's interview, September 2008.

150 *Today, any detectable level* National Toxicology Program, "Draft NTP Brief on Bisphenol A" (April 2008), http://cerhr.niehs.nih.gov/chemicals/bisphenol/bisphenol.html.

151 *No new, comprehensive laws* Cheryl Hogue, "The future of U.S. chemical regulation," *Chemical & Engineering News* 85 (January 2007), pp. 34–38.

151 *In 2006, the European Union* European Commission, "REACH in Brief" (October 2007), http://ec.europa.eu/environment/chemicals/reach/reach_intro.htm.

151 *The basis for REACH* Holger Breithaupt, "The costs of REACH," *European Molecular Biology Organization* 7 (October 2006), pp. 968–971.

152 *China and other fast-developing* DaeYoung Park, "REACHing Asia: Recent Trends in Chemical Regulations of China, Japan and Korea" (May 2008), http://papers.ssrn.com/sol3/papers.cfm?abstract_id=1121404.

152 *On a global scale* Jared Diamond, *Collapse: How Societies Choose to Fail or Succeed* (New York: Viking Adult, 2004), p. 6.

153 *Called the Comparative* The Comparative Toxicogenomics Database can be found at http://ctd.mdibl.org.

154 *The team also integrates* Carolyn J. Mattingly, Mount Desert Island Biological Laboratory, via e-mail, September 2008.

155 *This case is doubly horrific* Some of these meds in the family of drugs known as selective serotonin reuptake inhibitors are among the most widely used in the world, such as Prozac, Celexa, and Zoloft.

155 *The culprit DNA* Melanie Johns Cupp et al., "Cytochrome P450: New nomenclature and clinical implications," *American Family Physician* (January 1998).

157 *Only two animals* Comparative Toxicogenomics Database, http://ctd.mdibl.org/detail.go?view=ixn&type=chem&acc=C511295.

157 *Out of more than* M. L. Takacs et al., "Gene Expression profiling in the lung and liver of PFOA-exposed mouse fetuses," *Toxicology* 239, nos. 1–2 (September 24, 2007), pp. 15–33; also consult the Comparative Toxicogenomics Database Web site, http://ctd.mdibl.org/detail.go?view=ixn&type=chem&acc=C023036.

157　*The study concluded*　Thomas Rattenborg et al., "Inhibition of E2-induced expression of BRCA1 by persistent organochlorines," *Breast Cancer Research* 4 (2002), p. R12.

158　*One cause of cancer*　Helen Pilcher, "Cancer: The traitors within," *New Scientist* (November 2006), p. 48.

158　*One 2004 study in France*　Géraldine Lemaire et al., "A PXR reporter gene assay in a stable cell culture system: CYP3A4 and CYP2B6 induction by pesticides," *Biochemical Pharmacology* 68 (December 2004), pp. 2347–2358.

159　*These SNPs seem*　Karin Schläwicke Engström et al., "Genetic variation in glutathione-related genes and body burden of methylmercury," *Environmental Health Perspectives* 116 (June 2008), pp. 734–739.

159　*Scientists have long known*　Ibid.

159　*"These genetic variations"*　Ibid.

162　*Having a certain SNP*　National Human Genome Research Institute, "Learning about Factor V Leiden Thrombophilia" (September 2007), www.genome.gov/15015167.

164　*Eventually, they hope*　Francis S. Collins et al., "Transforming environmental health protection," *Science* 319 (February 2008), pp. 906–907.

166　*"While performing tests"*　NIH ethics consultation service, Consultation Report, internal memo (May 22, 2008).

168　*Huang and Xia*　Menghang Xia et al., "Compound cytotoxicity profiling using quantitative high-throughput screening," *Environmental Health Perspectives* 116 (March 2008), pp. 284–291.

169　*The fifty "assays" included*　Ibid.

169　*Huang shows me*　Ibid.

171　*An activist in global health*　World Health Organization, "Facts and Figures: Water, Sanitation and Hygiene Links to Health, Diarrhea Facts and Figures," Web site, www.who.int/water_sanitation_health/publications/factsfigures04/en.

171　*Some parents of autistic*　FDA, "Thimerosal in Vaccines" (June 2008), www.fda.gov/CBER/vaccine/thimerosal.htm.

172　*As a distinct field*　NIH, "The Genes, Environment and Health Initiative (GEI)" (September 2008), www.gei.nih.gov.

172　*Selenium is another*　Laura J. Raymond et al., "Mercury: selenium interactions and health implications," *SMDJ Seychelles Medical and Dental Journal* 7 (November 2004), pp. 72–77.

172　*This study will tap*　The Nurses' Health Study can be found at www.channing.harvard.edu/nhs.

173　*"Given the biologic relevance"*　Harvard School of Public Health, "Genes and Environment Initiative Launched at School," *Harvard Public Health NOW* (June 2008), www.hsph.harvard.edu/now/20080605/genes-and-environment-initiative-launched-at-school.html.

173　*This study would require*　Sarah Carr, NIH, via e-mail, September 15, 2008.

3. Brain

The epigraph to this section is taken from Emerson Pugh, quoted in Barry Gibb, *The Rough Guide to the Brain* (London: Rough Guides, 2007), p. 15.

181 *As science writer Steven Johnson* Recounted in Stephen Johnson, *Mind Wide Open* (New York: Scribner, 2004), p. 158.

182 *She tracked down this* Cathryn Ramin, *Carved in Sand: When Attention Fails and Memory Fades in Midlife* (New York: HarperCollins, 2007), p. 115.

182 *"It was a sobering thought"* Ibid., p. 166.

182 *I'm not going to spend* John Searle, *Mind, a Brief Introduction* (Oxford: Oxford University Press, 2004), pp. 47–49.

182 *I also have moments* Wallace Stevens, "Reality Is an Activity of the Most August Imagination," from *Opus Posthumous*, edited by Milton J. Bates (New York: Random House, 1989), pp. 135–136.

183 *This reminds me* Douglas Adams, *The Hitchhiker's Guide to the Galaxy* (New York: Del Rey, 1995).

183 *The late neuroscientist and psychiatrist* Paul MacLean, *The Triune Brain in Evolution* (New York: Plenum Press, 1990).

183 *These three brains* Thomas Lewis, Fari Amini, and Richard Mannon, *A General Theory of Love* (New York: Vintage, 2000), p. 31.

184 *Brewer tells me* The homepage of the James Brewer lab is http://hml.ucsd.edu/hml.

185 *For instance, patients with schizophrenia* F. X. Castellanos et al., "Developmental trajectories of brain volume abnormalities in children and adolescents with attention-deficit/hyperactivity disorder," *Journal of the American Medical Association* 288 (2002), pp. 1740–1748.

185 *And people who drink* Carol Anne Paul, "In a longitudinal analysis, continuous heavy drinking was also significantly negatively associated with brain volume," American Academy of Neurology 59th Annual Meeting: Abstract P05.030 (April 28–May 5, 2007); also Susan Jeffrey, "High alcohol consumption linked to reduced brain volume," *Medscape Medical News* (May 3, 2007), www.medscape.com/viewarticle/555978.

185 *As you can see* For complete results of the author's genetic tests and for citations and sources, go to www.experimentalman.com.

185 *I think of myself* "Right Brain vs. Left Brain," *Sydney Herald Sun*, www.news.com.au/heraldsun/story/0,21985,22556281-661,00.html; Sergio Della Sala, *Mind Myths: Exploring Popular Assumptions about the Mind and Brain* (New York: Wiley, 1999).

186 *Left- versus right-handedness* Insup Taylor with M. Martin Taylor, *Psycholinguistics: Learning and Using Language* (Englewood Cliffs, NJ: Prentice-Hall 1990), p. 362; Goulven Josse and Nathalie Tzourio-Mazoyer, "Review: Hemispheric specialization for language," *Brain Research Reviews* 44 (2003), pp. 1–12.

186 *Damage to either side* Michael V. Johnston, "Clinical disorders of brain plasticity," *Brain and Development* 26, no. 2 (March 2004), pp. 73–80.

187   *The first is a visual test*   "Right Brain vs. Left Brain," *Sydney Herald Sun.*

187   *I take one more*   "Hemispheric Dominance Inventory Test," from a Web site developed by Intelegen, Inc., www.web-us.com/brain/brain-dominance.htm.

187   *Arguably, my belief*   R. Cabeza, "Hemispheric asymmetry reduction in older adults: The HAROLD model," *Psychology and Aging* 17, no. 1 (March 2002), pp. 85–100.

188   *In 2000, researchers*   Eleanor A. Maguire et al., "Navigation-related structural change in the hippocampi of taxi drivers," *Proceedings of the National Academies of Sciences* 97, no. 8 (April 11, 2000), pp. 4398–4403.

189   *"We very much hope"*   Roger Dobson, "Taxi Drivers' Knowledge Helps Their Brains Grow," *The Independent*, December 17, 2006, www.independent.co.uk/life-style/health-and-wellbeing/health-news/taxi-drivers-knowledge-helps-their-brains-grow-428834.html.

191   *Neuroscientists call this*   Adam Gazzaley, University of California at San Francisco neurologist, author's interview, July 11, 2007.

192   *"Electroencephalography (EEG) was used"*   Adam Gazzaley et al., "Age-related top-down suppression deficit in the early stages of cortical visual memory processing," *Proceedings of the National Academies of Science*, in press (scheduled for March 2009), excerpt from abstract of study.

193   *For this test, researchers*   M. F. Gosso et al., "The SNAP-25 gene is associated with cognitive ability: Evidence from a family-based study in two independent Dutch cohorts," *Molecular Psychiatry* 11, no. 9 (September 2006), pp. 878–886.

193   *This last test was run*   G. Altmon et al., "Lipoprotein genotype and conserved pathway for exceptional longevity in humans," *PLoS Biology* 4, no. 4 (April 2006), p. e113.

194   *For this book, I took several*   The author took tests offered online by IQTest.com at www.iqtest.com and the International High IQ Society at www.highiqsociety.org/iq_tests; both Web sites were accessed from July to September 2008.

195   *For example, the hippocampus*   Mark F. Bear, Barry W. Connors, and Michael A. Paradiso, *Neuroscience, Exploring the Brain*, 2nd ed. (Baltimore: Lippincott, Williams and Wilkins, 2001), p. 752.

196   *Scientists believe*   For extensive information about memory and how memories are stored in the brain, see ibid., chapters 22 and 23, pp. 739–807.

199   *He is a director*   For information on this company and its tests, visit the Cognitive Drug Research, Ltd., Web site, www.cognitivedrugresearch.com.

199   *Besides drug trials*   Keith Wesnes, author's interview, August 2008; for a wide range of studies using CDR tests to assess cognitive function in patients suffering from diseases, on different diets, and much more, visit the CDR Web site, www.cognitivedrugresearch.com. CDR's Pat Turk also provided me with a nineteen-page list of study citations from peer-reviewed journals.

200   *Other tests include*   Wesnes, author's interview, August 2008.

200   *Tests have been run*   Keith Wesnes et al., "Breakfast reduces declines in attention and memory over the morning in school children," *Appetite* 41 (2003), pp. 329–331.

200   *and other tests were run*   Wesnes, author's interview, August 2008.

200   *A competitor of*   For more information on CogState, go to its Web site, www.cogstate.com.

200   *We ran a mini-experiment*   The main Web site address is www.portfolio .com. Also see David Ewing Duncan, "How Smart Are You? The Answer's Here," *Portfolio.com* (September 14, 2008), www.portfolio.com/views/ columns/natural-selection/2008/09/14/Cognition-and-Memory-Tests-Part-II.

203   *"Let's move on," he says*   Other CDR tests the author took were:

    Digit Vigilance: A target digit is randomly selected and constantly displayed to the right of the screen. A series of digits is then presented in the center of the screen at the rate of 150 per minute, and the patient is required to press the YES button as quickly as possible every time the digit in the series matches the target digit. There are forty-five targets in the series. The task lasts for three minutes.

    Choice Reaction Time: Either the word NO or the word YES is presented on the screen, and the patient is instructed to press the corresponding button as quickly as possible. There are fifty trials for which each stimulus word is chosen randomly with equal probability, and there is a varying inter-stimulus interval.

210   *The amygdala flashes signals*   Bear et al., *Neuroscience, Exploring the Brain*, pp. 589–591.

210   *The hypothalamus orchestrates*   Ibid., pp. 499–511.

210   *I assume "sex" is a*   Ibid., pp. 508–509.

210   *Yet I am aware*   Henry Greely, "On neuroethics," *Science* 318, no. 26 (October 2007), p. 533.

211   *Last year, a study*   Jordan W. Smoller, "Influence of RGS2 on anxiety-related temperament, personality, and brain function," *Archives of General Psychiatry* 65, no. 3 (2008), pp. 298–308.

211   *"We found that"*   "Gene Variants May Increase Risk of Anxiety Disorder," press release, Massachusetts General Hospital, March 3, 2008.

211   *I'm also including*   D. Denys et al., "Association between the dopamine D2 receptor TaqI A2 allele and low activity COMT allele with obsessive-compulsive disorder in males," *European Neuropsychopharmacology* 16, no. 6 (August 2006), pp. 446–450; R. Mössner et al., "Transmission disequilibrium of polymorphic variants in the tryptophan hydroxylase-2 gene in children and adolescents with obsessive-compulsive disorder," *International Journal of Neuropsychopharmacology* 9, no. 4 (August 2006), pp. 437–442.

213   *Goldin has published*   P. R. Golden et al., "The neural bases of amusement and sadness: A comparison of block contrast and subject-specific emotion intensity regression approaches," *Biological Psychiatry* 15:63, no. 6 (March 2007), pp. 577–586.

218   *I seldom go to church*   Sam Harris, *The End of Faith: Religion, Terror, and the Future of Reason* (New York: W. W. Norton, 2005); Christopher Hitchens, *God Is Not Great: How Religion Poisons Everything* (New York: Twelve Books, Hachette Book Group, 2007).

219   *"Objective evidence and certitude"*   William James, "The Will to Believe," quoted in Jonah Lehrer, *Proust Was a Neuroscientist* (Boston: Mariner-Houghton Mifflin, 2007), p. 17.

220   *In 2006, Grafman's lab*   Kristine M. Knutson et al., "Politics on the brain: An fMRI investigation," *Society of Neuroscience* 1, no. 1 (March 2006), pp. 25–40.

221   *Grafman is an affable*   Dan Gordon, ed., *Your Brain on Cubs: Inside the Heads of Players and Fans* (Washington, DC: Dana Press, 2008).

221   *"There is some evidence"*   Quotes from George Will, "Your Brain on Cubs," *Newsweek*, April 7, 2008, www.newsweek.com/id/129576.

221   *Grafman says that Cubs*   Jon Greenberg, "Loveable Losers? It's All in Your Head," MLB.com (March 27, 2008), http://mlb.mlb.com/content/printer_friendly/mlb/y2008/m03/d27/c2461052.jsp.

222   *In another experiment*   Matthew Alper, *The "God" Part of the Brain: A Scientific Interpretation of Human Spirituality and God* (Naperville, IL: Sourcebooks, 2006).

222   *David Wulff, a psychologist*   David Wulff, quoted in Sharon Begley, "Your Brain on Religion: Mystic Visions or Brain Circuits at Work?" *Newsweek*, May 7, 2001; this article is a good layperson's summary of the emerging field of neurotheology.

223   *Wulff, however, says*   Lee Kirkpatrick, *Attachment, Evolution, and the Psychology of Religion* (New York: Guilford Press, 2004).

223   *British evolutionary biologist*   Richard Dawkins, *The God Delusion* (London: Black Swan, 2006), p. 356.

223   *Studies comparing twins*   Nancy Segal, quoted in Michael Shermer, *How We Believe: Science, Skepticism, and the Search for God* (New York: Henry Holt, 2003).

223   *In another survey*   Shermer, *How We Believe*.

223   *In 2004, geneticist Dean Hamer*   Dean Hamer, *The God Gene: How Faith Is Hardwired into Our Genes* (New York: Doubleday, 2004).

223   *"The field of"*   Carl Zimmer, "Faith-boosting genes: A search for the genetic basis of spirituality," *Scientific American* (October 2004), www.sciam.com/article.cfm?id=faith-boosting-genes&ref=sciam.

231   *The glaring difference*   Searle, *Mind, a Brief Introduction*, pp. 13–18.

232   *Glimcher is a leading figure*   Glimcher's thoughts and ideas are collected in Paul Glimcher, *Decisions, Uncertainty, and the Brain: The Science of Neuroeconomics* (Cambridge, MA: MIT Press, 2004).

233   *"Many of us in neuroeconomics"*   P. W. Glimcher and A. Rustichini, "Neuroeconomics: The consilience of brain and decision," *Science* 306, no. 5695 (October 15, 2004), pp. 447–452, www.ncbi.nlm.nih.gov/pubmed/15486291.

234   *In a 2007 study*   Tali Sharot et al., "Neural mechanisms mediating optimism bias." *Nature* 450 (November 1, 2007), pp. 102–105.

235   *Optimal coding*   Glimcher Lab at New York University, lab Web site, www.cns.nyu.edu/~glimcher/people.html.

238   *"After about an hour"*   John Cassidy, "Mind Games: What Neuroeconomics Tells Us about Money and the Brain," *New Yorker* (September 18, 2006), www .newyorker.com/archive/2006/09/18/060918fa_fact.

239   *A famous experiment*   Samuel M. McClure et al., "Neural correlates of behavioral preference for culturally familiar drinks," *Neuron* 44 (September 19, 2004), pp. 379–387.

240   *"You're going to be"*   Stephanie Lazzaro first ran a Functional Localizer test on the author in the MRI, showing him a lottery in which he had a 50 percent chance of winning $2 and a 50 percent chance of losing $2. She looked for activation in the author's brain when he realized that he had won $2 versus when he realized that he had lost $2. This established a baseline for how his brain looks when he is feeling rewarded, a profile that Lazzaro and the researchers used to analyze the author's likes and dislikes of various products while viewing the products in the MRI.

245   *A magazine editor*   This section is based on David Ewing Duncan, "Brain Boosters: Our Reporter Enters the New World of Neuroenhancers," *Technology Review* (July 2007), www.technologyreview.com/biomedicine/ 18881.

246   *Last year, Cephalon*   "Ohio Sales Rep's Information Launched Massive Government Investigation of Cephalon," *Marketwatch* (September 29, 2008), www.marketwatch.com/news/story/ohio-sales-reps-information-launched/story.aspx?guid=%7BB373BB70-689A-4131-A5EE-DC00DE7 BB59E%7D&dist=hppr.

248   *In Cambridge, England*   D. C. Turner et al., "Cognitive enhancing effects of modafinil in healthy volunteers," *Psychopharmacology (Berl)* 165, no. 3 (2003), pp. 260–269.

248   *Other researchers saw*   "PROVIGIL Studies Demonstrate Improved Performance and Alertness under Conditions of Sleep Deprivation," press release, Cephalon (June 21, 2000); report on presentations delivered at the 14th Annual Meeting of the Associated Professional Sleep Societies, "New Findings in the Treatment of Performance Impairing Sleepiness Associated with Sleep Loss," http://cphln1.customedialabs.com/media/news-releases/ by-product/product/actiq/article/provigil-studies-demonstrate-improved-performance-and-alertness-under-conditions-of-sleep-deprivation.

249   *Neuroscientists are inserting*   H. Fisher, A. Aron, and L. L. Brown, "Romantic love: A mammalian brain system for mate choice," *Philosophical Transactions of the Royal Society: Biological Sciences* 361 (November 13, 2006), pp. 2173–2186.

250   *At the University of California*   "Creating an Atlas of the Human Brain, Neuroscientists Chart a New World in 3-D," *UCLA Today* (October 22, 2002), www.today.ucla.edu/2002/021022brain_atlas.html.

251   *"You can't just point"*   Ibid; and author's interview with John Mazziotta, June 12, 2008.

251   *"The idea of the unity"*   Edward O. Wilson, *Consilience, the Unity of Knowledge* (New York: Vintage, 1998), p. 5.

4. Body

The epigraph to this section is taken from *The Captive, Modern Library* Series 5 (New York: Random House, 1993), p. 343.

255   *Launched in 1996*   The Entelos, Inc., Web site is www.entelos.com; the Entelos description of the author's test is on the Entelos MyDigitalHealth Web site: www.mydigitalhealth.com/personal_story.htm.

255   *His number comes from*   Framingham Heart Study Web site, a project of the National Heart, Lung, and Blood Institute, and Boston University, official Web site, www.framinghamheartstudy.org.

258   *Created in the liver*   This information comes from Tom Paterson of Entelos.

261   *ARIC tracks factors*   A. E. Chambless et al., "Coronary heart disease risk prediction in the Atherosclerosis Risk in Communities (ARIC) study," *Journal of Clinical Epidemiology* 56, no. 9 (September 2003), pp. 880–890, www.ncbi.nlm.nih.gov/pubmed/14505774?dopt=Abstract.

266   *"A PhysioLab model"*   Entelos, "Physiolab Technology," www.entelos.com/physiolabModeler.php.

266   *Krauss's lab is studying*   R. M. Krauss, "Dietary and genetic probes of atherogenic dyslipidemia," *Arteriosclerosis, Thrombosis, and Vascular Biology* 25 (2005), pp. 2265–2272; R. M. Krauss et al., "Separate effects of reduced carbohydrate and weight loss on atherogenic dyslipidemia," *American Journal of Clinical Nutrition* (in press).

266   *His lab and others*   J. A. Simon et al., "Phenotypic predictors of response to simvastatin therapy among African Americans and Caucasians: The Cholesterol and Pharmacogenetics (CAP) Study," *American Journal of Cardiology* (in press).

267   *"These results have proven"*   Ronald Krauss, Children's Hospital Oakland Research Institute, www.chori.org/Principal_Investigators/Krauss_Ronald/krauss_overview.htm.

267   *A SNP variation*   E. S. Tai and J. M. Ordovas, "Clinical significance of apolipoprotein A5," *Current Opinion in Lipidology* 19, no. 4 (August 2008), pp. 349–354.

267   *Unfortunately, I am AA*   D. Corella et al., "APOA5 gene variation modulates the effects of dietary fat intake on body mass index and obesity risk in the Framingham Heart Study," *Journal of Molecular Medicine* 85, no. 2 (February 2007), pp. 119–128.

267   *Another SNP variation*   Len A. Pennacchio et al., "Two independent apolipoprotein A5 haplotypes influence human plasma triglyceride levels," *Human Molecular Genetics* 11, no. 24 (2002), pp. 3031–3038.

267 *Other markers in the APOA5*   Roberto Elosua et al., "Variants at the APOA5 locus, association with carotid atherosclerosis, and modification by obesity: The Framingham Study," *Journal of Lipid Research* 47 (May 2006), pp. 990–996.

267 *"They also suggest"*   Ronald Krauss, Lab Web site.

267 *A research team combined*   Sekar Kathiresan et al., "Polymorphisms associated with cholesterol and risk of cardiovascular events," *New England Journal of Medicine* 358 (March 20, 2008), pp. 1240–1249.

271 *"That, more or less, is"*   Michael Pollan, "Unhappy Meals," *New York Times*, January 28, 2007.

272 *Nonetheless, a slight panic*   See Robert Atkins, *Dr. Atkins New Diet Revolution*, rev. ed. (Blue Ridge Summit, PA: M. Evans, 2003).

273 *This nutrient is needed*   Ibid.; and "Definition of Vitamin B," Mednet .com, www.medterms.com/script/main/art.asp?articlekey=12865.

273 *The abnormal one*   "TCN2 Gene, Genecards," www.genecards.org/ cgi-bin/carddisp.pl?gene=TCN2.

275 *It's a "Tanita Body"*   Tanita, "How BIA Works," www.tanita.com/ HowBIAworks.shtml.

281 *People fifty years and older*   C. Daniel Johnson, et al., "Accuracy of CT colonography for detection of large adenomas and cancers," *New England Journal of Medicine* 359, no. 12 (September 18, 2008), pp. 1207–1217.

281 *"Seventy-five percent"*   Judy Yee, *Virtual Colonoscopy* (Philadelphia: Wolters Kluwer, 2008), p. 1.

281 *I ask whether*   MV 2006 CT Market Summary Report (Des Plaines, IL: IMV Medical Information Division, 2006).

284 *A recent paper by*   David Brenner, "Computed tomography—an increasing source of radiation exposure," *New England Journal of Medicine* 357 (November 29, 2007), pp. 2277–2284.

290 *To find out, I dispatch*   Sirtris can be found at www.sirtrispharma.com.

290 *In mice, these genes*   Raul Mostoslavsky et al., "Genomic instability and aging-like phenotype in the absence of mammalian SIRT6," *Cell* 124 (January 2006), pp. 315–329.

291 *When its enzyme*   Zachary Gerhart-Hines et al., "Metabolic control of muscle mitochondrial function and fatty acid oxidation through SIRT1/ PGC-1," *EMBO Journal* 26 (March 2007), pp. 1913–1923.

291 *Mitochondrial capacity*   Stéphanie Grandemange et al., "Stimulation of mitochondrial activity by p43 overexpression induces human dermal fibroblast transformation," *Cancer Research* 65 (May 2005), pp. 4282–4291.

291 *In rodents, boosting*   Haim Y. Cohen et al., "Calorie restriction promotes mammalian cell survival by inducing the SIRT1 deacetylase," *Science* 305 (July 2004), pp. 390–392.

291 *Perhaps more interesting*   Ralph R. Alcendor et al., "Sirt1 regulates aging and resistance to oxidative stress in the heart," *Circulation Research* 100 (May 2007), pp. 1512–1521.

291   *Much needs to be*   Hongying Yang et al., "Nutrient-sensitive mitochondrial NAD+ levels dictate cell survival," *Cell* 130 (September 2007), pp. 1095–1107.

291   *SIRT5 is still poorly*   Peter J. Elliott et al., "Sirtuins: Novel targets for metabolic disease," *Current Opinion in Investigational Drugs* 9 (April 2008), pp. 371–378.

291   *The first is caloric*   Cohen et al., "Calorie restriction promotes mammalian cell survival by inducing the SIRT1 deacetylase."

292   *Animals subjected to a*   Mayo Clinic, "Calorie restriction: Is this anti-aging diet worth a try?" (May 11, 2007), www.mayoclinic.com/health/anti-aging/HQ00233.

292   *A few hearty people*   Luigi Fontana et al., "Long-term calorie restriction is highly effective in reducing the risk for atherosclerosis in humans," *Proceedings of the National Academy of Sciences* 101 (April 2004), pp. 6659–6663.

292   *Other tests have*   Leonie K. Heilbronn et al., "Effect of 6-month calorie restriction on biomarkers of longevity, metabolic adaptation, and oxidative stress in overweight individuals," *Journal of the American Medical Association* 295 (April 2006), pp. 1539–1548.

292   *Discovered in 2003*   Konrad T. Howitz et al., "Small molecule activators of sirtuins extend *Saccharomyces cerevisiae* lifespan," *Nature* 425 (September 2003), pp. 191–196; Cohen et al., "Calorie restriction promotes mammalian cell survival by inducing the SIRT1 deacetylase."

292   *A cofounder of Sirtris*   Lenny Guarente, *Ageless Quest: One Scientist's Search for Genes That Prolong Youth* (Woodbury, CT: Cold Spring Harbor Laboratory Press, 2002).

293   *Sinclair and Konrad Howitz*   Howitz et al., "Small molecule activators of sirtuins extend *Saccharomyces cerevisiae* lifespan."

293   *In a 2004 interview*   Jennifer Couzin, "Aging research's family feud," *Science* 303 (February 2004), pp. 1276–1279.

293   *When mutant, DAF2*   Cynthia Kenyon et al., "A C. elegans mutant that lives twice as long as wild type," *Nature* 366 (December 1993), pp. 461–464.

293   *Naysayers, however*   Biologist Leonard Hayflick has been a leading critic.

293   *I first visited Sinclair's lab*   Joseph A. Baur et al., "Resveratrol increases health and survival of mice on a high calorie diet," *Nature* 444 (November 2006), pp. 337–342.

294   *Subsequent studies have*   Jamie L. Barger et al., "A low dose of dietary resveratrol partially mimics caloric restriction and retards aging parameters in mice," *PLoS ONE* 3 (June 2008), p. e2264.

294   *Sinclair's paper came*   Marie Lagouge et al., "Resveratrol improves mitochondrial function and protects against metabolic disease by activating SIRT1 and PGC-1a," *Cell* 127 (December 2006), pp. 1109–1122.

294   *In 2008, Sirtris began*   "Sirtris Announces Clinical Results from Phase 1 Trials of SRT501, a Sirtuin Therapeutic, Company Initiates Phase 1b Study in Type 2 Diabetes Patients" (October 2006), press release, www.sirtrispharma.com/press/2006–100406.html.

298   *In another study*   Maris Kuningas et al., "SIRT1 gene, age-related diseases, and mortality: The Leiden 85-Plus Study," *The Journals of Gerontology Series A: Biological Sciences and Medical Sciences* 62 (September 2007), pp. 960–965.

298   *The SNP, though*   Maris Kuningas, Leiden University Medical Center, via e-mail, October 2, 2008.

299   *The number of centenarians*   Danielle Conceicao, U.S. Census Bureau, via e-mail, October 20, 2008.

299   *One person, a supercentenarian*   Craig R. Whitney, "Jeanne Calment, World's Elder, Dies at 122," *New York Times*, August 5, 1997.

299   *In the year 2000*   Conceicao, via e-mail, October 20, 2008.

299   *We already have had*   Elizabeth Arias, "United States Life Tables, 2004," *National Vital Statistics Reports* 56 (December 2007), p. 35.

299   *Then came better hygiene*   Hsiang-Ching Kung et al., "Deaths: Final Data for 2005," *National Vital Statistics Reports* 56 (April 2008), p. 1.

301   *In ancient Greek mythology*   Mark P. O. Morford et al., *Classical Mythology* (New York: Oxford University Press, 2003).

301   *"The woods deca"*   Alfred Tennyson, *Idylls of the King and a Selection of Poems* (New York: Signet Classic, 2003), p. 325.

301   *More recent is the story*   David Ewing Duncan, *Hernando De Soto: A Savage Quest in the Americas* (New York: Crown, 1996).

301   *During the Enlightenment*   Nick Bostrom, "A history of transhumanist thought," *Journal of Evolution and Technology* 14 (April 2005).

301   *Mary Shelley's* Frankenstein   Mary Wollstonecraft Shelley, *Frankenstein, or, the Modern Prometheus, the 1818 Text* (Chicago: University of Chicago Press, 1982).

301   *In 1923, geneticist J. B. S. Haldane*   John Burdon Sanderson Haldane, *Daedalus or Science and the Future* (London: Kegan Paul, Trench, Trubner, 1924).

301   *Transhumanists have traced*   Cintra Wilson, "Droid Rage," *New York Times*, October 21, 2007.

302   *He has set a date*   Ray Kurzweil, Kurzweil Technologies, via e-mail, October 21, 2008.

302   *The emphasis is on sentient*   The Abolitionist Society can be found at www .abolitionist-society.com.

302   *Other derivations include*   World Transhumanist Association, "What Currents Are There within Transhumanism?" www.transhumanism. org/index.php/WTA/faq21/81.

302   *Kelly has suggested*   Kevin Kelly, "The Maes-Garreau Point" (March 14, 2007), www.kk.org/thetechnium/archives/2007/03/the_maesgarreau.php.

302   *My friend Gregory Stock*   Gregory Stock, *Redesigning Humans: Our Inevitable Genetic Future* (Boston: Houghton Mifflin, 2002).

Epilogue: Eternity

The epigraph to this section is taken from Chris Rodley, ed., *Cronenberg on Cronenberg* (London: Faber & Faber, 1992).

308   *My eyes were examined*   David Ewing Duncan, "Stem Cells Bring Sight to the Blind in India," *Fortune* (October 24, 2007).

308   *For taste, I met with*   U.-K. Kim et al., "Human Mutation" (in press), www.nidcd.nih.gov/research/scientists/draynad.asp.

309   *The test has twenty-six*   The Distorted Tunes Test can be found at www.nidcd.nih.gov/tunetest.

309   *NeuroVigil, which plans*   Philip S. Low, NeuroVigil, via e-mail, October 20, 2008.

310   *In 2007, he*   Philip S. Low, *A New Way to Look at Sleep* (San Diego: University of California Press, 2007).

310   *When I met him*   Henry Fountain, "In Sleep, We Are Birds of a Feather," *New York Times*, July 1, 2008.

312   *One of Agus's positions*   The Spielberg Family Center for Applied Proteomics can be found at http://sfcap.cshs.org/index.html.

314   *I had read a 2005*   Xeni Jardin, "Applied Minds Think Remarkably," *Wired* (June 21, 2005).

317   *I'm not sure how*   NIH, "Human Microbiome Project (HMP)," http://nihroadmap.nih.gov/hmp.

GLOSSARY

allele One side of the paired sequences of DNA that make up base pairs, genes, and chromosomes.

amino acids These small organic molecules—there are twenty different types—join together to form proteins based on instructions from a gene.

amygdala An almond-shaped structure in the brain that is thought to be a primary emotion center. It also plays a role in the development of memories.

atherogenic dyslipidemia A condition in which a patient has high triglycerides, low HDL, and an increase in small LDL particles (see the definitions of HDL, LDL, and triglycerides further on).

base pair DNA is a long strand of nucleotides (As, Ts, Cs, Gs) arranged on a double helix structure; the nucleotides come in pairs, such as AA—CT—TT—CC.

caloric restriction The practice of limiting the amount of food one eats in the hope that it will improve health and slow the aging process.

chromosome Genes and other DNA are arranged in groupings called chromosomes—humans have twenty-three chromosomes.

computer tomography (CT) An imaging technique that uses X-rays and computer analysis to provide a picture of body tissues and structures.

electroencephalogram An EEG is a procedure that records the brain's continuous electrical activity by means of electrodes attached to the scalp.

enzyme A protein in the body that helps a biological process go faster.

genetic code A person's genome is made up of long strands of nucleotides arranged by twos on a double helix. There are four nucleotides designated as A, G, C, and T. Only a small portion of human DNA is capable of coding for the creation of proteins.

genome/genomics All of the genetic information, the hereditary material, and the DNA in an organism; genomics is the study of everything related to genomes.

genotype An individual's DNA configuration that is passed down from his or her parents for a specific trait, which is different from genotypes of other individuals that carry different configurations. An individual can share the same genotype with other people.

haplotype A set of closely linked genetic markers present on one chromosome, which tend to be inherited together (they are not easily separable by the natural recombination that occurs when a mother and a father pass on their DNA).

heterozygous Having two different alleles (nucleotide letters—A, G, T, and C) at a given locus on a pair of chromosomes (e.g., GT or AC).

hippocampus An area buried deep in the forebrain that helps regulate emotion and memory.

homozygous Having two identical alleles in a SNP of a particular gene (e.g., GG or TT). See **SNP**.

hypothalamus Located at the bottom center of the middle brain, the hypothalamus is a neuro-thermostat that influences appetite, hormones, digestion, sexuality, circulation, emotions, and sleep.

LDL and HDL LDL means low-density lipoproteins. This is a form of cholesterol that circulates in the blood, commonly called "bad" cholesterol. High levels of LDLs in the blood stream increase the risk of developing heart disease. HDL means high-density lipoproteins. HDL transports cholesterol from the tissues to the liver, where it can be eliminated in bile. HDL cholesterol is considered "good" cholesterol, because higher blood levels of HDL cholesterol are associated with a lower risk of developing heart disease.

lipids Lipids function in the long-term storage of biochemical energy, insulation, structure, and control. Examples of lipids include fats and oils; an example is cholesterol.

lipoprotein (a) A compound of protein that carries fats and fatlike substances, such as cholesterol, in the blood.

mitochondria Rod-shaped structures within cells that are responsible for energy production. Mitochondria also contain a small amount of DNA.

protein A molecule composed of strings of amino acids often having highly complex structures; proteins are responsible for most biological activity and structure in a living organism.

polymorphism The difference in DNA sequences among individuals.

proteome/proteomics A proteome is the collection of all proteins in the body of an organism; proteomics is the study of proteins.

risk factor A characteristic (race, sex, age, obesity) or a variable (smoking, pesticide exposure, genetic variation) that is associated with an increased chance of developing a disease, a trait, or toxic effect.

SNP Single nucleotide polymorphism (pronounced "snips"); these are single-letter differences in DNA coding from one person to another. Changes from, say, an "A" in one person to a "G" in another can be responsible for an increased risk of developing a disease.

triglycerides These are lipids that are carried through the bloodstream to tissues. Most of the body's fat tissue is in the form of triglycerides, stored for use as energy.

EXPERIMENTAL MAN ONLINE

You are invited to join in a final experiment—the coupling of this book with a Web site: the Experimental Man project, located at www.experimentalman.com.

The Experimental Man project has produced far more data and information than can be contained in the confines of a printed book. To view the author's complete results, tests, and analyses, please visit the Experimental Man Project Web site. The site also features videos of tests, interviews, photographs, and updates.

The Web site, however, contains much more than the author's data. The site is a source of additional information about tests he took and some he didn't. It offers portals to many of the entities—labs, researchers, companies, and others—that tested him and assisted with the project.

The Experimental Man Web site also encourages you to participate in the project by enrolling as a wiki-contributor. Be your own Experimental Man or Experimental Woman. Share your data, thoughts, studies, and comments on your own wiki-page within the site. And researchers, please share your data, experiments, papers, and thoughts as well.

The Experimental Man Web site is a joint project with the Center for Life Science Policy at the University of California at Berkeley.

INDEX

ABCB1 gene, and arsenic resistance, 156
absolute risk, 46–47
ACE (angiotensin-I converting enzyme) gene, and high blood pressure, 26
Adams, Douglas, 183
addiction to heroin, genetic marker for, 95–96
additive effect of chemical exposure, 139
adjusted odds ratio, 46
Adler, Josh, 8, 15–20, 22, 33, 48, 63, 113, 173–74, 255, 281, 284–85, 320–21
ADRB2 gene, 162
Affymetrix Genome-Wide Human SNP Array, 32, 37
Agency for Toxic Substances and Disease Registry (ATSDR), 130, 133
age-related macular degeneration (AMD), genetic markers for, 37
aging

brain changes and, 179, 187, 198, 199–208
caloric restriction (CR) effect on, 292
SIRT1 gene and slowing of, 291, 294–95
Agus, David, 46, 103, 312–13, 314–15, 316, 317
alcohol, effects on brain of, 185
alcohol flush, DNA marker for, 97
alpha–1 antitrypsin deficiency gene, and lung and liver disease, 26
Altshuler, David, 47–48, 68, 69, 112
Alzheimer's disease, 67–69, 169, 290
American Chemistry Council (ACC), 150–51
American College of Medical Genetics, 110
amino acids, in collagen sequencing, 91–92
AmpliChip CYP450, 155
amygdala, and anxiety, 210, 211, 213, 215, 231, 234–35